K Eriksson D Estep P Hansbo

Computational Diff
Equations

CW00766086

CAMBRIDGE
UNIVERSITY PRESS

Published in Sweden, Finland, Norway, Denmark & Iceland by Studentlitteratur, Lund, Sweden.
Published elsewhere by the Press Syndicate of the University of Cambridge
The Pitt Building, Trumpington Street, Cambridge CB2 1RP
40 West 20th St, New York, NY 10011-4211, USA
10 Stamford Road, Oakleigh, Melbourne 3166, Australia

Cover designed by Mårten Levenstam

First published 1996

Printed in Sweden at Studentlitteratur, Lund

A catalogue record for this book is available from the British Library

Library of Congress cataloguing in publication data

ISBN   0 521 56312 7 (hardcovers)
         0 521 56738 6 (paperback)

To Anita, Floris, Ingrid, and Patty.

Contents

Preface

I admit that each and every thing remains in its state until there is reason for change. (Leibniz)

I'm sick and tired of this schism between earth and sky.
Idealism and realism sorely our reason try. (Gustaf Fröding)

This book, together with the companion volumes *Introduction to Computational Differential Equations* and *Advanced Computational Differential Equations*, presents a unified approach to computational mathematical modeling using differential equations based on a principle of a fusion of mathematics and computation. The book is motivated by the rapidly increasing dependence on numerical methods in mathematical modeling driven by the development of powerful computers accessible to everyone. Our goal is to provide a student with the essential theoretical and computational tools that make it possible to use differential equations in mathematical modeling in science and engineering effectively. The backbone of the book is a new unified presentation of numerical solution techniques for differential equations based on Galerkin methods.

Mathematical modeling using differential and integral equations has formed the basis of science and engineering since the creation of calculus by Leibniz and Newton. Mathematical modeling has two basic dual aspects: one symbolic and the other constructive-numerical, which reflect the duality between the infinite and the finite, or the continuum and the discrete. The two aspects have been closely intertwined throughout the development of modern science from the development of calculus in the work of Euler, Lagrange, Laplace and Gauss into the work of von Neumann in our time. For example, Laplace's monumental *Mécanique Céleste* in five volumes presents a symbolic calculus for a mathematical model of gravitation taking the form of Laplace's equation, together with massive numerical computations giving concrete information concerning the motion of the planets in our solar system.

However, beginning with the search for rigor in the foundations of calculus in the 19th century, a split between the symbolic and constructive aspects gradually developed. The split accelerated with the invention of the electronic computer in the 1940s, after which the constructive aspects were pursued in the new fields of numerical analysis and computing sciences, primarily developed outside departments of mathematics. The unfortunate result today is that symbolic mathematics and constructive-numerical mathematics by and large are separate disciplines and are rarely taught together. Typically, a student first meets calculus restricted to its symbolic form and then much later, in a different context, is confronted with the computational side. This state of affairs lacks a sound scientific motivation and causes severe difficulties in courses in physics, mechanics and applied sciences building on mathematical modeling. The difficulties are related to the following two basic questions: (i) How to get applications into mathematics education? (ii) How to use mathematics in applications? Since differential equations are so fundamental in mathematical modeling, these questions may be turned around as follows: (i) How can we teach differential equations in mathematics education? (ii) How can we use differential equations in applications?

Traditionally, the topic of differential equations in basic mathematics education is restricted to separable scalar first order ordinary differential equations and constant coefficient linear scalar n'th order equations for which explicit solution formulas are presented, together with some applications of separation of variables techniques for partial differential equations like the Poisson equation on a square. Even slightly more general problems have to be avoided because the symbolic solution methods quickly become so complex. Unfortunately, the presented tools are not sufficient for applications and as a result the student must be left with the impression that mathematical modeling based on symbolic mathematics is difficult and only seldom really useful. Furthermore, the numerical solution of differential equations, considered with disdain by many pure mathematicians, is often avoided altogether or left until later classes, where it is often taught in a "cookbook" style and not as an integral part of a mathematics education aimed at increasing understanding. The net result is that there seems to be no good answer to the first question in the traditional mathematics education.

The second question is related to the apparent principle of organization of a technical university with departments formed around particular differential equations: mechanics around Lagrange's equation, physics around Schrödinger's equation, electromagnetics around Maxwell's equations, fluid and gas dynamics around the Navier-Stokes equations, solid mechanics around Navier's elasticity equations, nuclear engineering around the transport equation, and so on. Each discipline has largely developed its own set of analytic and numerical tools for attacking its special differential equation independently and this set of tools forms the basic theoretical core of the discipline and its courses. The organization principle reflects both the importance of mathematical modeling using differential equations and the traditional difficulty of obtaining solutions.

Both of these questions would have completely different answers if it were possible to compute solutions of differential equations using a unified mathematical methodology simple enough to be introduced in the basic mathematics education and powerful enough to apply to real applications. In a natural way, mathematics education would then be opened to a wealth of applications and applied sciences could start from a more practical mathematical foundation. Moreover, establishing a common methodology opens the possibility of exploring "multi-physics" problems including the interaction of phenomena from solids, fluids, electromagnetics and chemical reactions, for example.

In this book and the companion volumes we seek to develop such a unified mathematical methodology for solving differential equations numerically. Our work is based on the conviction that it is possible to approach this area, which is traditionally considered to be difficult and advanced, in a way that is comparatively easy to understand. However, our goal has not been to write an easy text that can be covered in one term in an independent course. The material in this book takes time to digest, as much as the underlying mathematics itself. It appears to us that the optimal course will involve the gradual integration of the material into the traditional mathematics curriculum from the very beginning.

We emphasize that we are not advocating the study of computational algorithms over the mathematics of calculus and linear algebra; it is always a fusion of analysis and numerical computation that appears to be the most fruitful. The material that we would like to see included in the mathematics curriculum offers a concrete motivation for the development of analytic techniques and mathematical abstraction. Computation does not make analysis obsolete, but gives the analytical mind a focus. Furthermore, the role of symbolic methods changes. Instead of being the workhorse of analytical computations requiring a high degree of technical complexity, symbolic analysis may focus on analytical aspects of model problems in order to increase understanding and develop intuition.

How to use this book

This book begins with a chapter that recalls the close connection between integration and numerical quadrature and then proceeds through introductory material on calculus and linear algebra to linear ordinary and partial differential equations. The companion text *Advanced Computational Differential Equations* widens the scope to nonlinear differential equations modeling a variety of phenomena including reaction-diffusion, fluid flow and many-body dynamics as well as material on implementation, and reaches the frontiers of research. The companion text *Introduction to Computational Differential Equations* goes in the other direction, developing in detail the introductory material on calculus and linear algebra.

We have used the material that serves as the basis for these books in a variety of courses in engineering and science taught at the California Institute of Technology, Chalmers University of Technology, Georgia Institute of Technology, and the University of Michigan. These courses ranged from mathematically oriented courses on numerical methods for differential equations to applications oriented courses in engineering and science based on computation. Students in these kinds of courses tend to have a diverse preparation in mathematics and science and we have tried to handle this by making the material of this book as accessible as possible and including necessary background material from calculus, linear algebra, numerical analysis, mechanics, and physics.

In our experience, beginning a course about solving differential equations by discretizing Poisson's equation presents an overwhelming array of topics to students: approximation theory, linear algebra, numerical solution of systems, differential equations, function spaces, etc. The sheer number of topics introduced at one time in this approach gives rise to an almost insurmountable hurdle to understanding topics which taken one at a time are not so difficult. To overcome these difficulties, we have taken a different approach.

In the first part of this book, we begin by considering the numerical solution of the simplest differential equation by quadrature and we develop the themes of convergence of numerical methods by giving a constructive proof of the Fundamental Theorem of Calculus. We also show the close relationship between convergence and error estimation by studying adaptive quadrature briefly. Next, we present background material on linear algebra and polynomial approximation theory, following a natural line started with the first chapter by applying this material to quadrature. After this, we introduce Galerkin's method for more general differential equations by considering three specific examples. In this chapter, we also raise the important issues that are addressed in the rest of the book. This part concludes with an introduction to the numerical solution of linear systems.

In the second part of the book, we discuss the discretization of time or space dependent ordinary differential equations. The basic theme of this part is to develop an intuitive sense of the classification of differential equations into elliptic, parabolic, and hyperbolic. By discretizing model problems representing these basic types, we can clarify the issues in discretization and convergence. We also develop a sense of the kind of behavior to be expected of approximations and their errors for the different kinds of problems.

Finally in the third part of the book, we study the discretization of the classic linear partial differential equations. The material is centered around specific examples, with generalizations coming as additional material and worked out in exercises. We also introduce the complexities of multi-physics problems with two chapters on convection-diffusion-absorption problems.

While we advocate the arrangement of the material in this book on pedagogical grounds, we have also tried to be flexible. Thus, it is entirely possible to choose a line based on a particular application or type of problem, e.g. stationary problems, and start directly with the pertinent chapters, referring back to background material as needed.

This book is a substantial revision of Johnson ([10]) with changes made in several key ways. First, it includes additional material on the derivation of differential equations as models of physical phenomena and mathematical results on properties of the solutions. Next, the unification of computational methodology using Galerkin discretization begun in the precursor is brought to completion and is applied to a large variety of differential equations. Third, the essential topics of error estimation and adaptive error control are introduced at the start and developed consistently throughout the presentation. We believe that computational error estimation and adaptive error control are fundamentally important in scientific terms and this is where we have spent most of our research energy. Finally, this book starts at a more elementary level than the precursor and proceeds to a more advanced level in the advanced companion volume.

Throughout the book, we discuss both practical issues of implementation and present the error analysis that proves that the methods converge and which provides the means to estimate and control the error. As mathematicians, a careful explanation of this aspect is one of the most important subjects we can offer to students in science and engineering. However, we delay discussing certain technical mathematical issues underlying the Galerkin method for partial differential equations until the last chapter.

We believe that the students' work should involve a combination of mathematical analysis and computation in a problem and project-oriented approach with close connection to applications. The questions may be of mathematical or computational nature, and may concern mathematical modeling and directly relate to topics treated in courses in mechanics, physics and applied sciences. We have provided many problems of this nature that we have assigned in our own courses. Hints and answers for the problems as well as additional problems will be given in the introductory companion volume. The book is complemented by software for solving differential equations using adaptive error control called Femlab that is freely available through the Internet. Femlab implements the computational algorithms presented in the book, and can serve as a laboratory for experiments in mathematical modeling and numerical solution of differential equations. It can serve equally well as a model and toolbox for the development of codes for adaptive finite element methods.

Finally, we mention that we have implemented and tested a reform of the mathematics curriculum based on integrating mathematics and computation during the past year in the engineering physics program at Chalmers University. The new program follows a natural progression from calculus in one variable and ordinary differential equations to calculus in several variables and partial differential equations while developing the mathematical techniques in a natural interplay with applications. For course material, we used this book side-by-side with existing texts in calculus and linear algebra. Our experience has been very positive and gives clear evidence that the goals we have stated may indeed be achieved in practice. With the elementary companion text, we hope to ease the process of fusing the new and classical material at the elementary level and thereby help to promote the reform in a wider context.

Acknowledgements

We wish to thank the students at Chalmers and Georgia Tech who have patiently borne the burden of the development of this material and who have enthusiastically criticized versions of this book. We also thank our colleagues, including K. Brattkus, M. Knaap, M. Larson, S. Larsson, M. Levenstam, A. Ruhe, E. Süli, A. Szepessy, L. Wahlbin, and R. Williams, who read various parts of early versions of this book and made many useful, and necessary, suggestions and criticisms.

We would also like to thank M. Larson, M. Levenstam, A. Niklasson, and R. Williams for help with several computational examples and pictures, and the generous permission of the Gottfried-Wilhelm-Leibniz-Gesellschaft and Prof. E. Stein in Hannover to include pictures from the life and work of Leibniz.

D. Estep and C. Johnson wish to thank the Division of International Programs at the National Science Foundation for supporting D. Estep's visits to Sweden during which much of this book was written. D. Estep also wishes to thank the Computational Mathematics Program in the Division of Mathematics at the National Science Foundation and the Georgia Tech Foundation for the research support they have generously given.

Part I

Introduction

This first part has two main purposes. The first is to review some mathematical prerequisites needed for the numerical solution of differential equations, including material from calculus, linear algebra, numerical linear algebra, and approximation of functions by (piecewise) polynomials. The second purpose is to introduce the basic issues in the numerical solution of differential equations by discussing some concrete examples. We start by proving the Fundamental Theorem of Calculus by proving the convergence of a numerical method for computing an integral. We then introduce Galerkin's method for the numerical solution of differential equations in the context of two basic model problems from population dynamics and stationary heat conduction.

1

The Vision of Leibniz

> Knowing thus the Algorithm of this calculus, which I call Differential Calculus, all differential equations can be solved by a common method. (Leibniz)

> When, several years ago, I saw for the first time an instrument which, when carried, automatically records the number of steps taken by a pedestrian, it occurred to me at once that the entire arithmetic could be subjected to a similar kind of machinery so that not only addition and subtraction, but also multiplication and division could be accomplished by a suitably arranged machine easily, promptly and with sure results.... For it is unworthy of excellent men to lose hours like slaves in the labour of calculations, which could safely be left to anyone else if the machine was used.... And now that we may give final praise to the machine, we may say that it will be desirable to all who are engaged in computations which, as is well known, are the mangers of financial affairs, the administrators of others estates, merchants, surveyors, navigators, astronomers, and those connected with any of the crafts that use mathematics. (Leibniz)

Building on tradition going back to the ancient Greek philosophers, Leibniz and Newton invented calculus in the late 17th century and thereby laid the foundation for the revolutionary development of science and technology that is shaping the world today. Calculus is a method for modeling physical systems mathematically by relating the state of a system to its neighboring states in space-time using differential and integral equations. Because calculus is inherently computational, this revolution began to accelerate tremendously in the 1940s when the electronic computer was created. Today, we are seeing what is essentially a "marriage" of calculus and computation in the creation of the field of *computational mathematical modeling.*

Figure 1.1: Gottfried Wilhelm Leibniz, 1646-1716.

Actually, Leibniz himself sought to realize a unification of calculus and computation, but failed because the mechanical calculator he invented was not sufficiently powerful. The next serious effort was made in the 1830s by Babbage, who designed a steam powered mechanical computer he called the Analytical Engine. Again, technical difficulties and low speed stopped his ambitious plans.

The possibility of realizing Leibniz' and Babbage's visions of a universal computing machine came with the invention of the electronic valve in the 1930s, which enabled the construction of high speed digital computers. The development took a leap during the World War II spurred by the computing demands of the military. Until this time, large scale computations were performed by rooms of people using mechanical adding machines. The war provided an immediate pressure to speed up the process of scientific development by using mathematical modeling to hone a physical problem down to a manageable level, and mathematicians and physicists became interested in the invention of an electronic computing device. The logical design of programmable electronic computers was developed by the mathematician von Neumann, among others. By the late forties, von Neumann was using the first ENIAC (Electronic Numerical Integrator And Calculator) computer to address questions in fluid dynamics and aerodynamics.

The subsequent development of computer power that has resulted in desktop computers of far greater power than the ENIAC, has been paralleled by the rapid introduction of computational mathematical modeling into all areas of science and engineering. Questions routinely addressed computationally using a computer include: What is the weather going to do in three days? Will this airplane fly? Can this bridge carry a load of ten trucks? What happens during a car collision? How do we direct a rocket to pass close by Saturn? How can we create an image of the interior of the human body using very weak X-rays? What is the shape of a tennis racket that has the largest "sweet spot"? What is a design of a bicycle frame that combines low weight with rigidity? How can we create a sharp picture from a blurred picture? What will the deficit be in Sweden in the year 2000? How much would the mean temperature of the earth increase if the amount of carbon dioxide in the atmosphere increased by 20 percent?

The physical situations behind these kinds of questions are modeled by expressing the laws of mechanics and physics (or economics) in terms of equations that relate derivatives and integrals. Common variables in these models are time, position, velocity, acceleration, mass, density, momentum, energy, stress and force, and the basic laws express conservation of mass, momentum and energy, and balance of forces. Information about the physical process being modeled is gained by solving for some of the variables in the equation, i.e. by computing the *solution* of the differential/integral equation in terms of the others, which are assumed to be known data. Calculus is the basic study of differential/integral equations and their solutions.

Sometimes it is possible to find an exact formula for the solution of a differential/integral equation. For example, the solution might be expressed in terms of the data as a combination of elementary functions or as a trigonometric or power series. This is the classical mathematical method of solving a differential equation, which is now partially automated in mathematical software for symbolic computation such as Maple or Mathematica. However, this approach only works on a relatively small class of differential equations. In more realistic models, solutions of differential equations cannot be found explicitly in terms of known functions, and the alternative is to determine an approximate solution for given data through numerical computations on a computer. The basic idea is to *discretize* a given differential/integral equation to obtain a system of equations with a finite number of unknowns, which may be solved using a computer to produce an approximate solution. The finite-dimensional problem is referred to as a *discrete problem* and the corresponding differential/integral equation as a *continuous problem*. A good numerical method has the property that the error decreases as the number of unknowns, and thus the computational work, increases. Discrete problems derived from physical models are usually computationally intensive, and hence the rapid increase of computer power has opened entirely new possibilities for this approach. Using a desktop computer, we can often obtain more information about physical situations by numerically solving differential equations than was obtained over all the previous centuries of study using analytical methods.

Predicting the weather

The progress in weather prediction is a good example for this discussion. Historically, weather forecasting was based on studying previous patterns to predict future behavior. A farmer's almanac gives predictions based on the past behavior, but involves so many variables related to the weather that determining meaningful correlations is an overwhelming task. By modeling the atmosphere with a set of differential equations, the number of variables is reduced to a handful that can be measured closely, albeit at many locations. This was envisioned by the English pioneer of numerical weather prediction Richardson in the 1920s, who proposed the formation of a department of 64,000 employees working in shifts to perform the necessary calculations using mechanical calculators more quickly than the weather changed. After this proposal, the attitude toward numerical weather prediction became pessimistic. Not until the development of the modern computer, could the massive computations required be performed sufficiently rapidly to be useful. The first meaningful numerical forecasts were made by von Neumann and Charney in the late forties using the ENIAC, but of course the reliability was very low due to the extremely coarse discretization of the earth's system they had to use. The most recent model for the global weather uses a discretization grid with roughly 50,000 points horizontally and 31 layers vertically giving a total of five million equations that are solved in a couple of hours on a super-computer.

There are three sources of errors affecting the reliability of a numerical weather forecast: (i) measurement errors in data (or lack of data) (ii) approximation errors in modeling and (iii) approximation errors in computation. The initial data at the start of the computer simulation are always measured with some error; the set of differential equations in the computer model only approximately describes the evolution of the atmosphere; and finally the numerical solution of the differential equations is only an approximation of the true solution. These sources add up to form the total prediction error. It is essential to be able to estimate the total error by estimating individually the contributions from the sources (i)-(iii) and improve the precision where possible. This is a basic issue in all applications in computational mathematical modeling.

Our experience tells that forecasts of the daily weather become very unreliable in predictions for more than say a week. This was discussed in the 1960s by the meteorolgist Lorenz, who coined the phrase "the butterfly effect" to describe situations in which a small cause can have a large effect after some time. Lorenz gave a simple example displaying this phenomenon in the form of the *Lorenz system* of ordinary differential equations with only three unknowns. We plot a typical solution in Fig. 1.2, showing the trajectory of a "particle" being ejected away from the origin to be attracted into a slowly diverging orbit to the left, then making a loop on the right, returning to a few orbits to the left, then back to the right etc. The trajectory is very sensitive to perturbations as to the number of loops to the left or right, and thus is difficult to compute accurately over a longer time interval, just as the evolution of the weather may be difficult to predict for more than a week.

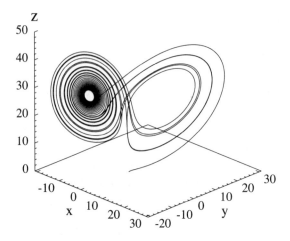

Figure 1.2: A solution of the Lorenz system computed with an error of
.1 or less over the time interval $(0, 30)$.

What is this book about?

If we summarize the Leibniz vision as a fusion of mathematical modeling,
mathematical analysis and computation, then there are three fundamental issues to be addressed:

- How are physical phenomena modeled using differential equations?

- What are the properties of solutions of differential equations?

- How are approximate solutions of differential equations computed
  and how can the accuracy of the approximations be controlled?

This book tries to answer these questions for a specific set of problems
and to provide a set of tools that can be used to tackle the large variety
of problems met in applications.

The book begins with some material directly from calculus. Partly
this is a review and partly a presentation of elementary material needed
to solve differential equations numerically. Next, we study the particular issues that arise in different classes of equations by studying a set of
simple model problems from physics, mechanics and biology. The scope
is then widened to cover basic linear models for heat conduction, wave
propagation, fluid flow and elastic structures. The companion volume

extends the scope further to nonlinear differential equations and systems of equations modeling a variety of phenomena including reaction-diffusion, fluid flow and many-body dynamics and reaches the frontiers of research.

Covering most of the material in this book would provide a good preparation and a flexible set of tools for many of the problems that are met in engineering and science undergraduate courses. It is essential to do a good portion of the problems given in the text in order to master the subject. We mark the more difficult and tedious problems (but they must be all the more rewarding, right?) with warnings and often give hints. A companion volume called *Advanced Computational Differential Equations* leads into graduate level, including material on nonlinear differential equations and implementation. Another companion book, *Introduction to Computational Differential Equations*, contains additional material on calculus and linear algebra, hints to problems in this volume and suggestions for project work.

The presentation is unified in the sense that it is always essentially the same set of basic tools that are put to use independently of the level of complexity of the underlying differential equation. The student will discover this gradually going through the material. The methodology is always presented in the simplest possible context to convey an essential idea, which later is applied to successively more complex problems just by "doing the same". This means that a thorough understanding of the simplest case is the best investment; for the student with limited time or energy this minimal preparation allows him or her to computationally address complex problems without necessarily having to go into all the details, because the main ideas have been grasped in a simple case. Thus, we seek to minimize the technical mathematical difficulties while keeping the essence of the ideas.

On the other hand, some ideas cannot be explained in just one application. As a result, the presentation in the simple cases may occasionally seem lengthy, like for instance the careful proof of the Fundamental Theorem of Calculus. But, the reader should feel confident that we have a carefully thought out plan in our minds and some reason for presenting the material in this way.

The book is supplemented by the software Cards and Femlab, where the algorithms presented in the book are implemented. This gives the possibility of a problem/project-oriented approach, where the student may test the performance of the algorithms and his own ideas of application and improvement, and get a direct experience of the possibilities of computational mathematical modeling. Femlab may also be used as a basis for code development in project-oriented work. Femlab is available through the World Wide Web by accessing http://www.math.chalmers.se/femlab. The software is presented in the introductory companion volume.

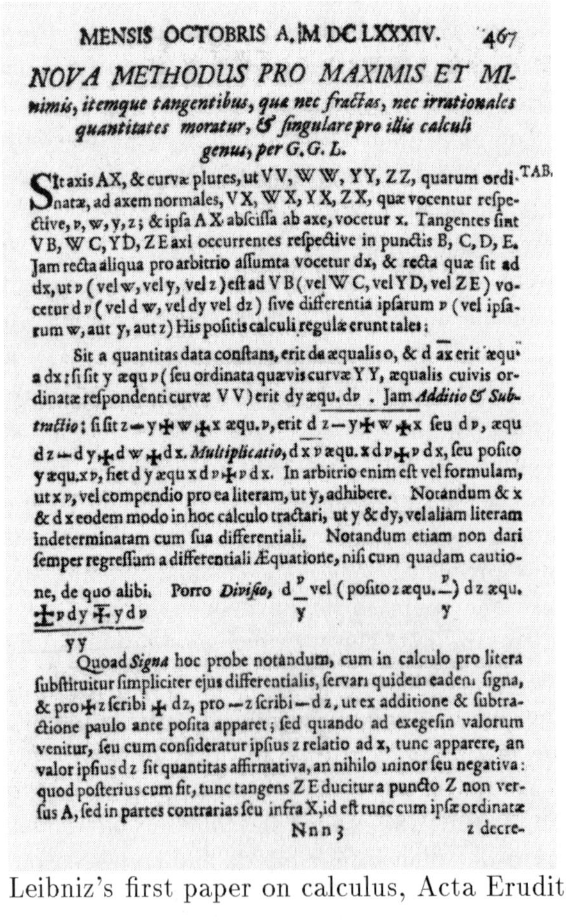

MENSIS OCTOBRIS A. M DC LXXXIV. 467

NOVA METHODUS PRO MAXIMIS ET MInimis, itemque tangentibus, quæ nec fractas, nec irrationales quantitates moratur, & singulare pro illis calculi genus, per G. G. L.

Sit axis AX, & curvæ plures, ut VV, WW, YY, ZZ, quarum ordinatæ, ad axem normales, VX, WX, YX, ZX, quæ vocentur respective, v, w, y, z; & ipsa AX abscissa ab axe, vocetur x. Tangentes sint VB, WC, YD, ZE axi occurrentes respective in punctis B, C, D, E. Jam recta aliqua pro arbitrio assumta vocetur dx, & recta quæ sit ad dx, ut v (vel w, vel y, vel z) est ad VB (vel WC, vel YD, vel ZE) vocetur dv (vel dw, vel dy vel dz) sive differentia ipsarum v (vel ipsarum w, aut y, aut z) His positis calculi regulæ erunt tales: TAB.

Sit a quantitas data constans, erit da æqualis 0, & d $\overline{ax}$ erit æqu. a dx: si sit y æqu v (seu ordinata quævis curvæ YY, æqualis cuivis ordinatæ respondenti curvæ VV) erit dy æqu. dv. Jam *Additio & Subtractio*; si sit z − y + w + x æqu. v, erit d $\overline{z - y + w + x}$ seu dv, æqu dz − dy + dw + dx. *Multiplicatio*, d $\overline{xv}$ æqu. x dv + v dx, seu posito y æqu. xv, fiet dy æqu x dv + v dx. In arbitrio enim est vel formulam, ut xv, vel compendio pro ea literam, ut y, adhibere. Notandum & x & dx eodem modo in hoc calculo tractari, ut y & dy, vel aliam literam indeterminatam cum sua differentiali. Notandum etiam non dari semper regressum a differentiali Æquatione, nisi cum quadam cautione, de quo alibi. Porro *Divisio*, d $\frac{v}{y}$ vel (posito z æqu. $\frac{v}{y}$) dz æqu. $\frac{\pm v\,dy \mp y\,dv}{yy}$

Quoad *Signa* hoc probe notandum, cum in calculo pro litera substituitur simpliciter ejus differentialis, servari quidem eadem signa, & pro + z scribi + dz, pro − z scribi − dz, ut ex additione & subtractione paulo ante posita apparet; sed quando ad exegesin valorum venitur, seu cum consideratur ipsius z relatio ad x, tunc apparere, an valor ipsius dz sit quantitas affirmativa, an nihilo minor seu negativa: quod posterius cum fit, tunc tangens ZE ducitur a puncto Z non versus A, sed in partes contrarias seu infra X, id est tunc cum ipsæ ordinatæ

Nnn 3 z decre-

Figure 1.3: Leibniz's first paper on calculus, Acta Eruditorum, 1684.

2

A Brief History

The quadrature of all figures follow from the inverse method of tangents, and thus the whole science of sums and quadratures can be reduced to analysis, a thing nobody even had any hopes of before. (Leibniz)

I would like to give a method of Speciosa Generalis, in which all truths of reason would be reduced to some kind of calculus. This could at the same time be a kind of language or universal script, but very different from all that have been projected hitherto, because the characters and even the words would guide reason, and the errors (except those of fact) would only be errors of computation. It would be very difficult to form or to invent this Language or Characteristic, but very easy to learn it without any Dictionaries. It would also serve to estimate the degree of likelihood (because we have not sufficient data to arrive at truths that are certain), and to see what is necessary to obtain them. And this estimate would be very important for its applications to life and for the deliberations of practice where by estimating the probabilities one miscalculates most often in more than half the cases. (Leibniz)

Before plunging into work, we make a brief digression into the history of mathematical modeling based on differential equations. The purpose is to give some perspective on the work that lies before us. We hope that the impressions gained by browsing through these pages will help when the material becomes difficult. In fact, it is no exaggeration to describe our subject as containing the finest creations of some of the most inventive minds of the last few centuries. Here, we can only give the briefest review. We strongly recommend the book Kline ([12]) as an interesting and readable background source for enriching the perspective.

From a large perspective the revolution in information technology now emerging, of which computational mathematical modeling is a part, could be viewed as the second major step after the first revolution in information technology created by the Babylonian culture around 2000 B.C. based on information-processing on clay tablets. The Babylonian culture had a high level of organization based on agriculture and trade and accordingly developed quite sophisticated mathematical tools for handling questions in astronomy, economy, and infrastructure such as dams and canals, using algebraic and geometric methods. In particular, algebraic methods for solving systems of equations involving the computation of square and cubic roots were developed.

2.1. A timeline for calculus

Calculus is the central mathematics underlying computational mathematical modeling and its development characterizes the modern scientific struggle to describe nature in quantitative terms.

17th century: Leibniz and Newton

The first and perhaps most stunning success of calculus was given by Newton (1642-1727) in his *Principia Mathematica* published in 1687. This summarized Newton's work in the period 1665-6, when Newton, having just finished his undergraduate studies, moved to his family home while Cambridge was struck by the plague. In the *Principia*, Newton gave a simple description of the motion of the planets in our solar system as solutions of differential equations that were derived under the assumption of a inverse square law for gravitational force. The precision of the predictions based on Newton's theory was impressive (and still is) and the success was boosted by the sheer size of the objects that seemed to be controlled by mathematics. In tackling this problem, Newton was following a long tradition. The origins of applications of calculus in astronomy goes back to Copernicus (1473-1543), who was the first person to seriously question the Aristotelian/Ptolemaic geocentric theory in which the planets and the Sun evolved around the earth in a very complex combination of circular motions, to Kepler (1571-1630), and to Galileo (1564-1642), who was condemned in Rome in 1632 for his heretical scientific views. All of these people formulated mathematical descriptions of the motion of the planets. Some of the discoveries of Galileo are now a standard topic in high school, such as the well known formulas $v = at$ and $s = at^2/2$ for the velocity v and traveled distance s at time t of an object starting at rest and having a constant acceleration a. Kepler formulated laws for the motion of the planets in elliptical orbits and also applied calculus to practical science on Earth: when married for a second time, he installed a wine cellar in his house and began to compute volumes of wine kegs to find optimal shapes and detect false volume specification. He did this by summing small pieces of volume. Cavalieri (1598-1647), a student of Galileo, actually formulated an early version of calculus. Fermat (1604-1665) and Descartes (1596-1650) continued this work, introducing analytical methods based on representing geometrical objects through equations for points in a coordinate system. Leibniz (1646-1716) and Newton, working independently, summarized and extended the previous work and created calculus in much the form we now use it. In particular, Leibniz laid the foundation of calculus as a formal symbolic "machine" with a surprising power and flexibility.

18th century: Euler and Lagrange

The development of calculus in the 18th century was largely advanced by Euler (1707-1783) and Lagrange (1736-1813) who used calculus to treat basic problems of mechanics. Euler's productivity is legendary; he authored over 800 books and articles and was the father of thirteen children. The *calculus of variations* was created by Euler and Lagrange to give a condensed formulation of problems in mechanics based on variational principles, where solutions are defined as minima or more generally stationary points of Lagrangian functions representing the total energy or the difference of kinetic and potential energy. The condition of stationarity could alternatively be expressed as a differential equation and thus variational formulations gave an alternative to description via differential equations. The finite element method is the modern realization of this deep relation.

19th and 20th centuries: Partial differential equations

In the 19th century and the beginning of the 20th century, the scope of calculus was widened by Gauss, Laplace, Poisson, Fourier, Cauchy, Riemann, Green, Stokes, Maxwell, Boltzmann, Einstein and Schrödinger among others, to phenomena such as heat conduction, fluid flow, mechanics of deformable bodies, electromagnetism, gas dynamics, relativity and quantum mechanics. Again, variational methods were often used and the corresponding Euler-Lagrange equations now took the form of partial differential equations. We name the basic such equations along with the approximate year of discovery or formulation: Laplace's equation in 1810, Poisson's equation in 1812, Cauchy-Navier's elasticity equations in 1828, Navier-Stokes equations in 1821/1845, Maxwell's equations in 1864, Boltzmann's equation in 1860, Einstein's equations in 1917, and Schrödinger's equation in 1925. For certain problems, the application of calculus again had a tremendous success: For example, building on the experiments of Faraday, Maxwell (1831-1879) formulated his famous model for the interaction of electric and magnetic fields known as Maxwell's equations for electromagnetism and predicted the possibility of propagation of electromagnetic waves and Einstein (1879-1955) created the theory of relativity and Schödinger quantum mechanics, based on calculus and, like Newton, opened a new perception of the world.

However, there was much less progress in other applications. The Navier-Stokes equations modeling fluid flow are virtually impossible to solve using analytical techniques and have become useful in applications only recently, when increased computer power has begun to allow accurate numerical solution. In fact, Newton's theoretical predictions of the impossibility of sustained motored flight, which was based on a crude model for fluid flow and overestimated the power necessary to create the necessary lift, were not corrected until the early 20th century after the actual flights by the Wright brothers, when Kutta and Jukowski discovered some approximate solutions of the Navier-Stokes equations that gave accurate predictions of the lift. In particular, the famous d'Alembert paradox (1752), which predicted both zero drag and lift in inviscid flow, posed a serious obstacle of getting theory and observation to agree. Even today, there are still many questions about the Navier-Stokes equations that we cannot answer.

2.2.   A timeline for the computer

The development of the computer has had a striking impact on mathematical modeling by extending our ability to compute information about the solutions of differential equations tremendously. In fact, this is becoming one of the main activities in science and engineering.

Before the 20th century

The idea of using a machine to handle tedious computations appears in several places. The abacus, for example, is a simple manually operated digital computer. Pascal (1623-1662) designed a machine to do additions and subtractions. Later in 1671, Leibniz designed a calculating machine that was capable of multiplication and division as well as addition and subtraction, and 1673 went to the Royal Society in London to get funding for its construction. In the 1830s, the English mathematician Babbage (1792-1871) was inspired by the punch-card programmable weaving machine of Jacquard to design the Analytical Engine; a mechanical steam powered programmable machine that was to be used to solve calculus problems from astronomy and navigation. Babbage built demonstration models with support from the British government, and the Swedes Georg and Edvard Scheutz in the 1850s designed a related "Difference Machine" built in 3 copies, but the complexity of the mechanics prevented a further spread of the technology. Babbage was aware of the generality of his concept and stated that "the whole of the conditions which enable a *finite machine* to make calculations of *unlimited* extent, are fulfilled in the Analytical Engine". By the "extent", he meant both the amount and accuracy of the data to be processed and the length and logical complexity of the algorithm to be performed. Interestingly, Babbage was a member of the Analytical Society in Cambridge that formed to study the calculus of Leibniz, or what was called the "d-ism" related to the du/dt notation of derivatives of Leibniz, as opposed to the "dot-age" notation $\dot{u}$ of Newton that dominated in England for many decades.

The 20th century

The ideas of Babbage were lost in science as the complete machine was never built. Then in the 1930s, Alan Turing (1912-1954) reopened the quest for a programmable "universal machine" in the form of the "Turing machine" studied in his dissertation *Computable numbers* in 1937 (apparently without being aware of the work by Babbage). Together with the development of the electronic valves, this opened the explosive development of the modern computer started during World War II. Turing was also interested in the ability of a computer to develop "intelligence" and devised the Turing Test. A person sits in a room with a teletype connected to two teletype machines located in one room with a human and one room with a computer. The person tries to determine which room holds the computer by asking questions using the teletype and studying the responses. If he can't find out, then the computer can be considered intelligent.

In fact, the theoretical basis of the computer revolution is rooted in the work on the foundations of mathematics in the 1920s and 1930s related to Hilbert's ambition of showing that any mathematical problem can be solved by a "fixed and mechanical process", or at least that the consistency (lack of contradictions) of mathematics can be established by "finite methods" that avoid reference to infinite numbers or an infinity of properties or manipulations. The work by Gödel, Church and Turing in the 1930s showing that certain questions in mathematics are "undecidable", or equivalently that certain quantities are "uncomputable", ruined Hilbert's hopes. But, this work did lay a foundation for the design of the modern computer. The mathematician von Neumann (1903-1957), initially related to Hilbert's program, played an important role in the development of the first generation of programmable electronic computers during and immediately after World War II, motivated by the need to solve certain differential equations of military importance. It is interesting to note the close connection between the abstract study of the very basis of mathematics and the concrete questions arising in the design of computers. This is an outstanding example of fruitful fusion of theory and practice.

John Atanassoff built a small-scale electronic digital computing device starting in 1939 at Iowa State University. In the same year, Howard Aiken began constructing an electro-mechanical computer called the Mark I (Automatic Sequence Controlled Calculator) at Harvard. Further development of early program-stored electronic computers was carried in parallel at several places during World War II: at the Institute of Advanced Study in Princeton and the University of Pennsylvania by von

Neumann and others, at the National Physical Laboratory in London by Turing, and also in Germany by Konrad Zuse. In the United States, the first general purpose electronic digital scientific computer was the ENIAC built during 1943-46 by Eckert and Mauchly at the University of Pennsylvania for the numerical computation of firing tables to replace the mechanical analog Differential Analyzer used there for the same purpose. The ENIAC was a big computer (100 feet long and 8 feet high), included 18,000 vacuum tubes, and was "programmed" manually by setting switches and plugging up cables. In 1944, von Neumann joined the ENIAC project and in a famous memo outlined the logical design of a program-stored computer, which was realized in the JOHNNIAC, MANIAC and the EDVAC (Electronic Discrete Variable Automatic Computer). In England, the computers developed during the war were the ACE (Analytical Computing Engine) and the Colossus, and in Germany, the Zuse 1-4. The UNIVAC I (Universal Automatic Computer) designed by Eckert and Mauchly and released in 1951 was the first computer to handle both numeric and alphabetic data equally well, and was also the first commercially available computer (48 systems were built).

Moving into the sixties, seventies, and eighties, new "generations" of computers have been introduced roughly every five to six years. In parallel, there has been a rapid development of computer-based methods in traditional areas of mathematics such as differential equations and optimization, in new areas of applications such as image processing and medical tomography, and in new branches of mathematics such as fractal geometry and complex "chaotic" systems. In the advanced companion volume, we continue this history with a focus on the development of the modern scientific computer and numerical analysis.

2.3. An important mathematical controversy

The logistic and contructivist schools

Leibniz considered calculus to be a formal theory for symbolic manipulation that was extremely useful, but which could not be given a completely rigorous justification because of the difficulties involved in defining concepts like "continuum" and "infinity". Leibniz was fascinated by the remarkable savings in computational effort that resulted when an integral could be computed using his formal symbolic methods instead of the tedious and time-consuming "method of exhaustion" of Eudoxus (408-355 B.C.). This method, also used by Euclid (approx 300 B.C.) and Archimedes (287-212 B.C.) computes the area of a figure by summing the individual areas of a large number of small pieces making up the figure. However, Leibniz was also aware of the fact that symbolic methods could not replace numerical calculation most of the time, and as we said, constructed a mechanical digital calculator in 1671.

During the 19th century, strong efforts were made by Cauchy (1789-1857), Weierstrass (1815-1897), Dedekind (1831-1916), and Cantor (1845-1918) to give calculus a theoretically satisfactory foundation which included precise definitions of the basic concepts of the continuum of real numbers, limit, continuity, derivative and integral. These efforts were largely successful in the sense that some basic concepts were clarified. Correspondingly, the "axiomatic formalistic", or "logistic", school of thought began to take a dominating role in mathematics education despite the strong criticisms from the "constructivists" in the beginning of the 20th century. The point of view of the constructivist school is that only mathematical objects that can be "constructed" by "finite methods" can be the subject of investigation, which effectively excludes infinite sets such as e.g. the set of real numbers used by the logistic school. Though many mathematicians had (and have) sympathy for the constructivist point of view, few worked actively in mainstream mathematics towards constructivist goals after the collapse of the Hilbert program in the 1930s. However, beginning in the 1940s, some constructivists, building on the positive aspects of the work by Gödel (1906-1978) and Turing, turned towards what has become computer science and developed the electronic computer. The dispute on the foundations of mathematics was left unsettled.

Symbolic and numerical computation

The use of computational methods boomed with the invention of the electronic computer in the 1940s. The basic problem in the constructivist point of view now appeared in the form of "computability" in computer science, which basically is the question of how much computational work is needed for a certain task. The question of computability also naturally comes up in the context of solving differential equations in the following form: how much work is required to compute a solution to a given differential equation to a certain accuracy? This in fact is the basic question of this book. If the computational work is too large, e.g. requiring more operations than the number of atoms in the universe, a constructivist would say that the solution does not exist in a concrete sense. A logicist could claim that the solution "exists" even though it is very difficult or impossible to construct.

The controversy between logicists and constructivists reflects a fundamental difference in nature between symbolic and numerical computation, where symbolic computation represents the logicist approach and the numerical computation the constructive one. One may argue, that nature itself is "constructive" and builds on "numerical computation" which is massive but finite, while symbolic computation, involving "infinite" sets such as the set of real numbers or Hilbert spaces of functions, may be used by mathematicians or scientists in descriptions of natural phenomena. Solving differential equations using numerical computation thus may be viewed as a more or less direct simulation of natural processes of finite dimension, while symbolic computation may have a different, less concrete, character. It appears that both symbolic and numerical computation are indispensable, reflecting in a sense a duality of soul and matter, and we may think of the good combination in science, which always remains to be found, as a Leibnizian synthesis of calculus and computation.

2.4.   Some important personalities

Leibniz

> If I were to choose a patron for cybernetics, I should have to choose Leibniz. (Wiener)

Gottfried Wilhelm Leibniz was maybe the last Universal Genius incessantly active in the fields of theology, philosophy, mathematics, physics, engineering, history, diplomacy, philology and many others. He was born in 1646 at Leipzig into an academic family. His father died in 1652 after which Leibniz took charge of his own education by using his father's large library to read works by poets, orators, historians, jurists, philosophers, mathematicians, and theologians. At the age of 8, he taught himself Latin and decided, when 15, to convert from the classical scholastic philosophy of Aristotle to the new philosophy of Bacon,

Pascal, and Descartes. His universal reading made him knowledgeable in almost every field. Leibniz attended university from the age of 14 to 22, finishing in 1667 with a doctorate in Jurisprudence based on a thesis entitled *On Difficult Cases in Law*, though there was some trouble in graduating because of his youth. The year after he developed a new method of teaching law including a new system seeking to define all legal concepts in terms of a few basic ones and deducing all specific laws from a small set of principles. He also wrote a dissertation in 1666 in philosophy entitled *The Art of Combinations*, in which he outlines a "universal language", which is a precursor to the development of the calculus of logic that Boole invented in the 19th century and formal languages underlying computer science of today. He pursued the idea of associating with each primitive concept a prime number and the composition of primitive concepts with the product; for example, if 3 represents "man", and 7 "rational", then 21 would represent "rational man". He then tried to translate the usual rules for reasoning into such a system, but was unsuccessful. The *Art of Combinations* also contains the Fundamental Theorem of Calculus in "difference form".

In 1670, Leibniz published his first philosophical work *On true principles, and the true method of philosophizing against false philosophers* and continued the year after with the two-volume treatise *Hypothesis Physica Nova* dedicated to the Academie des Sciences de Paris and the Royal Society of London. This contained, among other things, discussions on the principles of conservation of quantity of motion or momentum in its correct vector form and conservation of energy. Leibniz took up a job as advisor to the Duke of Mainz and came up with a plan to distract Louis XIV from Europe into a conquest of Egypt, which was eventually followed by Napoleon, and was sent in 1672 to Paris to present his ingenious ideas. In this context, Leibniz published anonymously a biting satire on Louis XIV called *Mars Christianissimus* (Most Christian War God) referring to the king's imperialism.

During his stay in Paris 1672-76, which included two visits to London, Leibniz ("the most teachable of mortals" in his own words), plunged himself into studies of the "pre-calculus" of Cavalieri, Pascal, Fermat, Descartes, Gregory and Barrow (Newton's teacher), with guidance in particular from Christian Huygens. With only little preparation in geometry and algebra, Leibniz quickly created a synthesis of calculus including the Fundamental Theorem of Calculus, and the basic notation and analytical tools still used today. This work was summarized in 1684 in six pages in Leibniz' journal *Acta Eruditorum*, while Newton's first work on calculus was published in *Principa* in 1686 (but was conceived around 1666). Later, a bitter academic fight developed between Newton and Leibniz concerning "who invented calculus", Leibniz being accused of plagiarism. The truth is that Newton and Leibniz developed calculus independently in different forms building on a common tradition including the work of many.

Leibniz returned to Germany in 1676 and served different Dukes of Brunswick in Hannover as counselor and librarian among other duties. He was commissioned to write the history of the House of Brunswick, and bravely started, after intense studies including long trips to Austria and Italy 1687-90, with the volume *Protagea: A dissertation on the first formation of the globe and the oldest traces of history in the very monuments of nature*, which turned out as a work in geology and natural history. As concerns the history of the Dukes, he never got beyond year 1009, which affected his position at the court. At the end of his life he was often ridiculed and was treated as an old fossil in his enormous black wig and once-fashionable ornate clothes.

Leibniz also worked in other branches of mathematics: he introduced determinants to solve linear systems of equations and in *Characteristica Geometria* from 1679 envisioned a form of combinatorial topology, but as often he was too far ahead of his time to get any response. He also proposed the binary system as suitable for machine calculation and designed a binary calculator and a converter from decimal to binary numbers, in addition to his decimal calculator in 1671.

Leibniz wrote more than 15,000 letters including a long-lasting correspondence with Jacob and Johan Bernoulli, who popularized and spread Leibniz's work on calculus. His mature work on philosophy includes *Discourse on Metaphysics* 1686, *New Essays on Human Understanding* 1704, *Theodicy* 1710, *Monadology* and *Principles of Nature and Grace, based on Reason* 1714. His work on physics includes *Dynamics* 1689-91 and the essay *A specimen of Dynamics* 1695. Leibniz opposed Newton's idea of absolute space and time basing his dynamics on the concept of force, preparing the way for Einstein's theory of special relativity.

Leibniz' work was governed by a principle of synthesis by which he always sought to combine the best elements of truth. Even after his conversion to the new philosophy, he kept ties with Aristotle's idealism and rejected a pure materialism. He tried to find paths to unite Catholics and Protestants and proposed to create a United States of Europe in the form of a Holy Roman Empire with state and church united. His Monadology builds on the principle that body and soul are united and express dual aspects of existence, and has been viewed as a precursor to modern quantum mechanics. It also leads into modern systems and complexity theory with the whole being more than the sum of its parts in the same way as a state has a nature beyond that expressed through a list of the members of state. His idea of the "best possible world" rests on a synthesis of maximal variety and simplicity of laws, coupling to todays research on fractals and chaos. Many of his ideas were amazingly visionary, including the formation of the European union, and the notions of computers and complexity. Leibniz also had a strong interest in matters of direct practical importance. His different calculating machines were only some of Leibniz's many technological inventions including improvements on time-keeping, distance-reckoning, barometry, the design of lenses, carriages, windmills, suction pumps, gearing mechanisms and other devices.

Leibniz rejected an academic career and called universities "monkish", charging them with possessing learning but no judgement and being absorbed in trifles. Instead, he urged the pursuit of real knowledge, a combination of mathematics, physics, geography, chemistry, anatomy, botany, zoology and history.

Leibniz did not marry. He considered it once at the age of fifty, but the person to whom he proposed wanted time to think about it, which also gave Leibniz time to reconsider his proposal. He kept close ties with both Duchess Sophie of Hannover and also her daughter Queen Sophie Charlotte of Brandenburg, who considered herself to be a student of Leibniz. He wrote a beautiful ode to Queen Charlotte on the occasion of her funeral in 1705 (see below). He spent months at a time without leaving his study, sleeping a few hours in a chair to wake up early to continue his work. He was easily angered but composed himself quickly, reacted badly to criticism but accepted it soon. He had an excellent memory, being called a Living Dictionary by Duke George Ludwig. Leibniz died in 1716, lonely and neglected, with only his secretary attending the funeral. His work is a landmark in the development of science and philosophy and of great relevance even today.

Wiener

Norbert Wiener, the "20th century Leibniz of America", was born in 1894 into a Jewish Russian emigrant family, was tutored early by his father (who spoke forty languages), got a B.A. in 1909 and a Ph.D. in Philosophy at Harvard when aged 18. Wiener then worked on the foundations of mathematics with Russell in Cambridge, returned to the United States for military service and then turned to integration theory, potential theory and statistical mechanics in the 1920s. He took up a position at the Massachusetts Institute of Technology working on applications of harmonic analysis in signal processing, proposed an analog computer for the solution of partial differential equations, laid a foundation of cybernetics, developed anti-aircraft prediction during World War II, worked in mathematical physiology, pattern recognition, aspects of automatization and automata theory. Together with von Neumann, Wiener founded the Teleological Society or Cybernetical Circle directed to automata theory, self-reproducing machines, information theory, brain-computer interaction etc. Wiener died in 1954 in Stockholm (as did Descartes in 1650).

von Neumann

John von Neumann was born in 1903 in Budapest as the oldest of three sons of a banker. At the age of 6, he divided 8-digit numbers without pen and paper and by 8 he had mastered calculus. He got a Ph. D. in Mathematics in 1926 in Budapest after training as a chemical engineer until 1925. He then held posts in Göttingen, Berlin and Hamburg, working on the set theoretic and logical foundations of mathematics and the mathematical basis of quantum mechanics developing the theory of Hilbert spaces. He moved in 1930 to Princeton, where he became professor in 1933 at the Institute of Advanced Study, together with a group of famous scientists including Einstein. To start with his interest was directed towards different areas of pure mathematics including number theory, algebra, geometry, topology and functional analysis, but gradually through his engagement in the Manhattan project and the further development of the ENIAC during and after World War II, his interest shifted towards applications. Von Neumann laid the theoretical basis for the design of the first generation of programmable computers (EDVAC), including speculative work on self-reproducing computers. He developed basic numerical methods for problems in hydrodynamics and meteorology, and laid the foundation of game theory and mathematical economics. Von Neumann's personality was direct and open, he liked people, jokes, puns, bridge and poker, food and wine. In addition to science he was knowledgeable in literature and history. Von Neumann died in 1957. Wiener and von Neumann had largely parallel careers starting with the foundations of mathematics, continuing with applications of functional and harmonic analysis to quantum mechanics and information theory, further working on electronic computers, automata, and systems science.

> One evening I was sitting in the rooms of the Analytical Society at Cambridge, my head leaning forward on the table in a kind of dreamy mood, with a Table of logarithms lying open before me. Another member, coming into the room, and seeing me half asleep, called out: Well, Babbage, what are you dreaming about? to which I replied "I am thinking that all these Tables (pointing at the logarithms) might be calculated by machinery". (Babbage)

> Today we know that the program-controlled computer began during the last century with Babbage. But he was so far ahead his time that his machine was nearly completely forgotten. So in Germany when I started in 1934 nobody knew of his work. I was a student then in civil engineering in Berlin. Berlin is a nice town and there were many opportunities for a student to spend his time in an agreeable manner, for instance with nice girls. But instead of that we had to perform big and awful calculations. Also later as an engineer in the aircraft industry, I became aware of the tremendous number of monotonous calculations necessary for the design of static and aerodynamic structures. Therefore I decided to design and construct calculating machines suited to solving these problems automatically. (Konrad Zuse)

3

A Review of Calculus

The two questions, the first that of finding the description of the
curve from its elements, the second that of finding the figure from
the given differences, both reduce to the same thing. From this it
can be taken that the whole of the theory of the inverse method
of the tangents is reducible to quadratures. (Leibniz 1673)

Mathematics is the science of quantity. (Aristotle)

This chapter serves two purposes. First, it is a review of calculus of one
variable recalling the basic tools we need in order to study differential
equations. Calculus is essentially the study of differential and integral
equations and their solutions, and is important because the mathematical models in science usually take the form of differential and integral
equations. It turns out that it is generally impossible to solve these
equations analytically by finding a symbolic expression for the solution,
say in terms of elementary functions. Instead, we usually have to be
satisfied with approximate, or *numerical*, solutions. The second purpose
of this chapter is to introduce the important issues that arise when computing approximate solutions of differential and integral equations. We
do this by giving a detailed proof of the Fundamental Theorem of Calculus which states that a function is the derivative of its integral and that
the integral of a function on an interval gives the area underneath the
graph of the function on that interval. The proof of the Fundamental
Theorem shows the existence of the integral as the solution of a simple
differential equation and also gives a method for computing approximate
solutions of the differential equation.

3.1. Real numbers. Sequences of real numbers

Recall the way real numbers are classified: there are the *natural numbers* $i = 1, 2, ..$, the *integers* $p = 0, \pm 1, \pm 2, ...$, the set of *rational numbers* $r = p/q$, where p and $q \neq 0$ are integers, and finally the set of *irrational numbers* which are all the real numbers that are not rational. For example, $\sqrt{2}$ and π are irrational numbers. We use $\mathbb{R}$ to denote the set of real numbers.

Suppose for each natural number $i = 1, 2, 3, ...,$ a corresponding real number ξ_i is specified. Then we say that $\{\xi_i\} = \xi_1, \xi_2, \xi_3, ..$ is an *infinite sequence* of real numbers, or a *sequence* of real numbers for short. For example $1, 2^{-2}, 3^{-2}, 4^{-2}, ..$ is a sequence of real numbers ξ_i with $\xi_i = i^{-2}$. We can not actually reach the "end" of an infinite sequence, but as a practical alternative, we can try to determine if the numbers ξ_i in the sequence approach some particular value or limit as the index i increases. If this is the case, we may "extrapolate" to infinity without actually going there. To make this more precise, we say that a sequence $\{\xi_i\}$ *converges* to the *limit* $\xi \in \mathbb{R}$ if for any $\epsilon > 0$ there is an integer N_ϵ such that

$$|\xi - \xi_i| < \epsilon \quad \text{for all } i > N_\epsilon, \tag{3.1}$$

which we write in compact form as $\xi = \lim_{i \to \infty} \xi_i$. This definition says that ξ_i can be made "arbitrarily" close to ξ by taking i sufficiently large. For example,

$$\lim_{i \to \infty} \frac{1}{i} = 0,$$

since for any $\epsilon > 0$, $|0 - 1/i| < \epsilon$ for all $i > 1/\epsilon$. Similarly,

$$\lim_{i \to \infty} \frac{i}{i + 1} = 1.$$

However, $\lim_{i \to \infty} (-1)^i$ is undefined because $\xi_i = (-1)^i$ the sequence never gets close to a single number as i increases.

This definition of convergence is purely qualitative since it does not give any quantitative measure of the *rate* at which the numbers ξ_i in a convergent sequence approaches the limit. For example, the sequences $\{1/i\}$, $\{1/i^3\}$, and $\{1/2^i\}$ all converge to 0; however the rates are vastly different. We plot a few terms of the sequences in Fig. 3.1. Quantitative

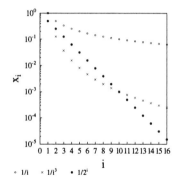

Figure 3.1: Three sequences that converge to zero plotted on a log scale.

measures of convergence play an important role in practical computation, and this is a central theme of this book.

In order to check if a sequence converges using the definition, we have to know a candidate for the limit. In the simple examples above, the limit is easy to find, but in general this is not true. For example, if the limit is a real number with an infinite, non-repeating decimal expansion, such as π, then we can't write down the limit explicitly. In such cases, it is natural to ask if there is a way to tell whether a sequence $\{\xi_i\}$ in $\mathbb{R}$ *has* a limit without actually producing the limit first. To answer this, we note that if a sequence $\{\xi_i\}$ actually has a limit, then the ξ_i must become close to each other for large indices since they get closer to the same value, i.e. the limit. It turns out that the converse is true for a sequence of real numbers. To make this precise, we define a sequence $\{\xi_i\}$ to be a *Cauchy sequence* if for any $\epsilon > 0$ there is an integer N_ϵ such that

$$|\xi_j - \xi_i| < \epsilon \quad \text{for all } i, j > N_\epsilon.$$

The set of real numbers $\mathbb{R}$ has the property that a Cauchy sequence $\{\xi_i\}$ of real numbers does converge to some real number ξ. This property is often expressed by saying that $\mathbb{R}$ is *complete*.

Problem 3.1. Show that a sequence that converges is a Cauchy sequence. Hint: write $\xi_i - \xi_j = (\xi_i - \xi) + (\xi - \xi_j)$ and use the triangle inequality.

Problem 3.2. Show that a sequence that converges cannot have two limits.

Verifying that a sequence of real numbers is a Cauchy sequence does not require any *a priori* (beforehand) knowledge of the limit, but requires only specific information about the sequence itself. This makes the verification of convergence a concrete matter. Moreover, though we do not know the limit, we have the means to determine approximations of it to any desired accuracy. We encounter this issue many more times in the course of this book.

3.1.1.   What is the square root of two?

We illustrate some of the issues related to the nature of the set of real numbers by examining the familiar real number $\sqrt{2}$. We learn in school two things about $\sqrt{2}$. First, it is the positive root of the equation $\xi^2 = 2$. Second, $\sqrt{2}$ is irrational with an infinitely long decimal expansion that never forms a regular pattern:

$$\sqrt{2} = 1.41421356237309504880168872420969807856967187537694...$$

Early in mathematical history, the Pythagoreans introduced $\sqrt{2}$ as the length of the diagonal in a square of side one and also discovered that $\sqrt{2}$ is not a rational number. Their proof is easy to understand: we assume that $\sqrt{2} = p/q$ with all common factors in p and q divided out. By squaring both sides, we get $2q^2 = p^2$. It follows that p must contain the factor 2, so that p^2 contains two factors of 2. But this means that q must also have a factor 2, which is a contradiction. The discovery was shocking to the Pythagoreans and had to be kept secret since the Pythagorean school was based on the principle that everything could be described in terms of natural numbers. It took more than 2000 years to resolve this dilemma.

The story continues in the late 19th century, when Cantor and Dedekind explained that $\sqrt{2}$ should be viewed as a limit of a sequence $\{\xi_i\}$ of rational numbers. Such a sequence is generated from the algorithm

$$\xi_{i+1} = \frac{\xi_i}{2} + \frac{1}{\xi_i}, \quad i = 1, 2, ..., \tag{3.2}$$

starting with $\xi_1 = 1$. We obtain $\xi_2 = 1.5$, $\xi_3 = 1.4167$, $\xi_4 = 1.4142157$, $\xi_5 = 1.41421356237469$, $\xi_6 = 1.414213562373095$, ... With a little work, we can show that this is a Cauchy sequence and therefore converges to a limit ξ. The limit must be $\xi = \sqrt{2}$ since replacing both ξ_{i+1} and ξ_i by ξ, we obtain the equation $\xi^2 = 2$ with the positive root $\xi = \sqrt{2}$. In fact, the sequence (3.2) is generated by *Newton's method* for solving the equation $\xi^2 = 2$ and variants are used in pocket calculators to compute square roots.

The resolution is that we use $\sqrt{2}$ in two different ways; as a symbol representing the positive root of the equation $\xi^2 = 2$, and as a limit of a sequence of numbers that can be computed to any desired accuracy. The symbol $\sqrt{2}$ may be used in symbolic calculations like $(\sqrt{2})^6 = 8$ without having to know its decimal expansion, whereas the constructive algorithm is used whenever we need the actual value to some precision, $\sqrt{2} = 1.4142135....$ To bring the two meanings together, one could try to identify the real number $\sqrt{2}$ with the algorithm (3.2) that gives approximations of $\sqrt{2}$. A mathematician from the constructivist school would like to do this. However, if we had to specify for each real number a convergent Cauchy sequence defining the number, then a discussion involving real numbers would get bogged down quickly. For this reason, it is often convenient to use a symbolic meaning of real numbers rather than a constructive meaning. The same dualism of symbolic and constructive aspects is useful for discussing solutions of differential equations.

Problem 3.3. Compute 8 iterations of (3.2) and compare the results to the value of $\sqrt{2}$ given by your calculator.

Problem 3.4. *(Pretty hard.)* Prove that (3.2) generates a Cauchy sequence. Hint: see Problem 3.14.

Problem 3.5. Is 0.9999.... equal to 1?

3.2. Functions

Ultimately, calculus is the study of functions. We say that $u = u(x)$ is a *function* of x defined on $\mathbb{R}$ if for each $x \in \mathbb{R}$ there is a number $u = u(x) \in \mathbb{R}$. We call x the *independent variable* and u the *dependent variable*. Functions may also be defined on subsets of $\mathbb{R}$ such as an *open* interval $(a, b) = \{x \in \mathbb{R} : a < x < b\}$ or *closed* interval $[a, b] = \{x \in \mathbb{R} : a \leq x \leq b\}$, where $a, b \in \mathbb{R}$. The set of points x for which a function $u(x)$ is defined is the *domain of definition* of u. The points $(x, u(x))$

for $a \le x \le b$ plotted in a (x, u) coordinate system is the *graph* of the function u on the interval $[a, b]$. We use x, y and z to denote independent variables, and typically u, v, w, f and g to denote functions. The concept of a function was introduced by Leibniz.

3.2.1. Limits and continuity

We say that $\bar{u}$ is the *limit* of $u(x)$ as x approaches $\bar{x}$, denoted

$$\bar{u} = \lim_{x \to \bar{x}} u(x),$$

if for any $\epsilon > 0$ there is a $\delta > 0$ such that

$$|\bar{u} - u(x)| < \epsilon \quad \text{for all } x \text{ with } 0 < |\bar{x} - x| < \delta. \tag{3.3}$$

This definition says that $u(x)$ comes arbitrarily close to $\bar{u}$ as x approaches $\bar{x}$. We emphasize that in the definition, we require $0 < |\bar{x} - x|$ and do not set $x = \bar{x}$. Thus, the limit tells about the behavior of $u(x)$ close to $\bar{x}$, but does not involve $u(\bar{x})$. For example, consider the three functions in Fig. 3.2. We clearly have $\lim_{x \to \bar{x}} u_1(x) = 1$, but also $\lim_{x \to \bar{x}} u_2(x) = 1$

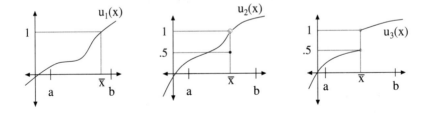

Figure 3.2: One continuous function and two discontinuous functions.

although $1 \neq u_2(\bar{x})$. In fact, the limit is independent of the value of u at $\bar{x}$, which needs not even be defined. Finally, $\lim_{x \to \bar{x}} u_3(x)$ does not exist, because u gets close to one value on the left of $\bar{x}$ and another value on the right. In this case we could speak of a left and right limit of $u_3(x)$ as x approaches $\bar{x}$ from left or right, but these limits are unequal so the limit itself is undefined.

Problem 3.6. (a) Compute, if possible, the limits: $\lim_{x \to 0} \sin(x)$ and $\lim_{x \to 0} \sin(1/x)$. (b) Write down a definition for $\lim_{x \to \bar{x}} u(x) = \infty$. (c) Determine $\lim_{x \to 0} 1/x^2$.

We distinguish the special class of functions $u(x)$ for which the limit $\lim_{x \to \bar{x}} u(x)$ can be evaluated by substituting $x = \bar{x}$. A function $u(x)$ is *continuous at a point* $\bar{x}$ if $\lim_{x \to \bar{x}} u(x) = u(\bar{x})$, or in other words if for any $\epsilon > 0$ there is a $\delta_\epsilon > 0$ such that

$$|u(x) - u(\bar{x})| < \epsilon \quad \text{if } |x - \bar{x}| < \delta_\epsilon.$$

For this to hold, u must be defined at $\bar{x}$, the limit must exist, and the limit must equal the value of u at $\bar{x}$. In the example above, u_1 is continuous at $\bar{x}$, while u_2 and u_3 are not. Notice that we may make u_2 continuous at $\bar{x}$ by changing the single function value $u_2(\bar{x})$ to be equal to 1, while this would not work for u_3.

We say that a function is *continuous in an interval* if it is continuous at every point in the interval. The function u_1 is continuous in $[a, b]$, while u_2 and u_3 are continuous at all points $x \neq \bar{x}$. We say that $u(x)$ is *uniformly continuous* on an interval $[a, b]$ if for any $\epsilon > 0$ there is a $\delta_\epsilon > 0$ such that for all $a \leq \bar{x} \leq b$,

$$|u(x) - u(\bar{x})| < \epsilon \quad \text{for all } a \leq x \leq b \text{ with } |x - \bar{x}| < \delta_\epsilon.$$

Comparing the definitions of continuity and uniform continuity, we see that uniform continuity means that the same δ_ϵ in the definition can be chosen simultaneously for every point in the interval. We use the concept of uniform continuity in the proof of the Fundamental Theorem below. It is possible to prove that a function that is continuous on a closed bounded interval is uniformly continuous on that interval.

Problem 3.7. Show that $f(x) = 1/x$ is continuous but not uniformly continuous on $(0, 1)$.

We recall some properties of continuous functions. A continuous function on a closed, bounded interval has both a maximum and a minimum value and takes on every value between its maximum and minimum value at least once in the interval. If we add two continuous functions $u(x)$ and $v(x)$ defined on a common domain, where their sum $u + v$ is defined by $(u + v)(x) = u(x) + v(x)$, then we get another continuous function. The product of two continuous functions is also continuous, and so is the ratio of two continuous functions $u(x)/v(x)$ at all points in the common domain where the denominator $v(x)$ is nonzero.

If $a < c < b$ and $u(x)$ is continuous separately on (a, c) and (c, b), then we say that $u(x)$ is *piecewise continuous* on (a, b). Such a function may have a jump at $x = c$. The function u_3 of Fig. 3.2 is piecewise continuous on (a, b).

3.2.2. Sequences of continuous functions

The solution of a differential or integral equation is a function and the main subject of this book is producing approximations of solutions of such equations. The approximations can be written as a sequence of functions where the index is related to the amount of work spent computing the approximation. The main mathematical issue is to show that an approximating sequence converges to the solution as the work increases, in other words, we get more accuracy with increased effort.

We start by discussing the convergence of sequences of continuous functions. A sequence of continuous functions $\{f_i\}$ *converges* to a function f at a point x if for every $\epsilon > 0$ there is an integer $N_\epsilon > 0$ such that

$$|f_i(x) - f(x)| < \epsilon \quad \text{for all } i \geq N_\epsilon.$$

$\{f_i\}$ converges to f on an interval I if $\{f_i\}$ converges to f at every point x in I. Note that the values of f_i at two different points may converge at different rates to the values of f at the points, i.e. we may have to choose different N_ϵ for different points in the interval. In contrast, $\{f_i\}$ *converges uniformly* to f on I if for every $\epsilon > 0$ there is an integer N_ϵ such that $i \geq N_\epsilon$ implies

$$|f_i(x) - f(x)| < \epsilon \quad \text{for all } x \text{ in } I.$$

Similarly, we define the sequence $\{f_i\}$ to be a *uniform Cauchy sequence* on the interval I if for every $\epsilon > 0$ there is an integer N_ϵ such that $i, j \geq N_\epsilon$ implies

$$|f_i(x) - f_j(x)| < \epsilon \quad \text{for all } x \text{ in } I.$$

A uniform Cauchy sequence of continuous functions $\{f_i\}$ on a closed, bounded interval I converges to a continuous function $f(x)$ on I. This follows from the convergence of the sequence of real numbers $\{f_i(x)\}$ for each x in I. For example, the sequence $\{x^i\}$ on $[0, 1/2]$ is a Cauchy sequence, since if $j \geq i$, then

$$|x^i - x^j| \leq x^i |1 - x^{j-i}| \leq 2^{-i} \cdot 1$$

for all $0 \leq x \leq 1/2$. Thus $\{x^i\}$ converges to a continuous function on $[0, 1/2]$, namely the (rather boring) function $f \equiv 0$.

Problem 3.8. Show that for any $0 < a < 1$ the sequence $\{x^i\}$ converges uniformly to 0 on $[0, a]$, but not for $a = 1$.

Problem 3.9. Let $\{f_i\}$ be a uniform Cauchy sequence consisting of uniformly continuous functions on a closed, bounded interval I. Show that the limit is uniformly continuous. Hint: use the definition of continuity and the triangle inequality as in Problem 3.1.

3.2.3. The derivative

The derivative is used to measure how much a function $u(x)$ varies for x close to a given point $\bar{x}$. The change in the function u from x to $\bar{x}$ is $u(\bar{x}) - u(x)$ which tends to zero as x approaches $\bar{x}$ if we assume that u is continuous. The average rate of change,

$$\frac{u(\bar{x}) - u(x)}{\bar{x} - x},$$

is defined for $x \neq \bar{x}$. If $u(x)$ represents the position of a car at time x then $u(x)$ has the units of length and x the units of time and the units of the average rate of change is distance/time, or speed. Recall that the average rate of change of u between x and $\bar{x}$ is the slope of the secant line to u passing through the points $(x, u(x))$ and $(\bar{x}, u(\bar{x}))$, see Fig. 3.4.

The average rate of change of a function indicates something about how the function changes with input, but it is not very precise. For a given x and $\bar{x}$, there are many functions with the same average rate of change, as we demonstrate in Fig. 3.3. The way to make the average rate more precise is to move x closer to $\bar{x}$. Of course, we can't take $x = \bar{x}$, and now we see why the concept of the limit is so useful in calculus. Using it, we can discuss the behavior of the average rate of change of a function over increasingly smaller intervals, or as $x \to \bar{x}$.

The *derivative* of u at $\bar{x}$, denoted $u'(\bar{x})$, is defined as the limit

$$u'(\bar{x}) = \lim_{x \to \bar{x}} \frac{u(\bar{x}) - u(x)}{\bar{x} - x} = \lim_{\Delta x \to 0} \frac{u(\bar{x}) - u(\bar{x} - \Delta x)}{\Delta x}, \qquad (3.4)$$

provided the limit exists. If the limit $u'(\bar{x})$ exists, then $u(x)$ is said to be *differentiable* at $\bar{x}$. Note that in (3.4), Δx tends to 0 from both directions, so that replacing Δx by $-\Delta x$ in the quotient we have

$$u'(\bar{x}) = \lim_{\Delta x \to 0} \frac{u(\bar{x} + \Delta x) - u(\bar{x})}{\Delta x},$$

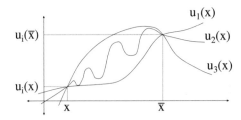

Figure 3.3: Three functions with the same average rate of change between x and $\bar{x}$.

which is equivalent to (3.4). Restricting Δx to be positive or negative
defines a *one-sided derivative*. If $u(x)$ is differentiable at x for all $x \in$
(a, b), then we say that $u(x)$ is differentiable on (a, b).

In Fig. 3.4, we plot secant lines for four points x_i that are successively
closer to $\bar{x}$. We recall, as confirmed in this plot, that the derivative of u
at $\bar{x}$ is the slope of the tangent line to u at $\bar{x}$. Note that in the definition
(3.4) we never actually substitute $x = \bar{x}$ since the denominator would
be zero.

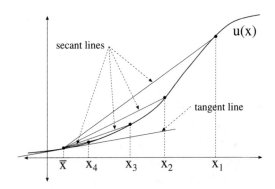

Figure 3.4: A sequence of secant lines approaching the tangent line at
$\bar{x}$.

The *derivative* $u'(x)$ of $u(x)$ is the function that takes the value of
the derivative of u at each point x. We also use the notation $Du(x) =$
$u'(x)$. Similarly, we write the nth order derivative as $D^n u(x)$, which is
computed recursively as the derivative of $D^{n-1}u$. The notation $\frac{du}{dx} = u'$,

$\frac{d^2u}{dx^2} = u''$ etc., and $f^{(q)} = d^q f / dx^q$ is also used.

The derivative of a constant function is zero while the derivative of a linear function $u(x) = ax + b$, where $a, b \in \mathbb{R}$, is the constant function $u'(x) = a$, or in other words, the slope of the line $u = ax + b$ is a. Further, the derivative of a polynomial $u(x) = x^m$, with $m \neq 0$, is $Dx^m = mx^{m-1}$, where we assume that $x \neq 0$ if $m < 0$. We recall the derivatives of some other elementary functions:

$$D \exp(x) = \exp(x), \quad D \sin(x) = \cos(x), \quad D \log(x) = x^{-1}.$$

We recall some of the properties of the derivative (for proofs, see Fig. 1.3):

$$(u + v)' = u' + v'$$
$$(uv)' = u'v + uv'$$
$$\left(\frac{u}{v}\right)' = \frac{u'v - uv'}{v^2} \quad (v \neq 0).$$

Another important rule of differentiation is the *chain rule* for differentiation of a function $w(x) = u(v(x))$ composed of two functions u and v that reads:

$$w'(x) = u'(v(x)) v'(x). \tag{3.5}$$

Using these formulas, it is possible to compute derivatives of functions that are made up of products, sums, quotients and compositions of the elementary functions such as polynomials, the exponential $\exp(x)$, and the trigonometric functions $\sin(x)$ and $\cos(x)$.

Problem 3.10. Compute the derivative of $1 + 2x^3 - \exp(x) \sin(5x + 1)$.

3.2.4. The mean value theorem and Taylor's theorem

We recall the *mean value theorem*.

Theorem 3.1. *Suppose that u is continuous on $[a, b]$ and differentiable on (a, b). Then there is a point ξ in (a, b) such that*

$$u(b) - u(a) = u'(\xi)(b - a).$$

Problem 3.11. Prove the mean value theorem by first reducing to the case $u(a) = u(b) = 0$ and then using the fact that $u(x)$ must take on a maximum or minimum value for some point $\bar{x}$ in (a, b). The case $u(a) = u(b) = 0$ is also referred to as *Rolle's theorem*.

Problem 3.12. Prove the chain rule (3.5) assuming u' is continuous at $v(x)$ and v is differentiable at x. Hint: use the mean value theorem.

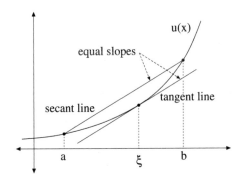

Figure 3.5: Illustration of the mean value theorem.

We also recall Taylor's theorem, which states that knowledge of the derivatives of a function $u(x)$ at a point $\bar{x}$ allows the function $u(x)$ to be approximated by a polynomial in an interval around $\bar{x}$. The mean value theorem is a particular case of Taylor's theorem with polynomial approximation of degree zero.

Theorem 3.2. *Assume that u has continuous derivatives of order $q+1$ in an interval (a,b) containing x_0. Then, for all $a < x < b$,*

$$u(x) = u(x_0) + u'(x_0)(x - x_0) + \frac{1}{2}u''(x_0)(x - x_0)^2 + \cdots$$
$$+ \frac{1}{q!}D^q u(x_0)(x - x_0)^q + \frac{1}{(q+1)!}D^{q+1}u(\xi)(x - x_0)^{q+1},$$

for some ξ between x and x_0.

The polynomial

$$u(x_0) + u'(x_0)(x - x_0) + \frac{1}{2}u''(x_0)(x - x_0)^2 + \cdots + \frac{1}{q!}D^q u(x_0)(x - x_0)^q$$

in x, is called the qth degree *Taylor polynomial* of the function $u(x)$ at x_0. If we use the Taylor polynomial to approximate the values of u, then the error is

$$\frac{1}{(q+1)!}D^{q+1}u(\xi)(x - x_0)^{q+1},$$

which is called the *remainder*. For example, the linear approximation to $u(x)$ is $u(x_0) + (x - x_0)u'(x_0)$ with the following estimate of the error:

$$|u(x) - u(x_0) - (x - x_0)u'(x_0)| \leq \frac{1}{2}(x - x_0)^2 \max_{a \leq z \leq b} |u''(z)| \tag{3.6}$$

for $a \leq x \leq b$. In Fig. 3.6, we plot the constant, linear and quadratic approximations to a function u. The size of the remainder depends on

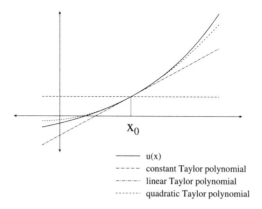

$$X_0$$

——— u(x)
- - - - constant Taylor polynomial
-·-·- linear Taylor polynomial
········ quadratic Taylor polynomial

Figure 3.6: Constant, linear, and quadratic Taylor approximations.

how close x is to x_0, the order of the Taylor polynomial, and on the size of $D^{q+1}u$ on (a, b).

Problem 3.13.   Plot the constant, linear, quadratic, and cubic Taylor polynomials for $\cos(x)$ computed at $x_0 = 0$ over the interval $[a, b] = [-\pi/2, \pi/2]$. In each case, compute a bound on the remainder.

Problem 3.14. *(Fixed point iteration.)* Let $u(x)$ be a differentiable function defined on $\mathbb{R}$. Consider the sequence $\{\xi_j\}$ generated by $\xi_j = u(\xi_{j-1})$, where x_0 is given. Suppose there is a constant $\theta < 1$ such that $|u'(x)| \leq \theta$ for $x \in \mathbb{R}$. Prove that $\{\xi_j\}$ is a Cauchy sequence converging to $\xi \in \mathbb{R}$ satisfying $\xi = u(\xi)$, where we refer to ξ as a *fixed point* of u. Draw a picture to illustrate. Apply the result to the function $u(x) = \frac{x}{2} + \frac{1}{x}$ related to the computation of $\sqrt{2}$ according to (3.2). Note that it suffices that the condition $|u'(x)| \leq \theta$ is satisfied close to the fixed point.

Problem 3.15. *(Zero of a function.)* Let $u(x)$ be a continuous function defined on $\mathbb{R}$. Consider the sequence $\{\xi_j\}$ generated by $\xi_j = \xi_{j-1} - \alpha u(\xi_{j-1})$, where ξ_0 is given and α is a constant. Find conditions that guarantee that $\{\xi_j\}$ converges to $\xi \in \mathbb{R}$ satisfying $u(\xi) = 0$.

Problem 3.16. *(Intermediate value theorem.)* Prove that if u is a continuous function on $[0, 1]$ such that $u(0) < 0$ and $u(1) > 0$, then there is point $0 < \xi < 1$ such that $u(\xi) = 0$. Hint: define a sequence of intervals I_j in $[0, 1]$ of length 2^{-j} such that $I_{j+1} \subset I_j$ and $u(x)$ changes sign in I_j. Choose a point ξ_j from each interval I_j and prove that $\{\xi_j\}$ is a Cauchy sequence converging to ξ satisfying $u(\xi) = 0$. Generalize to prove that a continuous function on an interval $[a, b]$ assumes all values between its maximum and minimum values.

3.3. Differential equations

A *differential equation* is an equation that relates derivatives of one or more functions over a specified domain of definition. Very often, the mathematical description of physical phenomena result in a differential equation, so that finding a *solution* of a differential equation, which is a function (or set of functions) that satisfies the differential equation at every point in the domain, is a centrally important problem in science. For example, when there is just one independent variable, the differential equation is called *ordinary*. The general form of a *first order* ordinary differential equation for one unknown function $u(x)$ is

$$\Psi(u'(x), u(x), x) = 0 \quad \text{for } a < x < b, \tag{3.7}$$

where $\Psi(y_1, y_2, y_3)$ is a given real-valued (continuous) function of the real variables y_1, y_2 and y_3, and (a, b) is a given interval. The general problem is to find all functions $u(x)$ that satisfy this equation at every point x in (a, b). To determine solutions uniquely a specification of the value of u at some point in (a, b) is generally required.

If we can solve for $u'(x)$ in terms of $u(x)$ and x in (3.7), then we obtain a problem of the form

$$\begin{cases} u'(x) = f(u(x), x) & \text{for } a < x \le b, \\ u(a) = u_0, \end{cases} \tag{3.8}$$

where $f(y_1, y_2)$ is a given function of the two variables y_1 and y_2, and we have specified the value $u(a) = u_0$. This problem is called an *initial value problem* and we may think of x as representing time so that $u(a)$

is the value of $u(x)$ given at the initial time $x = a$. We require the function $u(x)$ to be continuous on $[a, b]$ and satisfy the differential equation $u'(x) = f(u(x), x)$ for all x in $(a, b]$. We could also require the differential equation to be satisfied on (a, b) or $[a, b]$ with $u'(x)$ defined as a one-sided derivative at the end-points. The situation is a little different in the case of approximate numerical solution for which the specification $(a, b]$ is more natural. We may think of the initial value $u(a) = u_0$ as "replacing" the differential equation at $x = a$.

The general form of a *second order* ordinary differential equation for one unknown function $u(x)$ is

$$\Psi(u''(x), u'(x), u(x), x) = 0, \quad a < x < b, \tag{3.9}$$

where $\Psi(y_1, y_2, y_3, y_4)$ is a given function of $y_1, .., y_4$. One possibility of determining a unique solution is to specify both the values $u(a)$ and $u(b)$, in which case (3.9) is called a *two-point boundary value problem* where x usually represents a spatial coordinate. Alternatively, we can specify $u(a)$ and $u'(a)$ and obtain an initial value problem.

Much of mathematics is devoted to the study of differential equations, concentrating on such questions as whether a given differential equation has solutions (*existence*), whether or not there is at most one solution (*uniqueness*), properties of solutions such as regularity and stability to perturbations, and numerical computation of solutions.

3.3.1. The simplest differential equation

We begin by considering the simplest differential equation: given the uniformly continuous function $f(x)$ on $(a, b]$, find a continuous function $u(x)$ on $[a, b]$ such that

$$u'(x) = f(x) \quad \text{for } a < x \leq b. \tag{3.10}$$

Solving this equation, i.e. finding what is also called a *primitive function* $u(x)$ of $f(x)$, is a basic problem of calculus. This problem was called the "inverse method of tangents" by Leibniz, who thought of this problem in terms of finding a continuous curve given the slope of its tangent at every point.

We note that a primitive function is determined only up to a constant, because the derivative of a constant is zero; if $u'(x) = f(x)$, then also $(u(x) + c)' = f(x)$ for any constant c. For example, both $u(x) = x^2$

and $u(x) = x^2 + c$ satisfy $u'(x) = 2x$. The constant may be specified by specifying the value of the primitive function $u(x)$ at some point. Specification at the left-hand end point leads to the following *initial value problem*: given the function $f(x)$ on $(a, b]$ and the initial value u_0 at $x = a$, find a continuous function $u(x)$ such that

$$\begin{cases} u'(x) = f(x), & a < x \le b, \\ u(a) = u_0. \end{cases} \tag{3.11}$$

The Fundamental Theorem of Calculus states that this problem has a unique solution if $f(x)$ is continuous (or more generally piecewise continuous).

An example of a physical situation modeled in this way is a cyclist biking along a straight line with $u(x)$ representing the position at time x so that $u'(x)$ is the speed at time x. Supposing that the bike is equipped with a simple speedometer reporting the speed $f(x)$ of the bike at each time x, the problem is to determine the position $u(x)$ of the cyclist at time x after specifying a starting position u_0 at some time $x = a$. This problem is actually solved automatically by most speedometers which keep track of the total traveled distance by using an analog "integrator" that accumulates distance in increments based on a mechanical system. In fact, the mathematical method we use to generate an approximate solution is also based on a principle of accumulation of increments.

The process of solving (3.11), or in other words, producing a function whose derivative is a specified function on an interval and that has a specified value at one point, is called *definite integration*. Anticipating the Fundamental Theorem, we suppose (3.11) can be solved and we use the notation

$$u(x) = \int_a^x f(y)\, dy \quad \text{for } a \le x \le b, \tag{3.12}$$

to denote the solution $u(x)$ in the special case that $u(a) = 0$. We refer to $\int_a^x f(y)\, dy$ as the *integral* of f over the interval $[a, x]$. The notation was introduced by Leibniz in 1675 who thought of the integral sign $\int$ as representing "summation" and dy as the "increment" in y. We refer to a and x as the lower and upper limits respectively and y is the integration variable. The proof of the Fundamental Theorem explains the reason for this language. Right now, this is just a symbolic way of denoting the solution $u(x)$ of (3.11) in the case $u_0 = 0$. The solution $u(x)$ of (3.11) with $u_0 \neq 0$, can then be expressed as

$$u(x) = \int_a^x f(y)\,dy + u_0 \quad \text{for } a \le x \le b.$$

Recall that it is indeed sometimes possible to produce a function $u(x)$ with a specified derivative $u'(x) = f(x)$ on an interval. For example, if $f(x) = 3x^2$ then $u(x) = x^3 + c$, if $f(x) = \sin(x)$ then $u(x) = -\cos(x) + c$, and if $f(x) = e^x$, then $u(x) = e^x + c$. In general, however, we cannot solve $u'(x) = f(x)$ symbolically by expressing $u(x)$ in terms of elementary functions, so even the large number of examples that are worked out during a typical calculus course do not prove that every continuous function $f(x)$ has a primitive function. This is the motivation for studying the Fundamental Theorem of Calculus in detail.

Problem 3.17. Show that if u is differentiable in (a, b) and continuous on $[a, b]$ and $u'(x) = 0$ for all $a < x < b$, then u is constant on $[a, b]$. Hint: use the mean value theorem.

Problem 3.18. Solve $u'(x) = \sin(x)$, $x > \pi/4$, $u(\pi/4) = 2/3$.

3.4.   The Fundamental Theorem of Calculus

Utile erit scribit $\int$ pro omnia. (Leibniz, October 29 1675)

We prove the Fundamental Theorem of Calculus by constructing a sequence of approximate solutions of (3.11). The construction is based on approximating the uniformly continuous function $f(x)$ by simple functions $F_N(x)$ that converge to $f(x)$ as N increases and for which primitive functions $U_N(x)$ can be determined easily. We show that $\{U_N\}$ form a Cauchy sequence and converges to a limit u, which we prove is the solution of (3.11). This proof is the analog of defining $\sqrt{2}$ by constructing a sequence that converges to $\sqrt{2}$. The proof also shows that the integral of a function between two points gives the area underneath the graph of the function between the points. This couples the problem of finding a primitive function, or computing an integral, to that of computing an area, that is to *quadrature*. For simplicity's sake, we assume that $f(x) \ge 0$ for all $a \le x \le b$. The proof directly extends to the more general case of a piecewise continuous function $f(x)$ of variable sign.

3.4.1. A simple case

According to the discussion above, it suffices to consider (3.11) with $u_0 = 0$. To motivate the computational method, we first consider the problem:

$$\begin{cases} U'(x) = F, & x_0 < x \leq x_1, \\ U(x_0) = 0. \end{cases} \qquad (3.13)$$

where F is a constant on $[x_0, x_1]$, and where in anticipation, we rename $a = x_0$ and choose $a < x_1 \leq b$. Since F is constant, the primitive function U is simply the linear function

$$U(x) = \int_{x_0}^{x} F \, dy = F \cdot (x - x_0).$$

We see that in this case $U(x)$ gives the area underneath the curve F between x_0 and x, see Fig. 3.7.

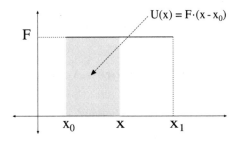

Figure 3.7: Integration of a constant function.

We choose F to be the maximum value of f on $[x_0, x_1]$, which exists because f is continuous. If the interval $[x_0, x_1]$ is small, then we expect $f(x)$ to be close to F for all $x_0 \leq x \leq x_1$, see Fig. 3.8, and therefore the solutions of $u' = f$ and $U' = F$ with $u(x_0) = U(x_0) = 0$ also to be close on $[x_0, x_1]$.

Problem 3.19. For $f(x) = e^x$, compare $\int_0^\delta f(x) \, dx$ and $\int_0^\delta F \, dx$ for small $\delta > 0$. (Hint: use Taylor's theorem.)

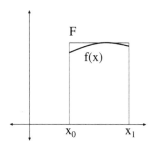

Figure 3.8: A constant approximation of a smooth function.

3.4.2. An approximate solution formula

In general, we have to deal with f over an interval $[a, b]$ that is not necessarily small and therefore we cannot expect f to be approximately constant over $[a, b]$. Instead, we use a piecewise constant approximation of f over $[a, b]$ based on dividing the interval up into sufficiently small pieces so that f is approximately constant on each. We choose an integer $N \geq 1$ and set $h = (b - a)/2^N$, then define $x_0 = a$ and $x_i = x_0 + i \cdot h$, for $i = 1, ..., 2^N$. In particular, $x_{2^N} = b$. The partition of $[a, b]$ induced by $x_0 < x_1 < \cdots < x_{2^N}$ is called a *mesh* and the x_i's are the *nodes*. We let $F_{N,i}$ denote the maximum value of f on $[x_{i-1}, x_i]$ and define the piecewise constant function $F_N(x) = F_{N,i}$ for $x_{i-1} < x \leq x_i$. We plot three examples in Fig. 3.9. The function $F_N(x)$ is piecewise constant on the partition $\{x_i\}$, and is our first example of a piecewise polynomial function.

Problem 3.20. Show that if $N \leq M$ are two integers, then the nodes corresponding to N are also nodes in the mesh corresponding to M. Meshes with this property are called *nested*.

Next, we compute the primitive function U_N corresponding to F_N. On the first interval $[x_0, x_1]$, we let U_N solve

$$\begin{cases} U_N'(x) = F_{N,1} & \text{for } x_0 < x \leq x_1, \\ U_N(x_0) = 0, \end{cases}$$

so that $U_N(x) = F_{N,1} \cdot (x - x_0)$ for $x_0 < x \leq x_1$. We repeat this process, interval by interval, solving

$$U_N'(x) = F_{N,i} \quad \text{for } x_{i-1} < x \leq x_i, \tag{3.14}$$

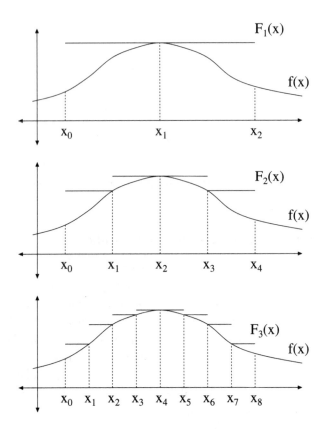

Figure 3.9: Three piecewise constant approximations to a smooth function.

with $U_N(x_{i-1})$ determined from the previous interval, to get

$$U_N(x) = U_N(x_{i-1}) + F_{N,i} \cdot (x - x_{i-1}) \quad \text{for } x_{i-1} < x \leq x_i.$$
(3.15)

Unwinding this formula back to the first interval, we obtain a formula for an approximate solution of the initial value problem (3.11):

$$U_N(x) = \sum_{j=1}^{i-1} F_{N,j} \, h + F_{N,i} \cdot (x - x_{i-1}) \quad \text{for } x_{i-1} < x \leq x_i.$$
(3.16)

By construction, $U_N(x)$ is the area underneath the graph of F_N from x_0 to x, written as the sum of the areas over each sub-interval. The next step is to prove that the sequence of functions $U_N(x)$ converges to a limit as N tends to infinity, which requires the uniform continuity of the function $f(x)$.

Problem 3.21. Prove (3.16).

3.4.3. Convergence

We shall prove that the sequence $\{U_N\}$ is a uniform Cauchy sequence on $[a, b]$ in the sense that for any $\epsilon > 0$ there is a N_ϵ such that

$$\max_{[a,b]} |U_N - U_M| < \epsilon \quad \text{for all } N, M > N_\epsilon.$$

We conclude that there is a limit $u(x) = \lim_{N \to \infty} U_N(x)$ that is also a continuous function on $[a, b]$. We then show that the construction of U_N implies that $u(x)$ solves (3.11) and the Fundamental Theorem follows.

To show that $\{U_N\}$ is a Cauchy sequence, we choose N and M with $N < M$ and let F_N and F_M denote the piecewise constant functions computed using the maximum values of f on intervals in the meshes with nodes $\{x_{N,i}\}_{i=0}^{2^N}$ and $\{x_{M,i}\}_{i=0}^{2^M}$ corresponding to N and M respectively. The values $U_N(x)$ and $U_M(x)$ represent the areas underneath the curves F_N and F_M between $x_{N,0} = x_{M,0} = a$ and x in $[a, b]$. We want to prove that these values are uniformly close for all x in $[a, b]$ if N and M are large.

To compare U_N and U_M, we introduce the piecewise constant function $\underline{F}_N$ defined using the minimum instead of the maximum value of $f(x)$ on each sub-interval and the primitive function $\underline{U}_N$ corresponding to $\underline{F}_N$.

Problem 3.22. Compute a formula for $\underline{U}_N$ analogous to (3.16).

The functions F_N, F_M and $\underline{F}_N$, as well as the areas given by U_N, U_M and $\underline{U}_N$ are displayed in Fig. 3.10. From the picture, it is clear that

$$\underline{F}_N(x) \leq F_M(x) \leq F_N(x) \quad \text{for } a \leq x \leq b,$$

which implies the analogous relation for the corresponding areas:

$$\underline{U}_N(x) \leq U_M(x) \leq U_N(x) \quad \text{for } a \leq x \leq b.$$

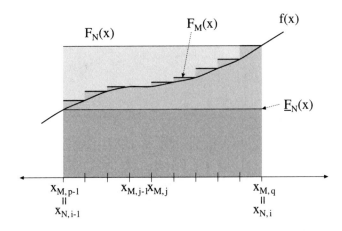

Figure 3.10: The area below F_M is between the areas below F_N and $\underline{F}_N$.

It follows that

$$\max_{[a,b]} |U_N - U_M| \leq \max_{[a,b]} |U_N - \underline{U}_N|. \qquad (3.17)$$

We estimate $\max_{[a,b]} |U_N - \underline{U}_N|$ in terms of $\max_{[a,b]} |F_N - \underline{F}_N|$ as follows: For $x_{N,i-1} \leq x \leq x_{N,i}$, we have with $h = (b-a)2^{-N}$,

$$0 \leq U_N(x) - \underline{U}_N(x) = \sum_{j=1}^{i-1} (F_{N,j} - \underline{F}_{N,j}) h + (x - x_{N,i-1})(F_{N,i} - \underline{F}_{N,i}),$$

which shows that

$$\max_{[a,b]} |U_N - \underline{U}_N| \leq (b-a) \max_{[a,b]} |F_N - \underline{F}_N|, \qquad (3.18)$$

because the sum of the lengths of the sub-intervals is less than or equal to $b - a$. To estimate $\max_{[a,b]} |F_N - \underline{F}_N|$, we use the assumption that $f(x)$ is uniformly continuous on $[a, b]$, i.e. for any $\epsilon > 0$ there is a $\delta_\epsilon > 0$ such that

$$|f(x) - f(y)| < \epsilon \quad \text{for all } a \leq x, y \leq b \text{ with } |x - y| < \delta_\epsilon.$$

This implies that for any $\epsilon > 0$ there is a N_ϵ such that,

$$\max_{[a,b]} |F_N - \underline{F}_N| < \epsilon \quad \text{if } N > N_\epsilon. \tag{3.19}$$

It follows from (3.18) that $\{U_N\}$ is a uniform Cauchy sequence of continuous functions on $[a, b]$ and therefore converges to a continuous function

$$u(x) = \lim_{N \to \infty} U_N(x) \quad \text{for } a \leq x \leq b. \tag{3.20}$$

Moreover,

$$\underline{U}_N(x) \leq u(x) \leq U_N(x) \quad \text{for all } a \leq x \leq b \tag{3.21}$$

and $\underline{U}_N(x)$ also converges to $u(x)$, see Fig. 3.11. We denote the limit $u(x)$ by

$$u(x) = \int_a^x f(y)\, dy,$$

where the motivation for this symbol comes from considering the limit of

$$U_N(x_i) = \sum_{j=1}^i F_{N,j} h$$

where $\int_a^x$ replaces the sum $\sum_j$, $f(y)$ replaces $F_{N,j}$ and dy replaces the stepsize h.

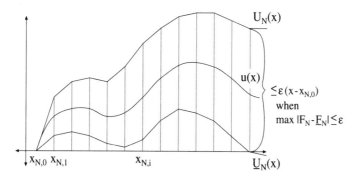

Figure 3.11: u is between U_N and $\underline{U}_N$.

Problem 3.23. Prove (3.18), (3.19), (3.21).

3.4.4.   The limit satisfies the differential equation

It remains to show that $u(x)$ satisfies (3.11). Obviously $u(a) = 0$ since $U_N(a) = 0$, so we have to show that $u(x)$ solves the differential equation $u'(x) = f(x)$ for $a < x \leq b$. We do this by using the definition of the derivative, so for $a < x \leq b$, let Δx denote a small positive number such that $a \leq x - \Delta x \leq b$. We assume that $x_{j-1} \leq x \leq x_j$ for some j, and that $x_{i-1} < x - \Delta x \leq x_i$ for some i, where we drop the sub-index N on the nodes to make the notation simpler, see Fig. 3.12. Using the formula

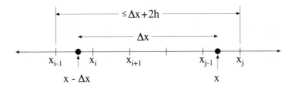

Figure 3.12: Locating x and $x - \Delta x$ in the mesh.

for $U_N(x)$, it follows that

$$|U_N(x) - U_N(x - \Delta x) - \Delta x f(x)| \leq \Delta x \max_{x_{i-1} \leq z, y \leq x_j} |f(y) - f(z)|.$$

Problem 3.24. Prove this.

This bound shows that U_N approximately satisfies the differential equation. To see this, note that by the uniform continuity of f, for any $\epsilon > 0$, there is a $\delta_\epsilon > 0$ such that

$$\max_{x_{i-1} \leq y, z \leq x_j} |f(y) - f(z)| < \epsilon, \qquad (3.22)$$

provided that Δx and h are so small that $\Delta x + 2h < \delta_\epsilon$; see Fig. 3.12. Dividing by Δx, we thus see that if Δx and h are small enough, then

$$\left| \frac{U_N(x) - U_N(x - \Delta x)}{\Delta x} - f(x) \right| < \epsilon.$$

Letting $N \to \infty$ and using the convergence of U_N to u, we conclude that

$$\left| \frac{u(x) - u(x - \Delta x)}{\Delta x} - f(x) \right| < \epsilon, \qquad (3.23)$$

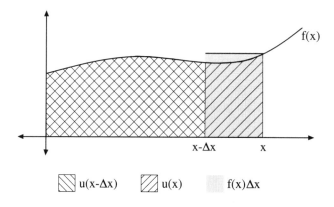

Figure 3.13: An illustration of (3.23). We see that that the difference
between the area under f from 0 to x, which is $u(x)$, and
the area under f from 0 to $x - \Delta x$, which is $u(x - \Delta x)$,
is approximately $f(x)\Delta x$.

if $\Delta x < \delta_\epsilon$. We illustrate this in Fig. 3.13. Using a very similar argument,
we see that (3.23) also holds for small $\Delta x < 0$. Finally, we let $\Delta x \to 0$
and conclude, using the definition of the derivative and the fact that ϵ is
arbitrarily small, that u is differentiable at x and $u'(x) = f(x)$. Finally, it
follows from Problem 3.17 that $u(x)$ is uniquely defined. This completes
the proof of the *Fundamental Theorem of Calculus*:

Theorem 3.3. *Suppose that f is continuous on the interval $[a, b]$. Then
there is a uniquely defined differentiable function $u(x) = \int_a^x f(y)\,dy$ on
$[a, b]$, which satisfies $u'(x) = f(x)$ for $a < x \leq b$, $u(a) = 0$. Further,
$u(x)$ gives the area below the graph of f from a to x.*

The theorem directly extends to the situation when f is piecewise
continuous and has variable sign. In this case, we have to interpret the
area with the proper sign.

Problem 3.25. Let $u(x)$ be continuous on $[a, b]$ and let $a = x_0 < x_1 <
\ldots < x_m = b$ be a partition of $[a, b]$. Prove that

$$\sum_{i=1}^{m} \frac{u(x_i) - u(x_{i-1})}{x_i - x_{i-1}}(x_i - x_{i-1}) = u(b) - u(a).$$

Study the relation between this formula and the Fundamental Theorem,
see Leibniz' *Art of Combinations* from 1666.

3.4.5. Properties of the integral

Properties of integration follow from the properties of differentiation and the proof of the Fundamental Theorem. We recall some of these now.

1. If f_1 and f_2 are piecewise continuous functions for $a \leq x \leq b$, x_0, x_1 are points in $[a, b]$ and c_1 and c_2 are constants, then

$$\int_{x_0}^{x_1} (c_1 f_1(x) + c_2 f_2(x))\, dx = c_1 \int_{x_0}^{x_1} f_1(x)\, dx + c_2 \int_{x_0}^{x_1} f_2(x)\, dx.$$

2. If u and v are two differentiable functions for $a \leq x \leq b$ and x_0 and x_1 are points in $[a, b]$, then

$$\int_{x_0}^{x_1} u'(x) v(x)\, dx = u(x_1) v(x_1) - u(x_0) v(x_0) - \int_{x_0}^{x_1} u(x) v'(x)\, dx.$$

3. If $f(x)$ is continuous for $a \leq x \leq b$ and x_0, x_1, and x_2 are points in $[a, b]$, then

$$\int_{x_0}^{x_2} f(x)\, dx = \int_{x_0}^{x_1} f(x)\, dx + \int_{x_1}^{x_2} f(x)\, dx.$$

4. If f_1 and f_2 are continuous functions and $f_1(x) \leq f_2(x)$ for all $a \leq x \leq b$ and x_0 and x_1 are points in $[a, b]$ with $x_0 < x_1$, then

$$\int_{x_0}^{x_1} f_1(x)\, dx \leq \int_{x_0}^{x_1} f_2(x)\, dx.$$

Problem 3.26. Prove #2. Hint: apply the Fundamental Theorem to the product uv.

Problem 3.27. (a) Prove #4. Hint: Problem 3.17. (b) Prove that $|\int_{x_0}^{x_1} f(x)\, dx| \leq \int_{x_0}^{x_1} |f(x)|\, dx$ if $x_0 < x_1$ by using the result from (a).

Problem 3.28. Prove Taylor's theorem. Hint: integrate successively by parts using the notation $k_n(x) = (x - z)^n / n!$ as follows:

$$u(z) = u(y) + \int_y^z Du(x) Dk_1(x)\, dx$$

$$= u(y) + Du(y)(z - y) - \int_y^z D^2 u(x) Dk_2(x)\, dx,$$

where in the last integral we used that $Dk_2 = k_1$. Continuing this way, we reach the conclusion using a variant of the mean value theorem.

Problem 3.29. Give an alternative proof of the fact that the limit $u(x)$ given by (3.20) satisfies $u'(x) = f(x)$ by starting from the fact that $u(x) - u(x - \Delta x) = \int_{x-\Delta x}^{x} f(y)\, dy$; see Fig. 3.13.

3.4.6. October 29, 1675

On October 29, 1675, Leibniz got a bright idea sitting at his desk in Paris. He writes: "Utile erit scribit $\int$ pro omnia" which translates to "It is useful to write $\int$ instead of omnia". This is the moment when the modern notation of calculus is created. Since 1673, Leibniz had been working with a notation based on a, l and "omnia" which represented in modern notation dx, dy and $\int$ respectively. He had been using formulas like

$$\text{omn.}l = y, \quad \text{omn.}yl = \frac{y^2}{2}, \quad \text{omn.}xl = x\text{omn.}l - \text{omn.omn.}la,$$

where "omn.", short for omnia, indicated a discrete sum and l and a denoted increments of finite size (often $a = 1$). In the new notation, these formulas became

$$\int dy = y, \quad \int y\, dy = \frac{y^2}{2}, \quad \int x\, dy = xy - \int y\, dx.$$

This opened up the possibility of dx and dy being arbitrarily small and the sum being replaced by the "integral".

Problem 3.30. Make sense out of the above formulas. Prove, as did Leibniz, the second from a geometrical argument based on computing the area of a right-angled triangle by summing thin slices of variable height y and thickness dy, and the third from computing similarly the area of a rectangle as the sum of the two parts below and above a curve joining two opposite corners of the rectangle.

3.4.7. Estimating the accuracy of the approximation

The proof of the Fundamental Theorem also gives a way to estimate the accuracy of the approximation computed on a particular mesh. By the construction of U_N, it follows that for $x_{m-1} < x \leq x_m$,

$$|u(x) - U_N(x)| \leq \sum_{j=1}^{m} \max_{x_{j-1} \leq y, z \leq x_j} |f(y) - f(z)|\, h,$$

$$(3.24)$$

where $h = (b - a)2^{-N}$. Using the estimate

$$\max_{x_{j-1} \le y, z \le x_j} |f(y) - f(z)| \le \max_{[x_{j-1}, x_j]} |f'| |x_j - x_{j-1}|, \tag{3.25}$$

which is a consequence of the mean value theorem, we obtain from (3.24)

$$|u(x) - U_N(x)| \le \sum_{j=1}^{m} \left(\max_{[x_{j-1}, x_j]} |f'| h \right) h. \tag{3.26}$$

In particular, (3.26) can be simplified to

$$|u(x) - U_N(x)| \le 2^{-N}(b - a) \max_{[a, x_m]} |f'|, \tag{3.27}$$

from which we conclude that any desired level of accuracy on $[a, b]$ can be reached by choosing N large enough to balance the size of $|f'|$ and the length of $[a, b]$.

Problem 3.31. Prove (3.24), (3.25), (3.26) and (3.27)

We illustrate by estimating the error in the approximation of $\log(x)$ using numerical integration. Recall that the natural logarithm $\log(x)$ solves the initial value problem

$$\begin{cases} u'(x) = x^{-1} & \text{for } x > 1 \text{ or } 0 < x < 1, \\ u(1) = 0, \end{cases}$$

or in other words

$$\log(x) = \int_1^x \frac{1}{y} \, dy. \tag{3.28}$$

Problem 3.32. Prove that $\log(e^x) = x$ for all x. Hint: use the chain rule to show that

$$\frac{d}{dx} \log(e^x) = 1,$$

and observe that $\log(e^0) = 0$.

We now apply the procedure presented above to compute an approximation of the integral $\log(x)$ given by (3.28) and bound the error using the estimate (3.26). Note that because $|f''(x)| = x^{-2}$ is decreasing,

$\max_{[x_{j-1}, x_j]} |f''| = x_{j-1}^{-2}$. Computing $\log(4) \approx 1.38629$ for a variety of h, we find that

h	Error Bound	True Error
.1	.0799	.0367
.01	.00755	.00374
.001	.000750	.000375
.0001	.0000750	.0000375

We see that (3.26) predicts the dependence of the error on h well in the sense that the ratio of the true error to the error bound is more or less constant as h decreases. Since the ratio is approximately .5, this means that (3.26) overestimates the error by a factor of 2 in this problem. In contrast, if we use the less accurate error bound (3.27), we obtain the estimates .3, .03, .003, and .0003 for $h = .1, .01, .001,$ and .0001 respectively, which is about 8 times too large for any given h.

Problem 3.33. Explain why (3.27) is less accurate than (3.26) in general.

Problem 3.34. Estimate the error using (3.26) and (3.27) for the following integrals: (a) $\int_0^2 2s \, ds$, (b) $\int_0^2 s^3 \, ds$, and (c) $\int_0^2 \exp(-s) \, ds$ using $h = .1$, .01, .001 and .0001. Discuss the results.

3.4.8. Adaptive quadrature

Once we can estimate the error of approximations of $u(x) = \int_a^x f(y) \, dy$, it is natural to pose the problem of computing an approximation of a specified accuracy. Using (3.26) it follows that if h is so small that

$$\sum_{j=1}^{m} \left(\max_{[x_{j-1}, x_j]} |f'| h \right) h \leq \text{TOL}, \qquad (3.29)$$

where TOL is a given *error tolerance*, then

$$|u(x) - U_N(x)| \leq \text{TOL}, \qquad (3.30)$$

if $x_{m-1} < x \leq x_m$. Since in general the amount of computer time, or *cost*, is a major concern, the practical problem is to compute an approximation of a given accuracy using the least amount of work, which in this case means the largest possible step size h, or equivalently, the smallest number of intervals.

However, the form of (3.29) carries an inherent inefficiency in terms of reaching a desired accuracy with minimum cost if $|f'|$ varies over the interval $[a, x_m]$. More precisely, we have to adjust h depending on the size of $|f'|$ to achieve (3.29), and in particular, the size of h in fact depends on the largest value of $|f'|$ in the entire interval $[a, x_m]$. This is the appropriate size of h for intervals $[x_{j-1}, x_j]$, $j \leq m$, where the largest value is achieved, but in intervals where $|f'|$ is significantly smaller, this value of h is unnecessarily small. Consequently, the number of intervals may be far from minimal.

A better procedure is to *adapt* the step size to compensate for the size of $|f'|$ in each sub-interval. Such a mesh is called an *adapted mesh*. We choose a non-uniform partition of $[a, x]$, $a = x_0 < x_1 < ... < x_m = x$, with sub-intervals $I_j = (x_{j-1}, x_j)$ of lengths $h_j = x_j - x_{j-1}$ and approximate $\int_a^x f(s)\, ds$ by the sum

$$U(x) = \sum_{j=1}^{m} f(x_j) h_j, \tag{3.31}$$

which is referred to as a *Riemann sum* or the *rectangle rule quadrature formula*. Using a straightforward modification of the proof above, it is possible to show that

$$\left| \int_a^{x_m} f(y)\, dy - \sum_{j=1}^{m} f(x_j) h_j \right| \leq \sum_{j=1}^{m} \left(\max_{[x_{j-1}, x_j]} |f'| h_j \right) h_j. \tag{3.32}$$

Problem 3.35. Prove (3.32).

By (3.32), we can guarantee that

$$\left| \int_a^{x_m} f(y)\, dy - \sum_{j=1}^{m} f(x_j)\, h_j \right| \leq \text{TOL}, \tag{3.33}$$

by choosing the step sizes h_j so that

$$\sum_{j=1}^{m} \left(\max_{[x_{j-1}, x_j]} |f'| h_j \right) h_j \leq \text{TOL}. \tag{3.34}$$

We refer to (3.34) as a *stopping criterion*; if the mesh sizes h_j satisfy this criterion, then the quadrature error is bounded by the tolerance TOL and we are satisfied.

Choosing the steps to achieve (3.34) is not straightforward because of the global aspect of the inequality, i.e. the size of one step h_j affects the possible sizes of all the remaining steps. In practice, we simplify the problem by devising a step-wise strategy, or *adaptive algorithm*, for choosing each step h_j in order to achieve (3.34). There are many possible adaptive algorithms and the choice depends partly on the accuracy requirements. We describe two.

In the first case, we estimate the sum in (3.34) as

$$\sum_{j=1}^{m} \left(\max_{[x_{j-1},x_j]} |f'| \, h_j \right) h_j \leq (x_m - a) \max_{1 \leq j \leq m} \left(\max_{[x_{j-1},x_j]} |f'| \, h_j \right),$$

where we use the fact that $\sum_{j=1}^{m} h_j = x_m - a$. It follows that (3.34) is satisfied if the steps are chosen by

$$h_j = \frac{\text{TOL}}{(x_m - a) \max_{[x_{j-1},x_j]} |f'|} \quad \text{for } j = 1, ..., m. \tag{3.35}$$

In general, this is a nonlinear equation for h_j since x_j depends on h_j.

We apply this adaptive algorithm to the computation of $\log(4)$ and obtain the following results

TOL	x_m	Steps	Approximate Area	Error
.1	4.077	24	1.36	.046
.01	3.98	226	1.376	.0049
.001	3.998	2251	1.38528	.0005
.0001	3.9998	22501	1.3861928	.00005

The reason x_m varies slightly in these results is due to the strategy we use to implement (3.35). Namely, we specify the tolerance and then search for the value of m that gives the closest x_m to 4. This procedure is easy to carry out by hand, but it is not the most efficient way to achieve (3.35) and we discuss more efficient methods later in the book.

We plot the sequence of mesh sizes for TOL $= .01$ in Fig. 3.14, where the adaptivity is plainly visible. In contrast, if we compute with a fixed mesh, we find using (3.29) that we need $N = 9/\text{TOL}$ points to guarantee an accuracy of TOL. For example, this means using 900 points to guarantee an accuracy of .01, which is significantly more than needed for the adapted mesh.

The second adaptive algorithm is based on an *equidistribution of error* in which the steps h_j are chosen so that the contribution to the error from each sub-interval is roughly equal. Intuitively, this should lead to the least number of intervals since the largest error reduction is gained if we subdivide the interval with largest contribution to the error. In this case, we estimate the sum on the left-hand side of (3.34) by

$$\sum_{j=1}^{m} \left(\max_{[x_{j-1},x_j]} |f'| \, h_j \right) h_j \leq m \max_{1 \leq j \leq m} \left(\max_{[x_{j-1},x_j]} |f'| \, h_j^2 \right)$$

and determine the steps h_j by

$$h_j^2 = \frac{\text{TOL}}{m \max\limits_{[x_{j-1},x_j]} |f'|} \quad \text{for } j = 1, ..., m. \qquad (3.36)$$

As above, we have to solve a nonlinear equation for h_j, now with the additional complication of the explicit presence of the total number of steps m.

We implement (3.36) using the same strategy used for the first adaptive algorithm for the computation of $\log(4)$ and obtain the following results:

TOL	x_m	Steps	Approximate Area	Error
.1	4.061	21	1.36	.046
.01	4.0063	194	1.383	.005
.001	3.9997	1923	1.3857	.0005
.0001	4.00007	19220	1.38626	.00005

We plot the sequence of step sizes for TOL $= .01$ in (3.14). We see that at every tolerance level, the second adaptive strategy (3.36) gives the same accuracy at $x_m \approx 4$ as (3.35) while using fewer steps. We may compare the efficiency of the two algorithms by estimating in each case the total number of steps m required to compute $\log(x)$ to a given accuracy TOL. We begin by noting that the equality

$$m = \frac{h_1}{h_1} + \frac{h_2}{h_2} + \cdots + \frac{h_m}{h_m},$$

implies that, assuming $x_m > 1$,

$$m = \int_1^{x_m} \frac{dy}{h(y)},$$

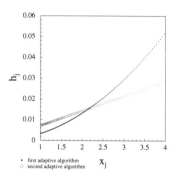
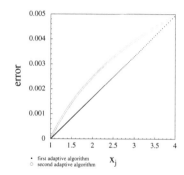

Figure 3.14: On the left, we plot the step sizes generated by two adaptive algorithms for the integration of $\log(4)$ using $\mathrm{TOL} = .01$. On the right, we plot the errors of the same computations versus x.

where $h(y)$ is the piecewise constant *mesh function* with the value $h(s) = h_j$ for $x_{j-1} < s \leq x_j$. In the case of the second algorithm, we substitute the value of h given by (3.36) into the integral to get, recalling that $f(y) = 1/y$ so that $f'(y) = -1/y^2$,

$$m \approx \frac{\sqrt{m}}{\sqrt{\mathrm{TOL}}} \int_1^{x_m} \frac{dy}{y},$$

or

$$m \approx \frac{1}{\mathrm{TOL}} \left(\log(x_m)\right)^2 . \tag{3.37}$$

Making a similar analysis of the first adaptive algorithm, we get

$$m \approx \frac{x_m - 1}{\mathrm{TOL}} \left(1 - \frac{1}{x_m}\right). \tag{3.38}$$

In both cases, m is inversely proportional to TOL. However, the number of steps needed to reach the desired accuracy using the first adaptive algorithm increases much more quickly as x_m increases than the number needed by the second algorithm, i.e. at a linear rate as opposed to a logarithmic rate. Note that the case $0 < x_m < 1$ may be reduced to the case $x_m > 1$ by replacing x_m by $1/x_m$ since $\log(x) = -\log(1/x)$.

   If we use (3.35) or (3.36) to choose the steps h_j over the interval $[a, x_m]$, then of course the error of the approximate integration over any smaller interval $[a, x_i]$ with $i \leq m$, is also smaller than TOL. For the first algorithm (3.35), we can actually show the stronger estimate

$$\left| \int_a^{x_i} f(y)\, dy - \sum_{j=1}^i f(x_j)\, h_j \right| \leq \frac{x_i - a}{x_m - a} TOL, \quad 1 \leq i \leq m, \tag{3.39}$$

i.e., the error grows at most linearly with x_i as i increases. However, this does not hold in general for the second adaptive algorithm. In Fig. 3.14, we plot the errors versus x_i for $x_i \leq x_m$ resulting from the two adaptive algorithms with TOL $= .01$. We see the linear growth predicted for the first algorithm (3.35) while the error from the second algorithm (3.36) is larger for $1 < x_i < x_m$.

Problem 3.36. Approximate the following integrals using the adaptive algorithms (3.35) and (3.36) to choose the steps: (a) $\int_0^2 2s\, ds$, (b) *(Harder)* $\int_0^2 s^3\, ds$, and (c) $\int_0^2 \exp(-s)\, ds$ using e.g. TOL $= .1, .01, .001$ and $.0001$. Discuss the results.

Problem 3.37. Compare theoretically and experimentally the number of steps of (3.35) and (3.36) for the computation of integrals of the form $\int_x^1 f(y)\, dy$ for $x > 0$, where $f(y) \sim y^{-\alpha}$ with $\alpha > 1$.

3.5. A look ahead

We proved above the existence of a solution $u(x)$ to the initial value problem

$$\begin{cases} u'(x) = f(u(x), x) & \text{for } a < x \leq b, \\ u(a) = u_0, \end{cases} \tag{3.40}$$

in the case that $f(u(x), x) = f(x)$ depends only on x, by studying the convergence of a numerical method of the form

$$U_j = U_{j-1} + h_j f(x_j) \quad \text{for } j = 1, 2, ..., m, \tag{3.41}$$

with $U_0 = u_0$, where $a = x_0 < x_1 < ... < x_m$ is an increasing sequence with step size $h_j = x_j - x_{j-1}$ and U_j is an approximation of $u(x_j)$. Such numerical methods extend to the more general case (3.40), where for instance f can also depend on $u(x)$, in the form

$$U_j = U_{j-1} + h_j f(U_{j-1}, x_j) \quad \text{for } j = 1, 2, ..., m, \tag{3.42}$$

with $U_0 = u_0$. Moreover, a variation of the argument used above proves that as the step sizes tend to zero, this method converges to a solution of (3.40). In this volume, we will study scalar and vector problems of the form (3.40) where f is linear in u, and then in the advanced companion volume, we will consider nonlinear f. This study is all the more important because once f is allowed to depend on u, then in general there is no equivalent of the definite integral that we can use to define the solution of the initial value problem.

The *elementary functions* in mathematics such as the exponential function $\exp(x) = e^x$ or the trigonometric functions $\sin(x)$, $\cos(x)$, are solutions to specific initial or boundary value problems. For instance, for $\lambda \in \mathbb{R}$ the function $\exp(\lambda x)$ solves the initial value problem

$$u'(x) = \lambda u(x) \quad \text{for } x > 0, \quad u(0) = 1,$$

corresponding to $f(u(x), x) = \lambda u(x)$. We will meet this problem in Chapters 6 and 9. For $\omega \in \mathbb{R}$, the function $\sin(\omega x)$ solves the initial value problem $u''(x) + \omega^2 u(x) = 0$ for $x > 0$, $u(0) = 0$, $u'(0) = \omega$, a problem we will meet in Chapter 10. Further, for $n = 1, 2, ...$, the function $\sin(nx)$ solves the boundary value problem $u''(x) + n^2 u(x) = 0$ for $0 < x < \pi$, $u(0) = u(\pi) = 0$, a problem we will meet in Chapters 6 and 8. Values of the elementary functions in general are computed by approximate solution of the corresponding differential equation. The computational methods presented in the book in particular gives methods for computing values of the elementary functions.

Problem 3.38. Verify that the functions $\exp(\lambda x)$, $\sin(\omega x)$ and $\sin(nx)$ solve the indicated initial or boundary value problems.

Problem 3.39. Find a correspondence between the set of departments at your university and the set of differential equations.

> In the direction of largeness it is always possible to think of a larger number....Hence this infinite is potential... and is not a permanent actuality but consists in a process of coming to be, like time... (Aristotle)

> The neglect of mathematics for thirty or forty years has nearly destroyed the entire learning of Latin Christendom. For he who does not know mathematics cannot know any of the other sciences; what is more, he cannot discover his own ignorance or find

proper remedies. So it is that the knowledge of this science pre-
pares the mind and elevates it to a well-authenticated knowledge
of all things. For without mathematics neither what is antecedent
nor consequent to it can be known; they perfect and regulate the
former, and dispose and prepare the way for that which succeeds.
(Roger Bacon, 1267)

A poor head, having subsidiary advantages,.... can beat the best,
just as a child can draw a line with a ruler better than the greatest
master by hand. (Leibniz)

Figure 3.15: An important Leibniz manuscript from October 29, 1675
that contains the origins of calculus.

4

A Short Review of Linear Algebra

Language is a lens that necessarily intervenes between the mind
and the world and that can, depending of the skill of the optician,
either distort or magnify our aprehension of the world. (Leibniz)

Linear algebra deals with vectors and linear transformations of vectors
and together with calculus forms the basis for the study of differential
equations. In this chapter, we give a speedy review of some parts of
linear algebra, concentrating on the Euclidean vector space $\mathbb{R}^d$. We also
give a first example of a vector space consisting of piecewise polynomial
functions, of which we will meet more examples in the next chapter.

The most familiar vector space is the Euclidean space $\mathbb{R}^d$ consisting
of the set of all *column vectors* $x = (x_1, ..., x_d)^\top$, where $\top$ indicates the
transpose and each x_j is a real number. Recall that $(x_1, ..., x_d)$ denotes
a *row vector*. When $x \in \mathbb{R}^1$, we call x a *scalar*. Note that we do not
distinguish vectors from scalars using an $\to$ or boldface for example. It
will be clear from the context whether a particular variable is a vector
or scalar.

Two vectors $x = (x_1, ..., x_d)^\top$ and $y = (y_1, ..., y_d)^\top$ may be added by
componentwise scalar addition

$$x + y = (x_1, ..., x_d)^\top + (y_1, ..., y_d)^\top = (x_1 + y_1, ..., x_d + y_d)^\top,$$

and a vector x may be multiplied by a scalar $\alpha \in \mathbb{R}$ componentwise
as well $\alpha x = (\alpha x_1, ..., \alpha x_d)^\top$. Thus, vectors in $\mathbb{R}^d$ may be added and
multiplied by scalars, which are the two basic operations in a vector

space. More generally, a set V is a *vector space*, if there are two operations defined on elements or "vectors" in V, namely addition of vectors denoted by $+$ and multiplication of a vectors by a scalar in $\mathbb{R}$, so that $x, y \in V$ implies $x + y \in V$, and $\alpha x \in V$ for any $\alpha \in \mathbb{R}$ and $x \in V$. Moreover, the operations should satisfy the rules satsified by vector addition and scalar multiplication $\mathbb{R}^d$. For example, the addition of vectors should be commutative and associative, i.e., the order in which vectors are added is irrelevant. A *subspace* of a vector space is a subset with the same properties, so that it forms a vector space in its own right. For example, $\{(x_1, x_2, 0)^\top \in \mathbb{R}^3 : x_i \in \mathbb{R}, i = 1, 2\}$ and $\{(x_1, x_2, x_3)^\top \in \mathbb{R}^3 : x_1 - 2x_2 + x_3 = 0\}$ are subspaces of $\mathbb{R}^3$ (verify!).

4.1.   Linear combinations, linear independency, and basis.

A *linear combination* of a set of vectors $\{v_i\}_{i=1}^d$ is a sum of the form $\sum_{i=1}^d \alpha_i v_i$ for some scalars $\alpha_i \in \mathbb{R}$. If $\{v_i\}_{i=1}^d$ is a set of vectors in a vector space V, then the set S of all possible linear combinations of the v_i,

$$ S = \Big\{ v : v = \sum_{i=1}^d \alpha_i v_i, \ \alpha_i \in \mathbb{R} \Big\}, $$

is a subspace of the vector space.

Problem 4.1.   Prove this.

A set of vectors $\{v_i\}_{i=1}^d$ is said to be *linearly independent* if the only linear combination of the vectors that sums to zero has coefficients equal to zero, i.e., if $\sum_{i=1}^d \alpha_i v_i = 0$ implies $\alpha_i = 0$ for all i. A *basis* for a vector space V is a set of linearly independent vectors $\{v_i\}_{i=1}^d$ such that any vector $v \in V$ can be written as a linear combination of the basis vectors v_i, i.e. $v = \sum_{i=1}^d \alpha_i v_i$, where the α_i are called the *coordinates* of v with respect to the basis $\{v_i\}_{i=1}^d$. The requirement that the vectors in a basis be linearly independent means that the coordinates of a given vector are unique. The set of vectors in $\mathbb{R}^d$:

$$ \Big\{ (1, 0, 0, \cdots, 0, 0)^\top, (0, 1, 0, \cdots, 0, 0)^\top, \cdots, (0, 0, 0, \cdots, 0, 1)^\top \Big\}, $$

often denoted by $\{e_1, ..., e_d\}$, is the *standard basis* for $\mathbb{R}^d$. A vector $x = (x_1, ..., x_d)^\top \in \mathbb{R}^d$ can be written $x = \sum_{i=1}^d x_i e_i$. The *dimension* of a vector space is the number of vectors in any basis for the space (this number is the same for all bases).

Problem 4.2. If the dimension of the vector space is d and $\{v_i\}_{i=1}^d$ is a set of d linearly independent vectors, prove that $\{v_i\}_{i=1}^d$ forms a basis.

A vector space has many different bases (if the dimension $d > 1$), and the coordinates of a vector with respect to one basis are not equal to the coordinates with respect to another basis.

Problem 4.3. Prove that the set of vectors $\{(1,0)^\top, (1,1)^\top\}$ is a basis for $\mathbb{R}^2$. If the coordinates of x are (x_1, x_2) in the standard basis, find the coordinates with respect to the new basis.

4.2. Norms, inner products, and orthogonality.

A *norm* $\|\cdot\|$ is a real-valued function of vectors with the following properties:

$$\|x\| \geq 0 \text{ for all vectors } x, \quad \|x\| = 0 \iff x = 0,$$
$$\|\alpha x\| = |\alpha|\,\|x\| \text{ for all scalars } \alpha \text{ and all vectors } x,$$
$$\|x + y\| \leq \|x\| + \|y\| \text{ for all vectors } x, y,$$

where the last inequality is referred to as the *triangle inequality*. A norm is used to measure the size of a vector. The most familiar norm is the *Euclidean norm* $\|x\| = \|x\|_2$ or "length" of a vector $x = (x_1, ..., x_d)^\top$ defined by

$$\|x\|_2 = \left(x_1^2 + \cdots + x_d^2\right)^{1/2},$$

which fits our geometric intuition about length. The Euclidean norm is closely related to the *Euclidean inner product* $(x, y)_2$ of two vectors $x = (x_1, ..., x_d)$ and $y = (y_1, ..., y_d)$, also denoted by $x \cdot y$ and referred to as the "dot-product", defined by

$$(x, y)_2 = x \cdot y = \sum_{i=1}^d x_i y_i.$$

The relation between the norm and the inner product is $\|x\|_2 = \sqrt{(x, x)_2}$. The Euclidean inner product $(x, y)_2$ is is The Euclidean norm is also called the l_2 norm. There are other ways of defining a norm of a vector $x = (x_1, ..., x_d)^\top$, such as the l_1 and l_∞ norms:

$$\|x\|_1 = |x_1| + \cdots + |x_d|,$$
$$\|x\|_\infty = \max_{1 \le i \le d} |x_i|.$$

Problem 4.4. Prove this estimate.

Problem 4.5. Plot the "unit circle" $\{x \in \mathbb{R}^2 : \|x\|_p = 1\}$ for the three norms $\|\cdot\|_p$, $p = 1, 2, \infty$.

Recall that an *inner product* or *scalar product* is a real-valued function of pairs of vectors denoted by $(\cdot, \cdot)$ with the following properties: if x, y, and z, are vectors and α, $\beta \in \mathbb{R}$, then

$$(\alpha x + \beta y, z) = \alpha(x, z) + \beta(y, z),$$
$$(x, \alpha y + \beta z) = \alpha(x, y) + \beta(x, z),$$
$$(x, y) = (y, x).$$

These rules may be summarized by saying that the inner product is *bilinear* and *symmetric*. An inner product also satisfies the Cauchy-Schwarz (or Cauchy) inequality

$$|(x, y)| \le \|x\| \, \|y\|,$$

for any two vectors x and y, where $\|x\| = \sqrt{(x, x)}$ is the associated *scalar product norm*. This inequality follows by noting that for all $s \in \mathbb{R}$,

$$0 \le \|x + sy\|^2 = (x + sy, x + sy) = \|x\|^2 + 2s(x, y) + s^2 \|y\|^2$$

and then choosing $s = -(x, y)/\|y\|^2$, which minimizes the right-hand side.

The basic example of an inner product is the Euclidean inner product $(\cdot, \cdot)_2$. Below, we will often suppress the index 2 in the Euclidean scalar product and norm. Thus, if nothing else is indicated, $\|x\| = \|x\|_2$ and $(x, y) = (x, y)_2$. Note that the l_1 and l_∞ norms are not connected to scalar products as the Euclidean norm is, but they do satisfy the relation: $|(x, y)_2| \le \|x\|_1 \, \|y\|_\infty$.

Problem 4.6. Prove that if $a_1, ..., a_d$, are given positive weights, then $(x, y) = \sum_{i=1}^{d} a_i x_i y_i$ is a scalar product on $\mathbb{R}^d$. Write down the corresponding norm and Cauchy inequality.

Problem 4.7. Prove the triangle inequality for a scalar product norm.

There is a helpful interpretation of the Euclidean inner product of two vectors x and v as a *generalized average* or *weighted average* of the coefficients of x. If x_1, x_2, ..., x_d are d numbers, then their average is $(x_1 + x_2 + ... + x_d)/d = \frac{1}{d} x_1 + \cdots + \frac{1}{d} x_d$. A generalized average is $v_1 x_1 + \cdots + v_d x_d$ where the v_i are called the *weights*. Note that this is the same as $(x, v)_2$ with $v = (v_1, ..., v_d)^\mathsf{T}$. Weighted averages might be familiar from a course where a test was given several times more weight than a quiz when the final grade was computed.

If x and y are two vectors, the *projection* of x in the direction of y is the vector αy, where $\alpha = (x, y)/\|y\|^2$. This vector has the property that $(x - \alpha y, y) = 0$ and is illustrated in Fig. 4.1.

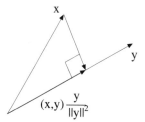

Figure 4.1: The projection of x in the direction of y.

Two vectors x and y are *orthogonal* if $(x, y) = 0$, i.e. if the projection of x onto y is the zero vector (and vica versa). Two non-zero orthogonal vectors are necessarily linearly independent, hence if the dimension of a vector space V is d and $\{v_1, ..., v_d\}$ are an orthogonal set of vectors, by which we mean the vectors are pairwise orthogonal, then they form a basis. Alternatively, it is possible, starting from any basis for a vector space V, to produce an orthogonal basis by successive subtraction of projections following the Gram-Schmidt algorithm.

To test whether a particular vector x in a vector space V is the zero vector, we can use orthogonality in the following way:

$$x = 0 \iff (x, v) = 0 \quad \text{for all } v \in V. \tag{4.1}$$

Interpreted in terms of averages, this means that x is zero if and only if every possible weighted average of its components is zero. We note that it suffices to check that (4.1) holds for all vectors v in a basis for V. For, if $\{v_i\}$ is a basis for V and $(x, v_i) = 0$ for all i and v is any vector in V, then $v = \sum_i \alpha_i v_i$ for some scalars α_i and $(x, v) = \sum_i \alpha_i (x, v_i) = 0$.

Problem 4.8. Prove (4.1).

The concept of orthogonality also extends to subspaces. A vector x in a vector space V is orthogonal to a subspace $S \subset V$ if x is orthogonal to all vectors $s \in S$. For example, $(0, 0, 1)^\top$ is orthogonal to the plane generated by the two vectors $(1, 0, 0)^\top$ and $(0, 1, 0)^\top$.

Problem 4.9. Plot the vectors and prove this claim.

A vector being orthogonal to a subspace means that certain weighted averages of its coefficients are zero. Note that it suffices to check that x is orthogonal to all the vectors in a basis for a subspace S to determine that x is orthogonal to S. The *orthogonal complement* of a subspace, denoted $S^\perp$, is the set of vectors in V that are orthogonal to S.

Problem 4.10. Prove that $S^\perp$ is a subspace of V. Show that the only vector both in S and in $S^\perp$ is 0. Show that $(S^\perp)^\perp = S$.

Similarly, the *projection* of a vector v onto a subspace S is the vector $v_s \in S$ such that

$$(v - v_s, s) = 0 \quad \text{for all vectors } s \in S. \tag{4.2}$$

The projection of a vector v is the best approximation of v in S in the following sense.

Lemma 4.1. *Let v_s denote the projection of v into the subspace S of a vector space. Then,*

$$\|v - v_s\| \leq \|v - s\| \quad \text{for all } s \in S.$$

Proof. Using the orthogonality (4.2), we have for any $s \in S$

$$(v - v_s, v - v_s) = (v - v_s, v - s) + (v - v_s, s - v_s)$$
$$= (v - v_s, v - s),$$

since $s - v_s \in S$. Taking absolute values and using the Cauchy-Schwarz inequality gives

$$\|v - v_s\|^2 \leq \|v - v_s\| \, \|v - s\|,$$

from which the claim follows.   ∎

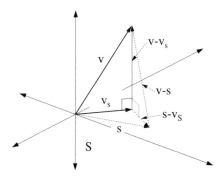

Figure 4.2: The projection of v onto the subspace S.

This basic argument occurs again in various forms, so it is worth taking the time to understand it now. We illustrate it in Fig. 4.2.

The above applies to the Euclidean inner product, but directly generalizes to an arbitrary scalar product.

Problem 4.11. Prove that the projection of a vector x in the direction of a vector y defined above is equal to the projection of x onto the subspace $\{v : v = \alpha y, \; \alpha \in \mathbb{R}\}$ consisting of all scalar multiples of y.

4.3. Linear transformations and matrices

A *transformation* or *map* $f(x)$ from $\mathbb{R}^d$ to $\mathbb{R}^d$ is a function that associates a vector $y = f(x) \in \mathbb{R}^d$ to each vector $x \in \mathbb{R}^d$. In component form in the standard basis, the transformation can be written $y_i = f_i(x)$, $i = 1, ..., d$, where each coordinate function f_i is a map from $\mathbb{R}^d$ to $\mathbb{R}$. The transformation $y = f(x)$ is a *linear transformation* if f has the property that $f(\alpha x + z) = \alpha f(x) + f(z)$ for all vectors x and z and scalars α. A linear transformation $y = f(x)$ can be written in component form

$$y_1 = a_{11}x_1 + a_{12}x_2 + \cdots + a_{1d}x_d$$
$$y_2 = a_{21}x_1 + a_{22}x_2 + \cdots + a_{2d}x_d$$
$$\vdots$$
$$y_d = a_{d1}x_1 + a_{d2}x_2 + \cdots + a_{dd}x_d$$

where $a_{ij} = f_i(e_j)$. Introducing the $d \times d$ *matrix* $A = (a_{ij})$ with elements a_{ij}, $i, j = 1, ..., d$, where i is the *row index* and j the *column index* this is written

$$y = Ax.$$

The matrix of a transformation depends on the choice of basis. Recall that the transpose $A^\top = (b_{ij})$ of a matrix $A = (a_{ij})$ is the matrix with elements b_{ij} where $b_{ij} = a_{ji}$. The transpose satisfies $(Ax, y) = (x, A^\top y)$.

The properties of linear transformations are reflected by properties of matrices. If $f(x)$ and $g(x)$ are two linear transformations on $\mathbb{R}^d$ with associated matrices A and B, then the composition $g \circ f(x) = g(f(x))$ has matrix BA with the usual definition of matrix multiplication. The inverse of a transformation exists if and only if the inverse of the associated matrix exists, where the inverse of a matrix A is the unique matrix A^{-1} such that $A^{-1}A = AA^{-1} = I$, where I is the $d \times d$ identity matrix. Recall that that the determinant of a matrix A, denoted by $\det A$, is a real number computable from the elements of A, which is non-zero if and only if A^{-1} exists.

Suppose that A is the matrix of a linear transformation f. Recall that the *range* $R(A)$ of the matrix A is the set of vectors $\{Ax, x \in \mathbb{R}^d\}$. If there is a solution of $Ax = b$, then b must be in the range of A. The set of vectors x such that $Ax = 0$ is the *null space* $N(A)$ of A. A basic result states that $Ax = b$ has a unique solution if and only if $Ay = 0$ implies $y = 0$. In particular $R(A) = \mathbb{R}^d$ if and only if $N(A) = \{0\}$. The range of A and the null space of $A^\top$ are related by the fact that they form a *decomposition* of $\mathbb{R}^d$ in the sense that any vector $z \in \mathbb{R}^d$ can be written uniquely in the form $z = x + y$ where $x \in R(A)$ and $y \in N(A^\top)$. In particular $\dim R(A) + \dim N(A^\top) = d$. Further, $\dim N(A) = \dim N(A^\top)$ so that also $\dim R(A) + \dim N(A) = d$.

4.4. Eigenvalues and eigenvectors

In general, a linear transformation can rotate and rescale vectors and it does this in different degrees to different vectors. In order to establish some structure on a given transformation, it is natural to ask if there are vectors on which the transformation acts only to scale the vector. An *eigenvector* v of a transformation represented by the matrix A, or an eigenvector of A for short, is a non-zero vector that has the property

$$Av = \lambda v,$$

for a scalar λ, which is called the *eigenvalue* of A associated to v. Since an eigenvector v solves $(A - \lambda I)v = 0$, where I is the identity matrix, λ is eigenvalue if and only if the determinant of $A - \lambda I$ vanishes. In other words, the eigenvalues are the roots of the polynomial equation $\det(A - \lambda I) = 0$. Since the roots may be complex, it may be convenient to assume that the scalars are the set of complex numbers. In this case the transpose $A^\top$ of a matrix $A = (a_{ij})$ is replaced by the adjoint A^* with elements b_{ij} where $b_{ij} = \bar{a}_{ji}$, where the bar denotes complex conjugate. The adjoint satisfies $(Ax, y) = (x, A^* y)$, where $(v, w) = \sum_i v_i \bar{w}_i$ is the scalar product with complex scalars.

Eigenvectors associated to distinct eigenvalues are linearly independent. On the other hand, an eigenvalue of A can be associated to several eigenvectors. The *eigenspace* of A corresponding to an eigenvalue is the subspace formed by taking all linear combinations of the eigenvectors associated to the eigenvalue.

We recall that the eigenvalues of a diagonal matrix with non-zero elements only in the diagonal are just the diagonal entries.

Problem 4.12. Prove this claim.

Thus, computing the eigenvalues of a matrix is greatly simplified if the matrix can be transformed into a diagonal matrix without changing the eigenvalues. Two matrices A and B are *similar* if there is an invertible matrix P such that $P^{-1}AP = B$. If v is an eigenvector of A corresponding to λ, then $w = P^{-1}v$ is an eigenvector of B corresponding to the same eigenvalue λ. Thus, similar matrices have the same eigenvalues. A matrix is *diagonalizable* if there is an invertible matrix P such that $P^{-1}AP$ is a diagonal matrix. Recall that a $d \times d$ matrix is diagonalizable if and only if it has d linearly independent eigenvectors forming a basis for $\mathbb{R}^d$. If P diagonalizes A, then the columns of P are just d linearly independent eigenvectors of A.

It is possible to show that certain types of matrices are diagonalizable. If A is symmetric, i.e. $A = A^\top$, then A is diagonalizable with real eigenvalues, and moreover the (real) matrix P may be chosen to be *orthogonal* which means that $P^{-1} = P^\top$ and the columns of P, which are the eigenvectors of A, are orthogonal to each other. We say that A is *positive definite* if $(Ax, x) > 0$ for all vectors $x \neq 0$. If A is positive definite and symmetric, then the eigenvalues of A are positive and moreover $\|x\|_A = \sqrt{(x, Ax)}$ is a vector norm. We will encounter several positive-definite symmetric matrices when computing on differential equations.

More generally, assuming the scalars to be complex, if A has d distinct (possibly complex) eigenvalues, then A is diagonalizable: $P^{-1}AP = \Lambda$, where Λ is diagonal with the eigenvalues on the diagonal and the columns of P are the corresponding eigenvectors. If A is diagonalizable with P orthogonal so that $P^{-1} = P^*$, then A is said to be *normal*. A basic fact of linear algebra states that A is normal if $A^*A = AA^*$. If A is real and symmetric so that $A = A^\top = A^*$, then A is normal.

4.5. Norms of matrices

It is often necessary to measure the "size" of a linear transformation, or equivalently the size of a corresponding matrix. We measure the size by measuring the maximum possible factor by which the transformation can rescale a vector. We define the *matrix norm* for a matrix A by

$$\|A\|_p = \max_{\substack{v \in \mathbb{R}^d \\ v \neq 0}} \frac{\|Av\|_p}{\|v\|_p}, \quad p = 1, 2, \text{ or } \infty.$$

Note that the matrix norm is defined in terms of a particular choice of vector norm for $\mathbb{R}^d$. Generally, matrix norms are harder to compute than vector norms. It is easiest for the $\|\cdot\|_1$ and the $\|\cdot\|_\infty$ norm:

$$\|A\|_1 = \max_{1 \leq j \leq n} \sum_{i=1}^{n} |a_{ij}| \quad \text{(maximum absolute column sum)}$$

$$\|A\|_\infty = \max_{1 \leq i \leq n} \sum_{j=1}^{n} |a_{ij}| \quad \text{(maximum absolute row sum)}.$$

The $\|\cdot\|_2$ norm is harder to compute, see Problem 4.15. If λ is an eigenvalue of A, then $\|A\| \geq |\lambda|$ for any matrix norm.

Problem 4.13. Prove the last claim.

In the case of a symmetric matrix and the l_2 norm, the following lemma can be used to give a converse to this result.

Lemma 4.2. *Let A be a $d \times d$ symmetric matrix with eigenvalues $\{\lambda_i\}$. Then*

$$\|Av\|_2 \leq \max_{1 \leq i \leq d} |\lambda_i| \|v\|_2 \tag{4.3}$$

Proof. We first show that if P is an orthogonal matrix, then $\|P\|_2 = \|P^\top\|_2 = 1$, because for any vector v,

$$\|Pv\|_2^2 = (Pv, Pv)_2 = (v, P^\top Pv)_2 = (v, Iv)_2 = (v, v)_2 = \|v\|_2^2,$$

hence

$$\frac{\|Pv\|_2}{\|v\|_2} = 1 \quad \text{for all } v \neq 0.$$

The proof for $P^\top$ is the same. Let now $\Lambda = P^\top AP$ be a diagonal matrix of eigenvalues of A obtained choosing the columns of P to be orthogonal eigenvectors of A, so that $A = P\Lambda P^\top$. We can now compute:

$$\|Av\|_2 = \|P\Lambda P^\top v\|_2 = \|\Lambda P^\top v\|_2 \leq \max_{1 \leq i \leq d} |\lambda_i| \|P^\top v\|_2 = \max_{1 \leq i \leq d} |\lambda_i| \|v\|_2.$$

∎

It follows immediately that if A is symmetric, then $\|A\|_2 = \max_{1 \leq i \leq n} |\lambda_i|$.

Problem 4.14. Prove the last statement.

Problem 4.15. Prove that for any matrix A, we have $\|A\|_2 = \sqrt{\rho(A^\top A)}$, where $\rho(A^\top A)$ is the *spectral radius* of $A^\top A$, that is the largest modulus of the eigenvalues of $A^\top A$.

4.6. Vector spaces of functions

There are more general vector spaces than $\mathbb{R}^d$. Later on, we use vector spaces for which the "vectors" are functions defined on an interval (a, b). A vector space may consist of functions, because two functions f and g on (a, b) may be added to give a function $f + g$ on (a, b) defined by $(f + g)(x) = f(x) + g(x)$, that is the value of $f + g$ at x is the sum of the values of $f(x)$ and $g(x)$. Moreover, a function f may be multiplied by a real number α to give a function αf defined by $(\alpha f)(x) = \alpha f(x)$. In particular, we use vector spaces consisting of polynomials or piecewise polynomials on (a, b).

Problem 4.16. Prove that the set of functions v on $(0, 1)$ of the form $v(x) = ax + b$, where a and b are real numbers, is a vector space.

Problem 4.17. Prove that the set of piecewise constant functions on a partition of an interval, is a vector space of functions.

4.6.1. Scalar product, L_2 norm, and Cauchy's inequality

There are many parallels between vectors in $\mathbb{R}^d$, and functions in a vector space of functions defined on an interval (a, b). For example, for functions f and g on (a, b) we may define an analog of the usual scalar product as follows:

$$(f, g) = \int_a^b f(x)g(x)\,dx, \tag{4.4}$$

where we usually suppress the dependence on the interval (a, b). The associated norm is the $L_2(a, b)$ norm defined by

$$\|f\|_{L_2(a,b)} = \sqrt{(f, f)} = \left(\int_a^b f^2(x)\,dx \right)^{1/2}.$$

The $L_2(a, b)$ norm measures the "root mean square" area underneath the graph of f. The inner product $(\cdot, \cdot)$, which we refer to as the L_2 inner product, satisfies Cauchy's inequality

$$|(f, g)| \leq \|f\|_{L_2(a,b)} \|g\|_{L_2(a,b)}. \tag{4.5}$$

The terminology from $\mathbb{R}^d$ carries over to functions: we say that two functions f and g are orthogonal on (a, b) if

$$(f, g) = \int_a^b f(x)g(x)\,dx = 0. \tag{4.6}$$

This concept of orthogonality is central in this book, and we will become very familiar with it.

Problem 4.18. Prove the above Cauchy inequality.

Problem 4.19. Find a function that is orthogonal to $f(x) = x$ on $(0, 1)$.

Problem 4.20. Prove that $\sin(nx)$ and $\sin(mx)$ are orthogonal on $(0, \pi)$ if $n, m = 1, 2, 3, ..., n \neq m$.

4.6.2. Other norms of functions

Just as for vectors in $\mathbb{R}^d$, there is a need to measure the "size" of a function over an interval (a, b) in different ways or in different norms. As an example, consider the two functions f_1 and f_2 defined on the interval (a, b) plotted in Fig. 4.3. One could argue that f_1 should be considered

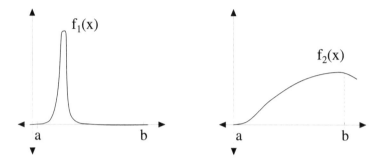

Figure 4.3: Two functions of different "sizes".

bigger because it reaches a higher maximum height. On the other hand, f_2 has bigger values over a greater portion of the interval $[a, b]$. We may compare the two functions using the $L_\infty(a, b)$ norm $\|\cdot\|_{L_\infty(a,b)}$ also denoted by $\|\cdot\|_{[a,b]}$, which measures a continuous function f on $[a, b]$ by its maximum absolute height over $[a, b]$:

$$\|f\|_{L_\infty(a,b)} = \|f\|_{[a,b]} = \max_{x\in[a,b]} |f(x)|,$$

Alternatively, we may compare the area underneath the graph of the absolute value of a function f, that is we may use the $L_1(a, b)$ norm defined by

$$\|f\|_{L_1(a,b)} = \int_a^b |f(x)| \, dx.$$

In the example above, f_2 has a larger L_1 norm than f_1 on (a, b), while f_1 has a larger L_∞ norm.

The L_1 and L_∞ norms are not associated to inner products, but they do satisfy the following variant of Cauchy's inequality

$$|(f, g)| \le \|f\|_{L_\infty(a,b)} \|g\|_{L_1(a,b)}.$$

Problem 4.21. Prove this.

Problem 4.22. Verify that the L_1, L_2 and L_∞ norms defined above satisfy the requirements of a norm.

Problem 4.23. Prove that if $0 < b-a \leq 1$, then $\|f\|_{L_1(a,b)} \leq \|f\|_{L_2(a,b)} \leq \|f\|_{L_\infty(a,b)}$. Generalize to $b > a + 1$.

Problem 4.24. Compute the L_∞, L_1, and L_2 norms of $\sin(\pi x)$, x^2, and e^x on $(0,1)$. Rank the functions in size in the different norms.

For $p = 1$, 2, or ∞, the set of functions on an interval (a,b) with finite $L_p(a,b)$ norm form a vector space which we denote by $L_p(a,b)$. Note that a function can belong to one of the spaces and not another. For example, $x^{-1/2}$ belongs to $L_1(0,1)$ but not $L_\infty(0,1)$.

Problem 4.25. (a) Determine r such that x^{-s} belongs to $L_1(0,1)$ for $s < r$. (b) Do the same for $L_2(0,1)$.

> It is, unfortunately, our destiny that, because of a certain aversion toward light, people love to be returned to darkness. We see this today, where the great ease for acquiring learning has brought forth contempt for the doctrines taught, and an abundance of truths of the highest clarity has led to a love for difficult nonsense... (Leibniz)

> "You see," he explained, "I consider that a man's brain originally is like a little empty attic, and you have to stock it with such furniture as you choose. A fool takes in all the lumber of every sort that he comes across, so that the knowledge which might be useful to him gets crowded out, or at best is jumbled up with a lot of other things, so that he has a difficulty laying his hands upon it. Now the skillful workman is very careful indeed as to what he takes into his brain-attic. He will have nothing but the tools which may help him in doing his work, but of these he has a large assortment, and all in the most perfect order. It is a mistake to think that the little room has elastic walls and can distend to any extent. Depend upon it there comes a time when for every addition of knowledge you forget something that you knew before. It is of the highest importance, therefore, not to have useless facts elbowing out the useful ones."

> "But the Solar System!" I protested.

> "What the deuce is it to me?" (A. C. Doyle)

5

Polynomial Approximation

> The best world has the greatest variety of phenomena regulated
> by the simplest laws. (Leibniz)

In Chapter 3 we computed approximations of the definite integral of a
function using piecewise constant approximation of the function. The
idea is that on one hand we know how to integrate a piecewise constant
analytically, and on the other hand, that a smooth function is approximately constant on small intervals. We will expand on this idea to
compute solutions of differential equations numerically using piecewise
polynomials.

In this chapter, we discuss different ways of approximating a function by polynomials on an interval or by piecewise polynomials on a
subdivision of an interval. We also derive different estimates for the
approximation error that are used in the rest of the book. We conclude
the chapter by applying the estimates to quadrature and making some
remarks on approximation by trigonometric polynomials.

5.1. Vector spaces of polynomials

We let $\mathcal{P}^q(a, b)$ denote the set of polynomials $p(x) = \sum_{i=0}^{q} c_i x^i$ of degree
at most q on an interval (a, b), where the $c_i \in \mathbb{R}$ are called the coefficients of $p(x)$. We recall that two polynomials $p(x)$ and $r(x)$ may be
added to give a polynomial $p + r$ defined by $(p + r)(x) = p(x) + r(x)$
and a polynomial $p(x)$ may be multiplied by a scalar α to give a polyno-
mial αp defined by $(\alpha p)(x) = \alpha p(x)$. Similarly, $\mathcal{P}^q(a, b)$ satisfies all the

requirements to be a vector space where each "vector" is a particular polynomial function $p(x)$.

Problem 5.1. Prove this claim.

A basis for $\mathcal{P}^q(a, b)$ consists of a special set of polynomials. A familiar basis is the set of monomials $\{1, x, ..., x^q\}$ which is a basis because this set is linearly independent since $a_0 \cdot 1 + a_1 \cdot x + \cdots + a_q x^q = 0$ for all $a < x < b$ implies $a_0 = a_1 = \cdots = a_q = 0$, and moreover any polynomial can be written as a linear combination of the monomials. It follows that the dimension of $\mathcal{P}^q(a, b)$ is $q+1$. Another basis is the set $\{(x - c)^i\}_{i=0}^q$, where c is a point in (a, b).

Problem 5.2. Find a formula for the coefficients of a polynomial $p \in \mathcal{P}^q(a, b)$ with respect to the basis $\{(x - c)^i\}_{i=0}^q$ in terms of the value of p and its derivatives at c.

Problem 5.3. Find the coefficients of $p(x) = c_0 + c_1 x + c_2 x^2$ with respect to the basis $\{1, (x - a), (x - a)^2\}$ in terms of the c_i and a.

Another basis for $\mathcal{P}^q(a, b)$ that is useful is the *Lagrange basis* $\{\lambda_i\}_{i=0}^q$ associated to the distinct $q + 1$ points $\xi_0 < \xi_1 < \cdots < \xi_q$ in (a, b), determined by the requirement that $\lambda_i(\xi_j) = 1$ if $i = j$ and 0 otherwise. The explicit expression for the basis function λ_i is

$$\lambda_i(x) = \frac{(x - \xi_0)(x - \xi_1) \cdots (x - \xi_{i-1})(x - \xi_{i+1}) \cdots (x - \xi_q)}{(\xi_i - \xi_0)(\xi_i - \xi_1) \cdots (\xi_i - \xi_{i-1})(\xi_i - \xi_{i+1}) \cdots (\xi_i - \xi_q)}$$

$$= \prod_{j \neq i} \frac{x - \xi_j}{\xi_i - \xi_j}.$$

For simplicity of notation, we suppress the dependence of the λ_i on q and the points ξ_i. We plot the Lagrange basis functions $\lambda_1(x) = (x - \xi_0)/(\xi_1 - \xi_0)$ and $\lambda_0(x) = (x - \xi_1)(\xi_0 - \xi_1)$ for $q = 1$ in Fig. 5.1. The polynomial $p \in \mathcal{P}^q(a, b)$ that has the value $p_i = p(x_i)$ at the *nodes* ξ_i, $i = 0, ..., q$, may be expressed in terms of the corresponding Lagrange basis as

$$p(x) = p_0 \lambda_0(x) + p_1 \lambda_1(x) + \cdots + p_q \lambda_q(x),$$

so that the values $\{p_i\}_{i=0}^q$ are the coefficients of $p(x)$ with respect to the Lagrange basis. We also refer to the Lagrange basis as a *nodal* basis.

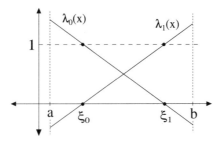

Figure 5.1: Linear Lagrange basis functions.

We extend the idea behind a nodal basis, i.e. defining a polynomial by giving its values at an appropriate number of distinct points, by determining a polynomial by specifying an appropriate number of values of the polynomial and its derivatives at a given set of points. A nodal basis with respect to a given set of values has the property that the coefficients of a general polynomial with respect to the basis are precisely these values. For example, a polynomial $p \in \mathcal{P}^2(a, b)$ may be specified by the values $p(a), p((a + b)/2)$, and $p(b)$ and a corresponding Lagrange basis λ_0, λ_1, and λ_2 related to the points $\xi_0 = a$, $\xi_1 = (a + b)/2$ and $\xi_2 = b$. We might also specify the values of $p(a)$, $p(b)$, and $p'(b)$. To determine a nodal basis with respect to these values, we seek polynomials μ_0, μ_1, and μ_2 such that

$$\mu_0(a) = 1, \ \mu_0(b) = 0, \ \mu_0'(b) = 0,$$
$$\mu_1(a) = 0, \ \mu_1(b) = 1, \ \mu_1'(b) = 0,$$
$$\mu_2(a) = 0, \ \mu_2(b) = 0, \ \mu_2'(b) = 1.$$

A polynomial $p \in \mathcal{P}^2(a, b)$ can then be written $p(x) = p(a)\mu_0(x) + p(b)\mu_1(x) + p'(b)\mu_2(x)$. Straightforward calculation shows that

$$\mu_0(x) = (b - x)^2/(b - a)^2,$$
$$\mu_1(x) = (x - a)(2b - x - a)/(b - a)^2,$$
$$\mu_2(x) = (x - a)(x - b)/(b - a).$$

Problem 5.4. Verify the formulas for μ_1, μ_2, and μ_3.

Problem 5.5. Find the nodal basis corresponding to specifying the values $p'(a)$, $p(b)$, and $p'(b)$.

Problem 5.6. *(Harder.)* (a) Prove that specifying the information $p(a)$, $p'(a)$, $p(b)$, and $p'(b)$ suffices to determine polynomials in $\mathcal{P}^3(a, b)$ uniquely. Determine the nodal basis for this set of values.

5.2.   Polynomial interpolation

One familiar example of a polynomial approximation of a function is the Taylor polynomial of degree q. This gives an approximation of a function $u(x)$ based on the values of $u(\bar{x})$ and the derivatives $D^r u(\bar{x})$ for $r = 1, ..., q$, at a specific point $\bar{x}$. Taylor's theorem gives a formula for the error between the function u and its Taylor polynomial. We consider an alternative approximation based on constructing a polynomial that agrees with a function at a set of distinct points.

We assume that f is continuous on $[a, b]$ and choose distinct *interpolation nodes* $a \leq \xi_0 < \xi_1 < \cdots < \xi_q \leq b$ and define a polynomial *interpolant* $\pi_q f \in \mathcal{P}^q(a, b)$, that *interpolates* $f(x)$ at the nodes $\{\xi_i\}$ by requiring that $\pi_q f$ take the same values as f at the nodes, i.e. $\pi_q f(\xi_i) = f(\xi_i)$ for $i = 0, ..., q$. Using the Lagrange basis corresponding to the ξ_i, we can express $\pi_q f$ using *Lagrange's formula:*

$$\pi_q f(x) = f(\xi_0)\lambda_0(x) + f(\xi_1)\lambda_1(x) + \cdots + f(\xi_q)\lambda_q(x) \quad \text{for } a \leq x \leq b. \tag{5.1}$$

We show two examples for $q = 0$ in Fig. 5.2, where $\pi_0(x) = f(\xi_0)$ with $\xi_0 = a$ and $\xi_0 = b$. In the case $q = 1$, choosing $\xi_0 = a$ and $\xi_1 = b$, see

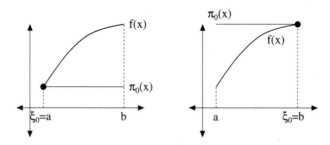

Figure 5.2: Two constant interpolants of a function.

Fig. 5.3, we get

$$\pi_1 f(x) = f(a)\frac{x - b}{a - b} + f(b)\frac{x - a}{b - a}. \tag{5.2}$$

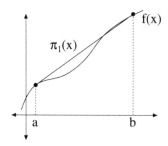

Figure 5.3: The linear interpolant $\pi_1 f$ of a function f.

Problem 5.7. Compute formulas for the linear interpolant of a continuous function f through the points a and $(b + a)/2$. Plot the corresponding Lagrange basis functions.

Problem 5.8. Write down the polynomial of degree 3 that interpolates $\sin(x)$ at $\xi_0 = 0$, $\xi_1 = \pi/6$, $\xi_2 = \pi/4$, and $\xi_3 = \pi/3$, and plot p_3 and sin on $[0, \pi/2]$.

5.2.1. A pointwise estimate of the interpolation error

The following theorem gives a pointwise error estimate for nodal interpolation.

Theorem 5.1. *Assume that f has $q + 1$ continuous derivatives in (a, b) and let $\pi_q f \in \mathcal{P}^q(a, b)$ interpolate f at the points $a \leq \xi_0 < \xi_1 < .. < \xi_q \leq b$. Then for $a \leq x \leq b$,*

$$|f(x) - \pi_q f(x)| \leq \left| \frac{(x - \xi_0) \cdots (x - \xi_q)}{(q + 1)!} \right| \max_{[a,b]} |D^{q+1} f|. \tag{5.3}$$

Proof. We present a proof for $q = 0$ and $q = 1$ that can be extended to functions of several variables; see Problem 5.12 for an alternative proof for one variable.

In the case $q = 0$, the interpolation error depends on the size of f' and the distance between x and ξ_0, see Fig. 5.4. Taylor's theorem (or the mean value theorem) implies that

$$f(x) = f(\xi_0) + f'(\eta)(x - \xi_0) = \pi_0 f(x) + f'(\eta)(x - \xi_0),$$

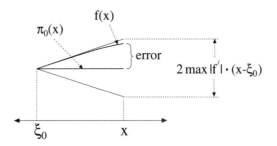

Figure 5.4: The error of a constant interpolant.

for some η between ξ_0 and x, so that

$$|f(x) - \pi_0 f(x)| \leq |x - \xi_0| \max_{[a,b]} |f'| \quad \text{for all } a \leq x \leq b,$$

proving the desired result.

The proof for $q = 1$ is a little more involved. In this case, the error estimate states that the error is proportional to $|f''|$, i.e. to the degree of concavity of f or the amount that f curves away from being linear, see Fig. 5.5. We start by recalling that

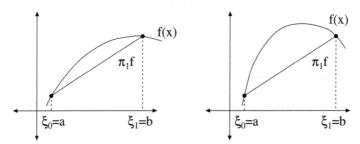

Figure 5.5: The error of a linear interpolant depends on the size of $|f''|$, which measures the degree that f curves away from being linear. Notice that the error of the linear interpolant of the function on the right is much larger than of the linear interpolant of the function on the left and the function on the right has a larger second derivative in magnitude.

$$\pi_1 f(x) = f(\xi_0)\lambda_0(x) + f(\xi_1)\lambda_1(x) = f(\xi_0)\frac{x - \xi_1}{\xi_0 - \xi_1} + f(\xi_1)\frac{x - \xi_0}{\xi_1 - \xi_0}. \tag{5.4}$$

Fixing x in (ξ_0, ξ_1) and using Taylor's theorem for $i = 0$ and 1, we get

$$f(\xi_i) = f(x) + f'(x)(\xi_i - x) + \frac{1}{2}f''(\eta_i)(\xi_i - x)^2, \tag{5.5}$$

where η_i lies between x and ξ_i. Substituting the Taylor expansions (5.5) into (5.4) and using the identities

$$\lambda_0(x) + \lambda_1(x) \equiv 1, \quad (\xi_0 - x)\lambda_0(x) + (\xi_1 - x)\lambda_1(x) \equiv 0, \tag{5.6}$$

we obtain the *error representation*

$$f(x) - \pi_1 f(x) = -\frac{1}{2}\left(f''(\eta_0)(\xi_0 - x)^2\lambda_0(x) + f''(\eta_1)(\xi_1 - x)^2\lambda_1(x)\right).$$

Note that expanding around the "common point" x, makes it possible to take into account the cancellation expressed by the second of the two identities in (5.6).

Problem 5.9. Verify (5.6) and the error representation formula.

From the formulas for λ_1 and λ_2, for $\xi_0 \leq x \leq \xi_1$, we get

$$|f(x) - \pi_1 f(x)|$$

$$\leq \frac{1}{2}\left(\frac{|\xi_0 - x|^2|x - \xi_1|}{|\xi_0 - \xi_1|}|f''(\eta_0)| + \frac{|\xi_1 - x|^2|x - \xi_0|}{|\xi_1 - \xi_0|}|f''(\eta_1)|\right)$$

$$\leq \frac{1}{2}\left(\frac{|\xi_0 - x|^2|x - \xi_1|}{|\xi_0 - \xi_1|} + \frac{|\xi_1 - x|^2|x - \xi_0|}{|\xi_1 - \xi_0|}\right) \max_{[a,b]}|f''|$$

$$= \frac{1}{2}\left(\frac{(x - \xi_0)^2(\xi_1 - x)}{\xi_1 - \xi_0} + \frac{(\xi_1 - x)^2(x - \xi_0)}{\xi_1 - \xi_0}\right) \max_{[a,b]}|f''|$$

$$= \frac{1}{2}|x - \xi_0||x - \xi_1| \max_{[a,b]}|f''|.$$

∎

Problem 5.10. Show that (5.3) holds for $q = 1$ and $a \leq x < \xi_0$ and $\xi_q < x \leq b$.

Problem 5.11. *(Hard.)* Prove the theorem for $q = 2$.

Problem 5.12. Give a different proof of Theorem 5.1 by considering the function

$$g(y) = f(y) - \pi_q f(y) - \gamma(x)(y - \xi_0)....(y - \xi_q),$$

where $\gamma(x)$ is chosen so that $g(x) = 0$. Hint: the function $g(y)$ vanishes at $\xi_0, \xi_1, ..., \xi_q$ and x. Show by repeated use of the mean value theorem that $D^{q+1}g$ must vanishes at some point ξ, from which it follows that $\gamma(x) = D^{q+1}f(\xi)/(q + 1)!$.

One advantage gained by using a linear interpolant of f is that it also approximates the derivative of f.

Theorem 5.2. *For $\xi_0 \le x \le \xi_1$,*

$$|f'(x) - (\pi_1 f)'(x)| \le \frac{(x - \xi_0)^2 + (x - \xi_1)^2}{2(\xi_1 - \xi_0)} \max_{[a,b]} |f''|.$$

We illustrate this in Fig. 5.6.

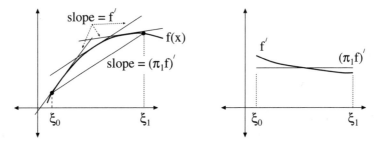

Figure 5.6: The derivative of a linear interpolant of f approximates the derivative of f. We show f and the linear interpolant $\pi_1 f$ on the left and their derivatives on the right.

Proof. Differentiating (5.4) with respect to x and using (5.5) together with the identities

$$\lambda_0'(x) + \lambda_1'(x) \equiv 0, \quad (\xi_0 - x)\lambda_0'(x) + (\xi_1 - x)\lambda_1'(x) \equiv 1,$$

$$(5.7)$$

we get the error representation

$$f'(x) - (\pi_1 f)'(x) = -\frac{1}{2}\left(f''(\eta_0)(\xi_0 - x)^2\lambda_0'(x) + f''(\eta_1)(\xi_1 - x)^2\lambda_1'(x)\right).$$

Taking absolute values, and noting that $|\lambda_i'(x)| = (\xi_1 - \xi_0)^{-1}$ proves the theorem.

Problem 5.13. Prove (5.7).

∎

5.2.2. Estimating interpolation errors in different norms

Theorems 5.1 and 5.2 state pointwise error estimates for the function values and the first derivative. In applications involving differential equations we use error bounds in various norms such as L_1, L_2 and L_∞ norms for function values or derivatives (recall the material on norms of functions in the previous chapter).

We first restate the estimates of Theorems 5.1 and 5.2 using the L_∞ norm. Choosing $a = \xi_0$ and $b = \xi_q$, (5.3) implies

$$\|f - \pi_q f\|_{L_\infty(a,b)} \leq \frac{1}{(q+1)!} \max_{x \in [a,b]} \|(x - \xi_0) \cdots (x - \xi_q)\| \| D^{(q+1)} f\|_{L_\infty(a,b)}.$$

In the case of $q = 0$, this reduces to

$$\|f - \pi_0 f\|_{L_\infty(a,b)} \leq (b - a)\|f'\|_{L_\infty(a,b)}, \tag{5.8}$$

while in the case $q = 1$, we get by Theorem 5.2,

$$\|f - \pi_1 f\|_{L_\infty(a,b)} \leq \frac{1}{8}(b - a)^2\|f''\|_{L_\infty(a,b)}, \tag{5.9}$$

and

$$\|f' - (\pi_1 f)'\|_{L_\infty(a,b)} \leq \frac{1}{2}(b - a)\|f''\|_{L_\infty(a,b)}. \tag{5.10}$$

These are basic maximum norm estimates for polynomial interpolation. Note the form of these estimates with the multiplicative factor $b - a$ to the power one for the function value when $q = 0$ and the derivative for $q = 1$ and the power two for the function value when $q = 1$. We call the constants 1, 1/8 and 1/2, *interpolation constants*.

Problem 5.14. Prove (5.9) and (5.10).

We will also use analogs of these estimates in the L_2 and L_1 norms. The interpolation constant, which we call C_i, changes depending on q and the norm that is used, and may also depend on the ratio $\min_i(\xi_i - \xi_{i-1})/(b-a)$, but does not depend on f or the interval (a,b). The values of the C_i in the case $p = 2$ and $q = 1$ with $\xi_0 = a$ and $\xi_1 = b$ are given in Problem 20.4 ($C_i = 1/\pi$ in (5.11) and $C_i = 1/\pi^2$ in (5.12)).

Theorem 5.3. *For* $p = 1, 2, \infty$, *there are constants* C_i *such that for* $q = 0, 1$,

$$\|f - \pi_q f\|_{L_p(a,b)} \leq C_i(b-a)\|f'\|_{L_p(a,b)}, \tag{5.11}$$

and for $q = 1$,

$$\|f - \pi_1 f\|_{L_p(a,b)} \leq C_i(b-a)^2 \|f''\|_{L_p(a,b)}, \tag{5.12}$$

$$\|f' - (\pi_1 f)'\|_{L_p(a,b)} \leq C_i(b-a) \|f''\|_{L_p(a,b)}. \tag{5.13}$$

Problem 5.15. Prove (5.11) for $q = 0$ with $C_i = 1$. Hint: use Taylor's theorem with the remainder in integral form. Prove (5.11) with $q = 1$ in the case $p = \infty$. *(Hard!)* Prove (5.12).

Problem 5.16. Compute approximate values of the interpolation constants C_i in Theorem 5.3 by explicit evaluation of the left- and right-hand sides assuming $f \in \mathcal{P}^{q+1}(a,b)$.

Problem 5.17. Prove by a change of variables that it suffices to prove the interpolation estimates of Theorem 5.3 in the case $(a,b) = (0,1)$.

Problem 5.18. Similar error bounds hold for interpolants of f that use values of f and its derivatives. *(Hard!)* Derive an error bound for the quadratic interpolating polynomial of f that interpolates the values $f(a)$, $f(b)$ and $f'(b)$.

5.2.3. Comparing accuracy

In this section, we compare the accuracy of different approximations of a function on the interval $[a, b]$. We caution that accuracy is not an absolute quantity; it must be measured with respect to the computational work used to produce the approximation. Computational work might be measured by the number of floating point operations or the amount of information about f that is required for example.

We first compare the quadratic Taylor polynomial, computed at a, with the quadratic interpolant computed using $\xi_0 = a$, $\xi_1 = (a+b)/2$, and $\xi_2 = b$. The error of the Taylor polynomial is bounded by $\frac{1}{6}(x-a)^3 \max_{[a,b]} |f'''|$ while the error of $p_2(x)$ is bounded by $\frac{1}{6}|x-\xi_0||x-\xi_1||x-\xi_2| \max_{[a,b]} |f'''|$. From this, we guess that the Taylor polynomial is more accurate in the immediate neighborhood of a, but not as accurate over the entire interval. In Fig. 5.7, we plot the errors of the two approximations of e^{-x} on $[0,1]$. In comparing the accuracy of the two

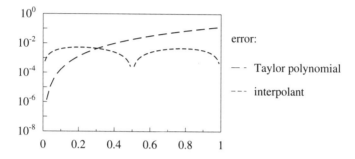

Figure 5.7: Errors of the quadratic Taylor polynomial and interpolant of e^{-x}.

approximations, we note that both use three pieces of information about f. If f, f', and f'' are known at a, then the Taylor polynomial is preferable if the approximation is needed in the immediate neighborhood of a. The interpolant is better for the purposes of approximating f over a larger interval.

We next compare error bounds for the constant and the linear interpolants. In this case, the linear interpolant uses two pieces of information about f, hence to make a fair comparison, we divide the interval $[a,b]$ into two equal pieces, and compare the linear interpolant $\pi_1 f$ to the piecewise constant interpolant $\tilde{\pi}_0 f(x)$ defined by

$$\tilde{\pi}_0 f(x) = \begin{cases} f(a), & a \le x \le (a+b)/2 \\ f(b), & (a+b)/2 < x \le b. \end{cases}$$

Recalling (5.8) and (5.9), we have for $a \le x \le b$,

$$|f(x) - \tilde{\pi}_0 f(x)| \le \frac{1}{2}(b-a) \max_{[a,b]} |f'|, \tag{5.14}$$

and

$$|f(x) - \pi_1 f(x)| \le \frac{1}{8}(b-a)^2 \max_{[a,b]} |f''|. \tag{5.15}$$

Comparing (5.14) and (5.15), we see that the bound of the error of the linear interpolant is smaller than that of the piecewise constant interpolant when

$$\max_{[a,b]} |f'| > \frac{b-a}{4} \max_{[a,b]} |f''|.$$

As $b - a$ becomes smaller, the error of the linear interpolant generally becomes smaller than the error of the piecewise constant interpolant. The size of $b - a$ for which the error of the linear interpolant is smaller depends on the relative size of $\max_{[a,b]} |f'|$ and $\frac{1}{4} \max_{[a,b]} |f''|$.

We have to be careful about drawing conclusions about the errors themselves, since the errors could be much smaller than the bounds. For example, if f is a constant, then the two approximations have the same error. However, it is possible to show that the error is about the size of the bound in (5.3) for almost all functions. We summarize by saying that the linear interpolant is *asymptotically* more accurate than a constant interpolant and that the piecewise constant approximation is *first order* accurate while the piecewise linear approximation is *second order* accurate. In general, increasing the degree of the interpolant increases the order of accuracy. But, we emphasize the asymptotic nature of this conclusion; to gain accuracy by increasing the degree, the interval may have to be decreased.

The final example we present shows that higher degree interpolants are not necessarily more accurate on a fixed interval. We set $f(x) = e^{-8x^2}$ and compute $\pi_8 f(x)$ using 9 equally spaced points in $[-2, 2]$. We plot f and $\pi_8 f$ in Fig. 5.8. While $\pi_8 f(x)$ agrees with $f(x)$ at the interpolation points, the error in between the points is terrible. If we use larger q, then the oscillations are even larger. The point is that the error bound depends both on the size of the interval and the derivatives of f, and both have to be taken into account in order to conclude that the error is small. In this case, e^{-8x^2} does not "act" like any polynomial on $[-2, 2]$ and as a result, high degree interpolants are not accurate.

Problem 5.19. Compute and graph $\pi_4(e^{-8x^2})$ on $[-2, 2]$, which interpolates e^{-8x^2} at 5 equally spaced points in $[-2, 2]$.

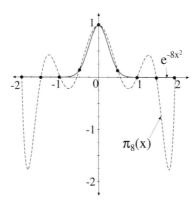

Figure 5.8: The interpolant of e^{-8x^2} using 9 points in $[-2, 2]$.

Problem 5.20. If you have access to a symbolic manipulating program, compute the error bound in (5.3) for $\pi_8(e^{-8x^2})$ on $[-2, 2]$. Graphing the derivatives of f is the easiest way to get the maximums.

5.3. Vector spaces of piecewise polynomials

As an alternative to increasing the order of the interpolating polynomial on a given interval, we may use piecewise polynomial approximations on increasingly finer partitions. Recall that we already made good use of this when proving the Fundamental Theorem of Calculus.

For a given interval $I = (a, b)$, we let $a = x_0 < x_1 < x_2 < \cdots < x_{m+1} = b$ be a *partition* or *triangulation* of (a, b) into sub-intervals $I_i = (x_{i-1}, x_i)$ of length $h_i = x_i - x_{i-1}$. We denote by $h(x)$ the *mesh function* defined by $h(x) = h_i$ for $x \in I_i$ and we use $\mathcal{T}_h = \{I_i\}$ to denote the set of intervals or *mesh* or triangulation. We illustrate the notation in Fig. 5.9.

We consider two vector spaces of piecewise polynomials of degree q on the mesh $\mathcal{T}_h$. The first is the space of discontinuous piecewise polynomials of degree q on (a, b), which we denote by

$$W_h^{(q)} = \{v : v|_{I_i} \in \mathcal{P}^q(I_i), \ i = 1, ..., m+1\}.$$

Here $v|_{I_i}$ denotes the *restriction* of v to the interval I_i, that is the function

Figure 5.9: A partition of an interval and the corresponding mesh function.

defined on I_i that agrees with v on I_i. The second is the set of continuous piecewise polynomials denoted by

$$V_h^{(q)} = \{v \in W_h^{(q)} : v \text{ is continuous on } I\}.$$

Note that these spaces depend on the underlying mesh, which is indicated with the subscript h, referring to the mesh function. In Fig. 5.12-Fig. 5.14, we show some typical functions in $V_h^{(1)}$ and $W_h^{(1)}$.

Problem 5.21. Prove that $W_h^{(q)}$ and $V_h^{(q)}$ are vector spaces.

A basis for $W_h^{(q)}$ can be constructed by combining for $i = 1, ..., m+1$, the set of functions that are equal to a basis function for $\mathcal{P}^q(I_i)$ for $x_{i-1} < x \le x_i$ and zero elsewhere. For example, to construct a basis for $W_h^{(1)}$, we choose the local basis functions for $\mathcal{P}^1(I_i)$,

$$\lambda_{i,0}(x) = \frac{x - x_i}{x_{i-1} - x_i} \quad \text{and} \quad \lambda_{i,1}(x) = \frac{x - x_{i-1}}{x_i - x_{i-1}},$$

and define the basis functions for $W_h^{(1)}$ by

$$\varphi_{i,j}(x) = \begin{cases} 0, & x \notin [x_{i-1}, x_i], \\ \lambda_{i,j}(x), & x \in [x_{i-1}, x_i], \end{cases} \quad i = 1, ..., m+1, \ j = 0, 1.$$

We plot such a basis function in Fig. 5.10. Generalizing this process, we see that the dimension of $W_h^{(q)}$ is $(m+1)(q+1)$.

Problem 5.22. Prove this formula.

Problem 5.23. Write down a basis for the set of piecewise quadratic polynomials $W_h^{(2)}$ on (a, b) and plot a sample of the functions.

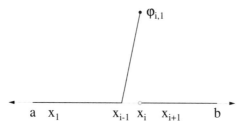

Figure 5.10: The basis function $\varphi_{i,1}$ associated to the interval $[x_{i-1}, x_i]$.

Constructing a basis for the space $V_h^{(q)}$ of continuous piecewise polynomials is a little more involved because the local Lagrange basis functions on each sub-interval have to be arranged so that the resulting basis functions are continuous on (a, b). In the case $q = 1$, we obtain the *hat functions* or *nodal basis functions* $\{\varphi_i\}$ for $V_h^{(1)}$ illustrated in Fig. 5.11. The formal definition is

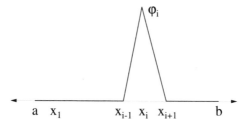

Figure 5.11: The hat function φ_i associated to node x_i.

$$\varphi_i(x) = \begin{cases} 0, & x \notin [x_{i-1}, x_{i+1}], \\ \dfrac{x - x_{i-1}}{x_i - x_{i-1}}, & x \in [x_{i-1}, x_i], \\ \dfrac{x - x_{i+1}}{x_i - x_{i+1}}, & x \in [x_i, x_{i+1}]. \end{cases}$$

The functions associated to the boundary nodes x_0 and x_{m+1} look like "half hats". Each hat function $\varphi_i(x)$ has the property that $\varphi_i \in V_h^{(1)}$ and $\varphi_i(x_j) = 1$ if $i = j$ and $\varphi_i(x_j) = 0$ otherwise. Observe that each hat function is defined on the whole of (a, b), but vanishes outside two intervals (or one interval if $i = 0$ or $m + 1$).

Problem 5.24. Write out equations for φ_0 and φ_{m+1}.

Problem 5.25. Prove that the set of functions $\{\varphi_i\}_{i=0}^{m+1}$ is a basis for $V_h^{(1)}$.

Since there is a hat function associated to each node, the dimension of the continuous piecewise linears $V_h^{(1)}$ is $m+2$ with one basis function for each node. Note that the hat functions are a *nodal basis* because the coefficients of $v \in V_h^{(1)}$ are simply the nodal values of v, i.e. each $v \in V_h^{(1)}$ has the representation

$$v(x) = \sum_{i=0}^{m+1} v(x_i)\varphi_i(x).$$

Problem 5.26. Prove this.

Problem 5.27. Determine a set of basis functions for the space of continuous piecewise quadratic functions $V_h^{(2)}$ on (a, b).

Problem 5.28. Determine a set of basis functions for the space of continuous piecewise cubic polynomials that have continuous first derivatives. Hint: see Problem 5.6.

5.4. Interpolation by piecewise polynomials

We now consider approximation using piecewise polynomials building on the above results for polynomial approximation. We let f be a given continuous function on an interval $[a, b]$ and consider interpolation into the spaces $V_h^{(1)}$ and $W_h^{(q)}$ for $q = 0, 1$, based on a partition $\{x_i\}_{i=0}^{m+1}$ of (a, b) with mesh function $h(x) = h_i = x_i - x_{i-1}$ on $I_i = (x_{i-1}, x_i)$.

5.4.1. Piecewise constant approximation

We define according to Fig. 5.12 a piecewise constant interpolant $\pi_h f \in W_h^{(0)}$ by

$$\pi_h f(x) = f(x_i) \quad \text{for } x_{i-1} < x \le x_i, \quad i = 1, ..., m+1.$$

Using Theorem 5.1 we obtain the following global error estimate:

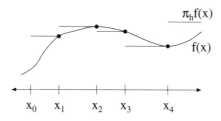

Figure 5.12: An example of a piecewise constant interpolant.

$$\|f - \pi_h f\|_{L_\infty(a,b)} \leq \max_{1 \leq i \leq m+1} h_i \|f'\|_{L_\infty(I_i)} = \|h\, f'\|_{L_\infty(a,b)}.$$
$$(5.16)$$

We remark that any value of f on a sub-interval I_i can be used to define $\pi_h f$ on I_i satisfying (5.16), not just one of the values at an endpoint as above. In particular, the mean value theorem for integrals asserts that there is a point $x_{i-1} \leq \xi_i \leq x_i$ such that $f(\xi_i)$ is equal to the average value of f on $[x_{i-1}, x_i]$, i.e.

$$f(\xi_i) = \frac{1}{x_i - x_{i-1}} \int_{x_{i-1}}^{x_i} f(x)\, dx,$$

and thus we may define $\pi_h f(x) = f(\xi_i)$ on I_i as the average of f over I_i.

Problem 5.29. Prove that any value of f on the sub-intervals can be used to define $\pi_h f$ satisfying the error bound (5.16). Prove that choosing the midpoint improves the bound by an extra factor $1/2$.

Problem 5.30. Prove the mean value theorem for integrals. Hint: use the mean value theorem.

5.4.2. Piecewise linear approximation

The continuous piecewise linear interpolant $\pi_h f \in V_h^{(1)}$ is defined by

$$\pi_h f(x) = f(x_{i-1}) \frac{x - x_i}{x_{i-1} - x_i} + f(x_i) \frac{x - x_{i-1}}{x_i - x_{i-1}} \quad \text{for } x_{i-1} \leq x \leq x_i,$$

for $i = 1, ..., m+1$; see Fig. 5.13.

Problem 5.31. Compute the continuous piecewise linear interpolant of e^{-8x^2} using 9 equally spaced nodes in $[0, 2]$ and compare to the polynomial interpolant on the same nodes.

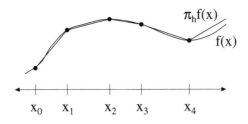

Figure 5.13: An example of a continuous piecewise linear interpolant.

A discontinuous linear interpolant $\pi_h f \in W_h^{(1)}$ may be defined by choosing for $i = 1, ..., m + 1$, two points $\xi_{i,0} \neq \xi_{i,1} \in [x_{i-1}, x_i]$ and setting

$$\pi_h f(x) = f(\xi_{i,0})\frac{x - \xi_{i,1}}{\xi_{i,0} - \xi_{i,1}} + f(\xi_{i,1})\frac{x - \xi_{i,0}}{\xi_{i,1} - \xi_{i,0}} \quad \text{for } x_{i-1} \leq x \leq x_i.$$

The resulting piecewise linear function is in general discontinuous across the nodes x_i, and we introduce some notation to account for this. We let $\pi_h f_i^+ = \lim_{x \downarrow x_i} \pi_h f(x)$ and $\pi_h f_i^- = \lim_{x \uparrow x_i} \pi_h f(x)$ denote the right- and left-hand limits of $\pi_h f(x)$ at x_i, and let $[\pi_h f]_i = \pi_h f_i^+ - \pi_h f_i^-$ denote the corresponding jump. We illustrate this in Fig. 5.14.

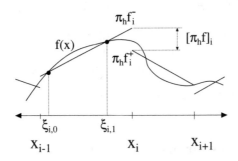

Figure 5.14: An example of a discontinuous piecewise linear interpolant.

The following error bound holds for the nodal interpolant for discontinuous and continuous piecewise linear approximation. The constants C_i depend on the choice of interpolation points in the case of discontinuous piecewise linears.

Theorem 5.4. *For $p = 1, 2$ and ∞, there are constants C_i such that*

$$\|f - \pi_h f\|_{L_p(a,b)} \le C_i \|h^2 f''\|_{L_p(a,b)}, \tag{5.17}$$

$$\|f - \pi_h f\|_{L_p(a,b)} \le C_i \|h f'\|_{L_p(a,b)}, \tag{5.18}$$

$$\|f' - (\pi_h f)'\|_{L_p(a,b)} \le C_i \|h f''\|_{L_p(a,b)}. \tag{5.19}$$

Problem 5.32. Prove (5.17) for approximations computed by interpolation and determine the constant C_i in the continuous and discontinuous cases.

Problem 5.33. Determine the continuous piecewise quadratic polynomial that interpolates a function and its derivative at the nodes of a partition. Note that the quadratic only may interpolate the function's derivative "on one side".

Problem 5.34. *(Hard.)* Construct a piecewise cubic polynomial function with a continuous first derivative that interpolates a function and its first derivative at the nodes of a partition.

5.4.3. Adapted meshes

Suppose $f(x)$ is a given function on an interval I and we want to compute a piecewise polynomial interpolant $\pi_h f$ of a specified degree such that the error in a specified norm is less than a given tolerance. It is natural to ask how to distribute the points efficiently in the sense of using the least number of points to achieve the desired accuracy. This is motivated because the cost (measured by computer time) of evaluating the interpolant is determined primarily by the number of interpolation points. The general answer to this question is that the mesh with the minimal number of mesh points is a mesh that *equidistributes* the error, so that the contribution to the total error from each sub-interval of the mesh is roughly equal. If we are using the maximum norm to measure the error, this means that the maximum error over each sub-interval is equal to the given tolerance. If the norm is an L_1 norm, equidistribution means that the integral of the absolute value of the error over each sub-interval is equal to the tolerance divided by the number of sub-intervals. Essentially, an equidistributed mesh has the minimal number of mesh points because dividing a sub-interval on which the error is largest reduces the total error maximally. We return to the principle of equidistribution of error below.

We illustrate this with the piecewise constant interpolation of e^x on $[0,1]$, that we require to be accurate within TOL $= .1$ in $\|\cdot\|_{L_\infty(0,1)}$, where TOL is the *error tolerance*. The mesh that equidistributes the error bound (5.16) on each interval I_i satisfies

$$(x_i - x_{i-1}) \max_{[x_{i-1},x_i]} e^x = \text{TOL}.$$

Since e^x is monotonically increasing, we have $\max_{[x_{i-1},x_i]} e^x = e^{x_i}$, and thus $x_i = x_{i-1} + .1e^{-x_i}$ for $i \geq 1$. This is a nonlinear equation, which we approximate by

$$x_i = x_{i-1} + .1e^{-x_{i-1}}, \quad i \geq 1,$$

where $x_0 = 0$. These equations are easily solved, in a step by step procedure, and we find that 17 nodes are required. In Fig. 5.15, we plot the approximation together with e^x on the left. On the right, we plot the error. Note that the error is slightly above the tolerance as a result of the

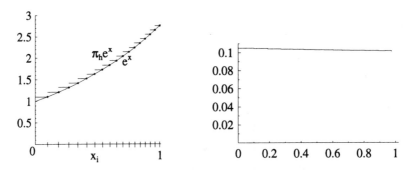

Figure 5.15: Interpolation on an approximately equidistributed mesh for e^x with TOL$= .1$ On the left, we show the mesh and approximation. On the right, we show the error.

approximation we made. If we use a uniform mesh, then we have to use the spacing of the smallest interval (the one containing 1) to achieve the same accuracy. This means using 26 nodes, about 50% more than in the equidistributed mesh. As the tolerance decreases, the relative number of points increases. Choosing a more wildly varying function can make the difference huge. In general, computation of an equidistributed mesh results in nonlinear equations which have to be solved approximately.

Problem 5.35. Determine the set of nodes $\{x_i\}$ in the above example.

Problem 5.36. Compute an equidistributed mesh for piecewise linear interpolation of e^x on $[0, 1]$ and compare to the mesh above.

Problem 5.37. Compute an equidistributed mesh for piecewise constant approximation of x^2 on $[0, 1]$. Plot the approximation and the error.

Problem 5.38. A related problem is to fix the number of nodes $m + 1$ (i.e. the amount of available work) and then compute the mesh that minimizes the error for that number of nodes. Once again, an equidistributed mesh provides the best approximation. This problem, however, results in a nonlinear system of equations. For piecewise constant interpolation of e^x on $[0, 1]$ using $m + 1$ nodes, write out the system of equations for the nodes that minimizes the error. Don't forget that the length of the sub-intervals has to sum to 1.

5.5. Quadrature

We apply the results in the previous section to the computation of the definite integral $\int_a^b f(x)\, dx$ by replacing f by different piecewise polynomial approximations on a mesh $a = x_0 < x_1 < \ldots < x_{m+1} = b$. This generalizes the strategy used in the proof of the Fundamental Theorem of Calculus in Chapter 3. We refer to the resulting quadratures as *composite* in the sense that they result from applying the same basic rule repeatedly over the sub-intervals. We consider a couple of quadrature rules based on piecewise constant approximation; the *rectangle rule*, the *endpoint rule*, and the *midpoint rule*. In the rectangle rule, the interpolation point is arbitrary; in the endpoint rule, it is one of the endpoints; and of course in the midpoint rule, it is the midpoint of each sub-interval. We also consider two rules based on piecewise linears; the trapezoidal rule based on continuous piecewise linear nodal interpolation, and the *Gauss rule* based on discontinuous piecewise linears.

5.5.1. The composite rectangle rule.

Approximating f by a piecewise constant interpolant interpolating at the points $x_{i-1} \le \xi_i \le x_i$, $i = 1, \ldots, m + 1$, we get the quadrature formula,

$$\int_a^b f(x)\,dx \approx \sum_{i=1}^{m+1} f(\xi_i)h_i, \qquad (5.20)$$

with the *quadrature points* $\{\xi_i\}$ and the *quadrature weights* $\{h_i\}$. This is the *composite rectangle rule*, that we used in the proof of the Fundamental Theorem.

We denote the absolute value of the quadrature error in computing $\int_a^b f(x)\,dx$ by the various quadrature formulas by $E(f)$, so in this case

$$E(f) = \left| \int_a^b f(x)\,dx - \sum_{i=1}^{m+1} f(\xi_i)h_i \right|.$$

Recalling (3.32), we have

$$E(f) \le \frac{1}{2}\sum_{i=1}^{m+1} h_i \max_{[x_{i-1},x_i]} |f'| h_i. \qquad (5.21)$$

5.5.2.   The composite midpoint rule

In the error analysis giving (5.21), the choice of quadrature points $x_{i-1} \le \xi_i \le x_i$ makes no visible difference to the accuracy. However, there is one choice, namely the midpoint $\xi_i = (x_{i-1} + x_i)/2$, that in general yields more accuracy than any other choice. The resulting quadrature formula is called the *composite midpoint rule* and it is the lowest order *composite Gauss rule*.

Problem 5.39. Write out the composite midpoint rule explicitly.

We can estimate the error in the composite midpoint rule in a way that takes into account a cancellation of errors that was lost in the proof of (5.21). We first consider the integration over one interval $[x_0, x_1]$, on which the midpoint quadrature rule is

$$\int_{x_0}^{x_1} f(x)\,dx \approx f\left(\frac{x_0 + x_1}{2}\right)(x_1 - x_0).$$

This formula gives the exact value of the integral for both constant and linear functions:

$$\int_{x_0}^{x_1} 1\,dx = (x_1 - x_0)$$

and

$$\int_{x_0}^{x_1} x\, dx = \frac{x_1^2}{2} - \frac{x_0^2}{2} = \frac{1}{2}(x_0 + x_1)(x_1 - x_0).$$

while, the endpoint rule in general is exact only if f is constant. We say that a quadrature rule has *precision* r if the rule gives the exact value for the integral of all polynomials of degree $r - 1$ or less, but there is some polynomial of degree r for which the quadrature rule is not exact. The rectangle rule has precision 1 in general, but the midpoint rule has precision 2.

The precision of a quadrature rule is related to the order of accuracy. To analyze the error of the midpoint rule we start from Taylor's theorem

$$f(x) = f\left(\frac{x_1 + x_0}{2}\right) + f'\left(\frac{x_1 + x_0}{2}\right)\left(x - \frac{x_1 + x_0}{2}\right) + \frac{1}{2}f''(\eta_x)\left(x - \frac{x_1 + x_0}{2}\right)^2,$$

for some η_x between x and $(x_1 + x_0)/2$, and integrate to get

$$\int_{x_0}^{x_1} f(x)\, dx = \int_{x_0}^{x_1} f\left(\frac{x_1 + x_0}{2}\right) dx + \int_{x_0}^{x_1} f'\left(\frac{x_1 + x_0}{2}\right)\left(x - \frac{x_1 + x_0}{2}\right) dx$$

$$+ \int_{x_0}^{x_1} \frac{1}{2}f''(\eta_x)\left(x - \frac{x_1 + x_0}{2}\right)^2 dx. \quad (5.22)$$

The first integral on the right-hand side of (5.22) is just the quadrature formula. The second integral on the right is zero because of cancellation.

Problem 5.40. Prove that

$$\int_{x_0}^{x_1} f'\left(\frac{x_1 + x_0}{2}\right)\left(x - \frac{x_1 + x_0}{2}\right) dx = 0.$$

We thus have

$$\left| \int_{x_0}^{x_1} f(x)\, dx - f\left(\frac{x_1 + x_0}{2}\right)(x_1 - x_0) \right|$$

$$\leq \frac{1}{2} \max_{[x_0, x_1]} |f''| \int_{x_0}^{x_1} \left(x - \frac{x_1 + x_0}{2}\right)^2 dx$$

$$\leq \frac{1}{24}(x_1 - x_0)^3 \max_{[x_0, x_1]} |f''|. \quad (5.23)$$

Problem 5.41. Verify (5.23).

For the error of the composite midpoint rule, accordingly

$$E(f) \le \sum_{i=1}^{m+1} \frac{1}{24} h_i^2 \max_{[x_{i-1}, x_i]} |f''| h_i. \tag{5.24}$$

Problem 5.42. Compare the accuracy of endpoint rule and the midpoint rule for the computation of $\int_0^1 e^x \, dx$ using $m = 10$.

5.5.3. The composite trapezoidal rule

Replacing f by a continuous piecewise linear nodal interpolant of f on a partition $\{x_i\}_{i=0}^{m+1}$ of $[a, b]$, we obtain the *composite trapezoidal rule*

$$\int_a^b f(x) \, dx \approx \sum_{i=1}^{m+1} \int_{x_{i-1}}^{x_i} \left(f(x_{i-1}) \frac{x - x_i}{x_{i-1} - x_i} + f(x_i) \frac{x - x_{i-1}}{x_i - x_{i-1}} \right) dx,$$

that is

$$\int_a^b f(x) \, dx \approx \sum_{i=1}^{m+1} \frac{f(x_{i-1}) + f(x_i)}{2} h_i$$

$$= f(x_0) \frac{h_1}{2} + \sum_{i=1}^{m} f(x_i) \frac{h_i + h_{i+1}}{2} + f(x_{m+1}) \frac{h_{m+1}}{2}. \tag{5.25}$$

Problem 5.43. Verify (5.25).

Problem 5.44. Show that the trapezoidal rule has precision 2.

To estimate the error in the trapezoidal rule, we use (5.3) to get

$$\left| \int_{x_{i-1}}^{x_i} f(x) \, dx - \int_{x_{i-1}}^{x_i} \pi_1 f(x) \, dx \right| \le \frac{1}{12} h_i^3 \max_{[x_{i-1}, x_i]} |f''|. \tag{5.26}$$

Problem 5.45. Prove (5.26).

Applying this estimate to each sub-interval, we find the following estimate for the composite trapezoidal rule

$$E(f) \le \sum_{i=1}^{m+1} \frac{1}{12} h_i^2 \max_{[x_{i-1}, x_i]} |f''| h_i. \tag{5.27}$$

Problem 5.46. (a) Derive a quadrature rule based on piecewise quadratic interpolation using the endpoints and midpoint of each sub-interval. Write down an explicit formula for the quadrature. This is called *Simpson's rule*. (b) Use (5.3) to prove the following error bound on the error $E(f)$ of Simpson's rule:

$$E(f) \leq \sum_{i=1}^{m+1} \frac{1}{192} h_i^3 \max_{[x_{i-1},x_i]} |f^{(3)}| h_i.$$

(c) Show that Simpson's rule has precision 4. This suggests that the error bound in (b) is too large. Show that in fact the error is bounded by

$$E(f) \leq \sum_{i=1}^{m+1} \frac{1}{2880} h_i^4 \max_{[x_{i-1},x_i]} |f^{(4)}| h_i.$$

5.5.4. The composite two-point Gauss rule

Comparing (5.24) and (5.27), the composite midpoint rule appears to be more accurate than the composite trapezoidal rule, though both approximations are second order accurate. It is natural to ask if there is a more accurate quadrature rule based on linear interpolation on the sub-intervals. This turns out to be the case if we use interpolation nodes inside, in which case the corresponding interpolant is discontinuous in general.

The interpolation nodes in $[x_{i-1}, x_i]$ are chosen as follows

$$\xi_{i,0} = \frac{x_i + x_{i-1}}{2} - \frac{\sqrt{3}}{6} h_i \quad \text{and} \quad \xi_{i,1} = \frac{x_i + x_{i-1}}{2} + \frac{\sqrt{3}}{6} h_i,$$

and the resulting quadrature formula is

$$\int_a^b \pi_h f(x)\, dx = \sum_{i=1}^{m+1} \left(\frac{h_i}{2} f(\xi_{i,0}) + \frac{h_i}{2} f(\xi_{i,1}) \right). \tag{5.28}$$

Problem 5.47. Prove that the two point Gauss rule has precision 4.

The quadrature error $E(f)$ in the corresponding composite rule is bounded by

$$E(f) \leq \sum_{i=1}^{m+1} \frac{1}{4320} h_i^4 \max_{[x_{i-1},x_i]} |f^{(4)}| h_i. \tag{5.29}$$

It is possible to derive higher order quadrature rules using piecewise polynomial interpolants of higher degree. The Newton formulas use equally spaced interpolation nodes in each sub-interval. It is also possible to derive higher order Gauss formulas. See Atkinson ([2]) for a complete discussion.

Problem 5.48. Compare the accuracies of the composite trapezoidal, Simpson's, and two-point Gauss rules for approximating $\int_0^2 (1 + x^2)^{-1}\, dx$ using $m = 10$ and a uniformly spaced mesh.

Problem 5.49. Develop adaptive algorithms for the above quadrature rules. Compare the performance on different functions $f(x)$.

5.6.   The L_2 projection into a space of polynomials

A polynomial interpolating a given function $f(x)$ agrees with point values of f or derivatives of f. An alternative is to determine the polynomial so that certain averages agree. These could include the usual average of f over an interval $[a, b]$ defined by

$$(b - a)^{-1} \int_a^b f(x)\, dx,$$

or a *generalized average* of f with respect to a *weight* g defined by

$$(f, g) = \int_a^b f(x) g(x)\, dx.$$

Introducing a weight is like changing scale in the average. Averages of functions are related naturally to inner products, just as for vectors in $\mathbb{R}^d$, since the generalized average is just the Euclidean inner product (f, g) of f and g on (a, b).

Problem 5.50. Prove that the space of continuous functions on an interval $[a, b]$ is a vector space and that the space of polynomials of degree q and less is a subspace.

The *orthogonal projection*, or L_2-projection, of f into $\mathcal{P}^q(a, b)$ is the polynomial $Pf \in \mathcal{P}^q(a, b)$ such that

$$(f - Pf, v) = 0 \quad \text{for all } v \in \mathcal{P}^q(a, b). \tag{5.30}$$

We see that Pf is defined so that certain average values of Pf are the same as those of f. Note that the orthogonality relation (5.30) defines Pf uniquely. Suppose that v and u are two polynomials in $\mathcal{P}^q(a, b)$ such that $(f - v, w) = (f - u, w) = 0$ for all $w \in \mathcal{P}^q(a, b)$. Subtracting, we conclude that $(u - v, w) = 0$ for all $w \in \mathcal{P}^q(a, b)$, and choosing $w = u - v \in \mathcal{P}^q(a, b)$ we find

$$\int_a^b |u - v|^2 \, dx = 0,$$

which shows that $u = v$ since $|u - v|^2$ is a non-negative polynomial with zero area underneath its graph. Since (5.30) is equivalent to an $(q + 1) \times (q + 1)$ system of equations, we conclude that Pf exists and is uniquely determined.

Problem 5.51. Prove the last statement.

As an example, the orthogonal projection of a function into the space of constant polynomials $\mathcal{P}^0(a, b)$ is simply the average value of f:

$$Pf = \frac{1}{b - a} \int_a^b f(x) \, dx.$$

For $q > 0$, computing the orthogonal projection involves solving a system of linear equations for the coefficients. For example in the case $q = 1$, seeking Pf on the form

$$Pf(x) = \alpha + \beta \left(x - \frac{a + b}{2} \right).$$

with the coefficients α and β to determine, we have for $v \in \mathcal{P}^1(a, b)$

$$(f - Pf, v) = \int_a^b \left(f(x) - \left(\alpha + \beta \left(x - \frac{a + b}{2} \right) \right) \right) v(x) \, dx.$$

Choosing first $v = 1$ and evaluating the integrals we find that

$$\alpha = \frac{1}{b - a} \int_a^b f(x) \, dx.$$

Next, we choose $v = x - \frac{a + b}{2}$ and compute to find that

$$\beta = \frac{12}{(b - a)^3} \int_a^b f(x) \left(x - \frac{a + b}{2} \right) dx.$$

We note that to compute the orthogonal projection defined by (5.30), it suffices to let v vary through a set of basis functions.

Problem 5.52. Prove versions of (5.8) and (5.11) for the L_2 projection of f into the constant polynomials.

Problem 5.53. (a) Compute the formula for the orthogonal projection of f into the space of quadratic polynomials on $[a, b]$. (b) Write down a system of equations that gives the orthogonal projection of a function into the space of polynomials of degree q.

Below we will use the L_2 projection $P_h f$ of a function f on (a, b) into a space of piecewise polynomials V_h, such as $W_h^{(q)}$ or $V_h^{(q)}$, defined by the requirement that $P_h f \in V_h$ and

$$(f - P_h f, v) = 0 \quad \text{for } v \in V_h.$$

For discontinuous piecewise polynomials this reduces to the L_2 projection on each sub-interval. For continuous piecewise linear polynomials, computing an L_2 projection requires the solution of a linear system of algebraic equations in the nodal degrees of freedom for V_h. We return to this issue at several occasions below, e.g. in Chapter 6, 8 and 15.

The L_2 projection of a function into a piecewise polynomial space V_h is the best approximation in the L_2 norm.

Lemma 5.5. *The L_2 projection $P_h f$ of a function f on an interval (a, b) into a vector space of piecewise polynomials V_h, satisfies*

$$\|f - P_h f\|_{L_2(a,b)} \leq \|f - v\|_{L_2(a,b)} \quad \text{for all } v \in V_h.$$
$$(5.31)$$

Problem 5.54. Prove this lemma. Hint: look at the proof of Lemma 4.1 in Chapter 4.

5.7. Approximation by trigonometric polynomials

Let f be defined on $\mathbb{R}$ and assume f is periodic with period 2π so that $f(x + 2n\pi) = f(x)$ for $n = \pm 1, \pm 2, \ldots$. If f is continuous and f' is continuous, except at a finite number of points in a period, then $f(x)$ can be represented by a convergent *Fourier series* as follows

$$f(x) = \frac{a_0}{2} + \sum_{n=1}^{\infty} (a_n \cos(nx) + b_n \sin(nx)),$$

where the *Fourier coefficients* a_n and b_n are defined by

$$a_n = \frac{1}{\pi} \int_0^{2\pi} \cos(nx) f(x)\, dx \text{ and } b_n = \frac{1}{\pi} \int_0^{2\pi} \sin(nx) f(x)\, dx.$$
$$(5.32)$$

It is thus natural to seek to approximate f by a finite sum

$$\pi_q f(x) = \frac{a_0}{2} + \sum_{n=1}^{q} \left(a_n \cos(nx) + b_n \sin(nx) \right),$$

where the a_n and b_n are the Fourier coefficients. This corresponds to choosing $\pi_q f$ as the $L_2(0, 2\pi)$ projection into the space spanned by $1, \cos(x), \sin(x), ..., \cos(qx), \sin(qx)$.

Problem 5.55. Prove this. Hint: use the fact that the functions $\{1, \cos(x), \sin(x), ..., \cos(qx), \sin(qx)\}$ are pairwise orthogonal.

Problem 5.56. Prove that if f has continuous periodic derivatives of order q, then there is a constant C such that $|a_n| + |b_n| \leq Cn^{-q}$ for $n = \pm 1, \pm 2, ...$

In principle we could try to use other coefficients a_n and b_n than the Fourier coefficients just discussed. One possibility is to require $\pi_q f$ to interpolate f at $2q + 1$ distinct nodes $\{\xi_i\}$ in the period.

Problem 5.57. Compute for $q = 2$ trigonometric interpolating polynomials for the periodic functions that reduce to (a) $f(x) = x$ (b) $f(x) = x(2\pi - x)$ on $[0, 2\pi)$.

In general, we have to use quadrature formulas to evaluate the integrals in (5.32). If we use the composite trapezoidal rule on $2q + 1$ equally spaced intervals in the period, then we obtain the same formulas for the coefficients obtained by requiring the trigonometric polynomial to interpolate f at the $2q + 1$ nodes.

If f is odd so that $f(-x) = f(x)$, then the a_n are all zero, and we may represent $f(x)$ as a sine series. In particular, if $f(x)$ is defined on $(0, \pi)$, then we may extend f as an odd 2π-periodic function and represent $f(x)$ as a sine series:

$$f(x) = \sum_{n=1}^{\infty} b_n \sin(nx), \quad \text{with} \quad b_n = \frac{2}{\pi} \int_0^{\pi} \sin(nx) f(x)\, dx$$
$$(5.33)$$

In order for the extended function to be continuous, it is necessary that $f(0) = f(\pi) = 0$.

6

Galerkin's Method

It is necessary to solve differential equations. (Newton)

Ideally, I'd like to be the eternal novice, for then only, the surprises would be endless. (Keith Jarret)

In Chapters 3 and 5, we discussed the numerical solution of the simple initial value problem $u'(x) = f(x)$ for $a < x \leq b$ and $u(a) = u_0$, using piecewise polynomial approximation. In this chapter, we introduce *Galerkin's method* for solving a general differential equation, which is based on seeking an (approximate) solution in a (finite-dimensional) space spanned by a set of basis functions which are easy to differentiate and integrate, together with an orthogonality condition determining the coefficients or coordinates in the given basis. With a finite number of basis functions, Galerkin's method leads to a system of equations with finitely many unknowns which may be solved using a computer, and which produces an approximate solution. Increasing the number of basis functions improves the approximation so that in the limit the exact solution may be expressed as an infinite series. In this book, we normally use Galerkin's method in the computational form with a finite number of basis functions. The basis functions may be global polynomials, piecewise polynomials, trigonometric polynomials or other functions. The finite element method in basic form is Galerkin's method with piecewise polynomial approximation. In this chapter, we apply Galerkin's method to two examples with a variety of basis functions. The first example is an initial value problem that models population growth and we use a global polynomial approximation. The second example is a boundary value problem that models the flow of heat in a wire and we use piecewise polynomial approximation, more precisely piecewise linear approximation. This is a classic example of the *finite element method*. For the second example, we also discuss the *spectral method* which is Galerkin's method with trigonometric polynomials.

The idea of seeking a solution of a differential equation as a linear combination of simpler basis functions, is old. Newton and Lagrange used power series with global polynomials and Fourier and Riemann used Fourier series based on trigonometric polynomials. These approaches work for certain differential equations posed on domains with simple geometry and may give valuable qualitative information, but cannot be used for most of the problems arising in applications. The finite element method based on piecewise polynomials opens the possibility of solving general differential equations in general geometry using a computer. For some problems, combinations of trigonometric and piecewise polynomials may be used.

6.1. Galerkin's method with global polynomials

6.1.1. A population model

In the simplest model for the growth of a population, like the population of rabbits in West Virginia, the rate of growth of the population is proportional to the population itself. In this model we ignore the effects of predators, overcrowding, and migration, for example, which might be okay for a short time provided the population of rabbits is relatively small in the beginning. We assume that the time unit is chosen so that the model is valid on the time interval $[0, 1]$. We will consider more realistic models valid for longer intervals later in the book. If $u(t)$ denotes the population at time t then the differential equation expressing the simple model is $\dot{u}(t) = \lambda u(t)$, where λ is a positive real constant and $\dot{u} = du/dt$. This equation is usually posed together with an initial condition $u(0) = u_0$ at time zero, in the form of an initial value problem:

$$\begin{cases} \dot{u}(t) = \lambda u(t) & \text{for } 0 < t \leq 1, \\ u(0) = u_0. \end{cases} \tag{6.1}$$

The solution of (6.1), $u(t) = u_0 \exp(\lambda t)$, is a smooth increasing function when $\lambda > 0$.

6.1.2. Galerkin's method

We now show how to compute a polynomial approximation U of u in the set of polynomials $V^{(q)} = \mathcal{P}^q(0, 1)$ on $[0, 1]$ of degree at most q using Galerkin's method. We know there are good approximations of the solution u in this set, for example the Taylor polynomial and interpolating polynomials, but these require knowledge of u or derivatives of u at certain points in $[0, 1]$. The goal here is to compute a polynomial approximation of u using only the information that u solves a specified differential equation and has a specified value at one point. We shall see that this is precisely what Galerkin's method achieves. Since we already know the analytic solution in this model case, we can use this knowledge to evaluate the accuracy of the approximations.

Because $\{t^j\}_{j=0}^q$ is a basis for $V^{(q)}$, we can write $U(t) = \sum_{j=0}^q \xi_j t^j$ where the coefficients $\xi_j \in \mathbb{R}$ are to be determined. It is natural to require that $U(0) = u_0$, that is $\xi_0 = u_0$, so we may write

$$U(t) = u_0 + \sum_{j=1}^{q} \xi_j t^j,$$

where the "unknown part" of U, namely $\sum_{j=1}^q \xi_j t^j$, is in the subspace $V_0^{(q)}$ of $V^{(q)}$ consisting of the functions in $V^{(q)}$ that are zero at $t = 0$, i.e. in $V_0^{(q)} = \{v : v \in V^{(q)}, v(0) = 0\}$.

Problem 6.1. Prove that $V_0^{(q)}$ is a subspace of $V^{(q)}$.

We determine the coefficients by requiring U to satisfy the differential equation in (6.1) in a suitable "average" sense. Of course U can't satisfy the differential equation at every point because the exact solution is not a polynomial. In Chapter 4, we gave a concrete meaning to the notion that a function be zero on average by requiring the function to be orthogonal to a chosen subspace of functions. The Galerkin method is based on this idea. We define the *residual error* of a function v for the equation (6.1) by

$$R(v(t)) = \dot{v}(t) - \lambda v(t).$$

The residual error $R(v(t))$ is a function of t once v is specified. $R(v(t))$ measures how well v satisfies the differential equation at time t. If the residual is identically zero, that is $R(v(t)) \equiv 0$ for all $0 \le t \le 1$, then the equation is satisfied and v is the solution. Since the exact solution u is not a polynomial, the residual error of a function in $V^{(q)}$ that satisfies the initial condition is never identically zero, though it can be zero at distinct points.

The Galerkin approximation U is the function in $V^{(q)}$ satisfying $U(0) = u_0$ such that its residual error $R(U(t))$ is orthogonal to all functions in $V_0^{(q)}$, i.e.,

$$\int_0^1 R(U(t))v(t)\, dt = \int_0^1 (\dot{U}(t) - \lambda U(t))v(t)\, dt = 0 \quad \text{for all } v \in V_0^{(q)}. \tag{6.2}$$

This is the *Galerkin orthogonality* property of U, or rather of the residual $R(U(t))$. Since the coefficient of U with respect to the basis function 1 for $V^{(q)}$ is already known ($\xi_0 = u_0$), we require (6.2) to hold only for functions v in $V_0^{(q)}$. By way of comparison, note that the true solution satisfies a stronger orthogonality condition, namely

$$\int_0^1 (\dot{u} - \lambda u)v\, dt = 0 \quad \text{for } \textit{all} \text{ functions } v. \tag{6.3}$$

We refer to the set of functions where we seek the Galerkin solution U, in this case the space $V^{(q)}$ of polynomials w satisfying $w(0) = u_0$, as the *trial space* and the space of the functions used for the orthogonality condition, which is $V_0^{(q)}$, as the *test space*. In this case, the trial and test space are different because of the *non-homogeneous* initial condition $w(0) = u_0$ (assuming $u_0 \ne 0$), satisfied by the trial functions and the homogeneous boundary condition $v(0) = 0$ satisfied by the test functions $v \in V_0^{(q)}$. In general, different methods are obtained choosing the trial and test spaces in different ways.

6.1.3.   The discrete system of equations

We now show that (6.2) gives an invertible system of linear algebraic equations for the coefficients of U. Substituting the expansion for U

into (6.2) gives

$$\int_0^1 \left(\sum_{j=1}^q j\xi_j t^{j-1} - \lambda u_0 - \lambda \sum_{j=1}^q \xi_j t^j \right) v(t)\, dt = 0 \quad \text{for all } v \in V_0^{(q)}.$$

It suffices to insure that this equation holds for every basis function for $V_0^{(q)}$, yielding the set of equations:

$$\sum_{j=1}^q j\xi_j \int_0^1 t^{j+i-1}\, dt - \lambda \sum_{j=1}^q \xi_j \int_0^1 t^{j+i}\, dt = \lambda u_0 \int_0^1 t^i\, dt, \quad i = 1, ..., q,$$

where we have moved the terms involving the initial data to the right-hand side. Computing the integrals gives

$$\sum_{j=1}^q \left(\frac{j}{j+i} - \frac{\lambda}{j+i+1} \right) \xi_j = \frac{\lambda}{i+1} u_0, \quad i = 1, ..., q. \tag{6.4}$$

This is a $q \times q$ system of equations that has a unique solution if the matrix $A = (a_{ij})$ with coefficients

$$a_{ij} = \frac{j}{j+i} - \frac{\lambda}{j+i+1}, \quad i, j = 1, ..., q,$$

is invertible. It is possible to prove that this is the case, though it is rather tedious and we skip the details. In the specific case $u_0 = \lambda = 1$ and $q = 3$, the approximation is

$$U(t) \approx 1 + 1.03448t + .38793t^2 + .301724t^3,$$

which we obtain solving a 3×3 system.

Problem 6.2. Compute the Galerkin approximation for $q = 1, 2, 3$, and 4 assuming that $u_0 = \lambda = 1$.

Plotting the solution and the approximation for $q = 3$ in Fig. 6.1, we see that the two essentially coincide.

Since we know the exact solution u in this case, it is natural to compare the accuracy of U to other approximations of u in $V^{(q)}$. In Fig. 6.2, we plot the errors of U, the third degree polynomial interpolating u at $0, 1/3, 2/3$, and 1, and the third degree Taylor polynomial

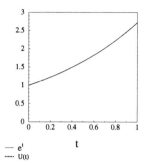

Figure 6.1: The solution of $\dot{u} = u$ and the third degree Galerkin approximation.

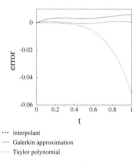

Figure 6.2: The errors of the third degree Galerkin approximation, a third degree interpolant of the solution, and the third degree Taylor polynomial of the solution.

of u computed at $t = 0$. The error of U compares favorably with the error of the interpolant of U and both of these are more accurate than the Taylor polynomial of u in the region near $t = 1$ as we would expect. We emphasize that the Galerkin approximation U attains this accuracy without any specific knowledge of the solution u except the initial data at the expense of solving a linear system of equations.

Problem 6.3. Compute the $L_2(0,1)$ projection into $\mathcal{P}^3(0,1)$ of the exact solution u and compare to U.

Problem 6.4. Determine the discrete equations if the test space is changed to $V^{(q-1)}$.

6.1.4. A surprise: ill-conditioning

Stimulated by the accuracy achieved with $q = 3$, we compute the approximation with $q = 9$. We solve the linear algebraic system in two ways: first exactly using a symbolic manipulation package and then approximately using Gaussian elimination on a computer that uses roughly 16 digits. In general, the systems that come from the discretization of a differential equation are too large to be solved exactly and we are forced to solve them numerically with Gaussian elimination for example.

We obtain the following coefficients ξ_i in the two computations:

exact coefficients   approximate coefficients

$$
\begin{pmatrix}
.14068... \\
.48864... \\
.71125... \\
.86937... \\
.98878... \\
1.0827... \\
1.1588... \\
1.2219... \\
1.2751...
\end{pmatrix}
\qquad
\begin{pmatrix}
152.72... \\
-3432.6... \\
32163.2... \\
-157267.8... \\
441485.8... \\
-737459.5... \\
723830.3... \\
-385203.7... \\
85733.4...
\end{pmatrix}
$$

We notice the huge difference, which makes the approximately computed U worthless. We shall now see that the difficulty is related to the fact that the system of equations (6.4) is *ill-conditioned* and this problem is exacerbated by using the standard polynomial basis $\{t^i\}_{i=0}^q$.

> **Problem 6.5.** If access to a symbolic manipulation program and to numerical software for solving linear algebraic systems is handy, then compare the coefficients of U computed exactly and approximately for $q = 1, 2, ...$ until significant differences are found.

It is not so surprising that solving a system of equations $A\xi = b$, which is theoretically equivalent to inverting A, is sensitive to errors in the coefficients of A and b. The errors result from the fact that the computer stores only a finite number of digits of real numbers. This sensitivity is easily demonstrated in the solution of the 1×1 "system" of equations $ax = 1$ corresponding to computing the inverse $x = 1/a$ of a given real number $a \neq 0$. In Fig. 6.3, we plot the inverses of two numbers a_1 and a_2 computed from two approximations $\tilde{a}_1$ and $\tilde{a}_2$ of the same accuracy. We see that the corresponding errors in the approximations $\tilde{x} = 1/\tilde{a}_i$ of the exact values $x = 1/a_i$ vary greatly in the two cases, since

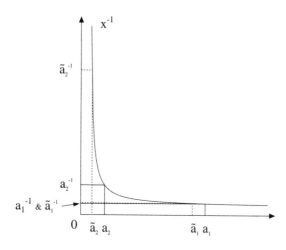

Figure 6.3: The sensitivity of the solution of $ax = 1$ to errors in a.

$1/a_i - 1/\tilde{a}_i = (\tilde{a}_i - a_i)/(a_i\tilde{a}_i)$. The closer a_i is to zero the more sensitive is the solution $1/a_i$ to errors in a_i. This expresses that computing $1/a$ is *ill-conditioned* when a is close to zero.

In general, the solution of $Ax = b$ is sensitive to errors in the entries of A when A is "close" to being non-invertible. Recall that a matrix is non-invertible if one row (or column) is a linear combination of the other rows (or columns). In the example of computing the coefficients of the Galerkin approximation with $q = 9$ above, we can see that there might be a problem if we look at the coefficient matrix A:

$$\begin{pmatrix}
0.167 & 0.417 & 0.550 & 0.633 & 0.690 & 0.732 & 0.764 & 0.789 & 0.809 \\
0.0833 & 0.300 & 0.433 & 0.524 & 0.589 & 0.639 & 0.678 & 0.709 & 0.735 \\
0.0500 & 0.233 & 0.357 & 0.446 & 0.514 & 0.567 & 0.609 & 0.644 & 0.673 \\
0.0333 & 0.190 & 0.304 & 0.389 & 0.456 & 0.509 & 0.553 & 0.590 & 0.621 \\
0.0238 & 0.161 & 0.264 & 0.344 & 0.409 & 0.462 & 0.506 & 0.544 & 0.576 \\
0.0179 & 0.139 & 0.233 & 0.309 & 0.371 & 0.423 & 0.467 & 0.505 & 0.538 \\
0.0139 & 0.122 & 0.209 & 0.280 & 0.340 & 0.390 & 0.433 & 0.471 & 0.504 \\
0.0111 & 0.109 & 0.190 & 0.256 & 0.313 & 0.360 & 0.404 & 0.441 & 0.474 \\
0.00909 & 0.0985 & 0.173 & 0.236 & 0.290 & 0.338 & 0.379 & 0.415 & 0.447
\end{pmatrix}$$

which is nearly singular since the entries in some rows and columns are quite close. On reflection, this is not surprising because the last two rows are given by $\int_0^1 R(U, t)t^8\, dt$ and $\int_0^1 R(U, t)t^9\, dt$, respectively, and t^8 and t^9 look very similar on $[0, 1]$. We plot the two basis functions in Fig. 6.4.

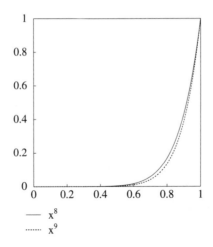

$$\begin{array}{ll} \text{———} & x^8 \\ \text{·······} & x^9 \end{array}$$

Figure 6.4: The basis functions t^8 and t^9.

In general, linear systems of algebraic equations obtained from the discretization of a differential equation tend to become ill-conditioned as the discretization is refined. This is understandable because refining the discretization and increasing the accuracy of the approximation makes it more likely that computing the residual error is influenced by the finite precision of the computer, for example. However, the degree of ill conditioning is influenced greatly by the differential equation and the choice of trial and test spaces, and even the choice of basis functions for these spaces. The standard monomial basis used above leads to an ill-conditioned system because the different monomials become very similar as the degree increases. This is related to the fact that the monomials are not an orthogonal basis. In general, the best results with respect to reducing the effects of ill-conditioning are obtained by using an orthogonal bases for the trial and test spaces. As an example, the Legendre polynomials, $\{\varphi_i(x)\}$, with $\varphi_0 \equiv 1$ and

$$\varphi_i(x) = (-1)^i \frac{\sqrt{2i+1}}{i!} \frac{d^i}{dx^i} \left(x^i (1-x)^i \right), \quad 1 \le i \le q,$$

form an orthonormal basis for $\mathcal{P}^q(0, 1)$ with respect to the L_2 inner product. It becomes more complicated to formulate the discrete equations using this basis, but the effects of finite precision are greatly reduced.

Another possibility, which we take up in the second section, is to use piecewise polynomials. In this case, the basis functions are "nearly orthogonal".

Problem 6.6. (a) Show that φ_3 and φ_4 are orthogonal.

6.2. Galerkin's method with piecewise polynomials

We start by deriving the basic model of stationary heat conduction and then formulate a finite element method based on piecewise linear approximation.

6.2.1. A model for stationary heat conduction

We model heat conduction a thin heat-conducting wire occupying the interval $[0, 1]$ that is heated by a *heat source* of intensity $f(x)$, see Fig. 6.5. We are interested in the stationary distribution of the temperature $u(x)$

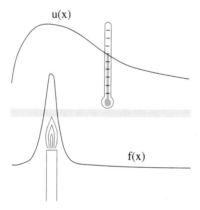

Figure 6.5: A heat conducting wire with a source $f(x)$.

in the wire. We let $q(x)$ denote the heat flux in the direction of the positive x-axis in the wire at $0 < x < 1$. Conservation of energy in a stationary case requires that the net heat flux through the endpoints of an arbitrary sub-interval (x_1, x_2) of $(0, 1)$ be equal to the heat produced in (x_1, x_2) per unit time:

$$q(x_2) - q(x_1) = \int_{x_1}^{x_2} f(x)\, dx.$$

By the Fundamental Theorem of Calculus,

$$q(x_2) - q(x_1) = \int_{x_1}^{x_2} q'(x)\,dx,$$

from which we conclude that

$$\int_{x_1}^{x_2} q'(x)\,dx = \int_{x_1}^{x_2} f(x)\,dx.$$

Since x_1 and x_2 are arbitrary, assuming that the integrands are continuous, we conclude that

$$q'(x) = f(x) \quad \text{for } 0 < x < 1, \tag{6.5}$$

which expresses conservation of energy in differential equation form. We need an additional equation that relates the heat flux q to the temperature gradient (derivative) u' called a *constitutive equation*. The simplest constitutive equation for heat flow is *Fourier's law*:

$$q(x) = -a(x)u'(x), \tag{6.6}$$

which states that heat flows from warm regions to cold regions at a rate proportional to the temperature gradient $u'(x)$. The constant of proportionality is the *coefficient of heat conductivity* $a(x)$, which we assume to be a positive function in $[0, 1]$. Combining (6.5) and (6.6) gives the *stationary heat equation* in one dimension:

$$-(a(x)u'(x))' = f(x) \quad \text{for } 0 < x < 1. \tag{6.7}$$

To define a solution u uniquely, the differential equation is complemented by *boundary conditions* imposed at the boundaries $x = 0$ and $x = 1$. A common example is the homogeneous *Dirichlet* conditions $u(0) = u(1) = 0$, corresponding to keeping the temperature zero at the endpoints of the wire. The result is a *two-point boundary value problem*:

$$\begin{cases} -(au')' = f & \text{in } (0, 1), \\ u(0) = u(1) = 0. \end{cases} \tag{6.8}$$

The boundary condition $u(0) = 0$ may be replaced by $-a(0)u'(0) = q(0) = 0$, corresponding to prescribing zero heat flux, or insulating the wire, at $x = 0$. Later, we aslo consider non-homogeneous boundary conditions of the from $u(0) = u_0$ or $q(0) = g$ where u_0 and g may be different from zero.

Problem 6.7. Determine the solution u of (6.9) with $a(x) = 1$ by symbolic computation by hand in the case $f(x) = 1$ and $f(x) = x$.

We want to determine the temperature u in the wire by solving the heat equation (6.8) with given f, boundary conditions, and heat conductivity $a(x)$. To compute the solution numerically we use the Galerkin finite element method.

6.2.2. The Galerkin finite element method

We consider the problem (6.8) in the case $a \equiv 1$, that is

$$\begin{cases} -u'' = f & \text{in } (0,1), \\ u(0) = u(1) = 0, \end{cases} \qquad (6.9)$$

and formulate the simplest finite element method for (6.9) based on continuous piecewise linear approximation.

We let $\mathcal{T}_h : 0 = x_0 < x_1 < \dots < x_{M+1} = 1$, be a *partition* or (*triangulation*) of $I = (0,1)$ into sub-intervals $I_j = (x_{j-1}, x_j)$ of length $h_j = x_j - x_{j-1}$ and let $V_h = V_h^{(1)}$ denote the set of continuous piecewise linear functions on $\mathcal{T}_h$ that are zero at $x = 0$ and $x = 1$. We show an example of such a function in Fig. 6.6. In Chapter 4, we saw that V_h is

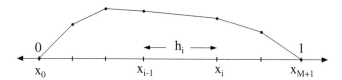

Figure 6.6: A continuous piecewise linear function in V_h.

a finite dimensional vector space of dimension M with a basis consisting of the hat functions $\{\varphi_j\}_{j=1}^M$ illustrated in Fig. 5.11. The coordinates of a function v in V_h in this basis are the values $v(x_j)$ at the interior nodes x_j, $j = 1, \dots, M$, and a function $v \in V_h$ can be written

$$v(x) = \sum_{j=1}^{M} v(x_j) \varphi_j(x).$$

Note that because $v \in V_h$ is zero at 0 and 1, we do not include φ_0 and φ_{M+1} in the set of basis functions for V_h.

As in the previous example, Galerkin's method is based on stating the differential equation $-u'' = f$ in the form

$$\int_0^1 (-u'' - f)v \, dx = 0 \quad \text{for all functions} v, \tag{6.10}$$

corresponding to the residual $-u'' - f$ being orthogonal to the test functions v, cf. (6.3). However, since the functions in V_h do not have second derivatives, we can't simply plug a candidate for an approximation of u in the space V_h directly into this equation. To get around this technical difficulty, we use integration by parts to move one derivative from u'' onto v assuming v is differentiable and $v(0) = v(1) = 0$:

$$-\int_0^1 u'' v \, dx = -u'(1)v(1) + u'(0)v(0) + \int_0^1 u'v' \, dx = \int_0^1 u'v' \, dx,$$

where we used the boundary conditions on v. We are thus led to the following *variational formulation* of (6.9): find the function u with $u(0) = u(1) = 0$ such that

$$\int_0^1 u'v' \, dx = \int_0^1 fv \, dx, \tag{6.11}$$

for all functions v such that $v(0) = v(1) = 0$. We refer to (6.11) as a *weak form* of (6.10).

The Galerkin finite element method for (6.9) is the following finite-dimensional analog of (6.11): find $U \in V_h$ such that

$$\int_0^1 U'v' \, dx = \int_0^1 fv \, dx \quad \text{for all } v \in V_h. \tag{6.12}$$

We note that the derivatives U' and v' of the functions U and $v \in V_h$ are piecewise constant functions of the form depicted in Fig. 6.7 and are not defined at the nodes x_i. However, the integral with integrand $U'v'$ is nevertheless uniquely defined as the sum of integrals over the subintervals. This is due to the basic fact of integration that two functions that are equal except at a finite number of points, have the same integral. We illustrate this in Fig. 6.8. By the same token, the value (or lack of

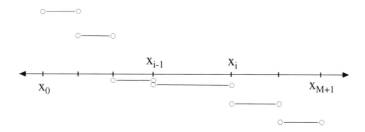

Figure 6.7: The derivative of the continuous piecewise linear function in Fig. 6.6.

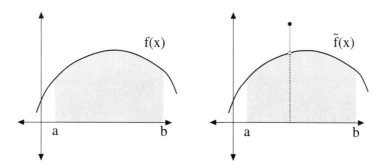

Figure 6.8: Two functions that have a different value at one point have the same area underneath their curves.

value) of U' and v' at the distinct node points x_i does not affect the value of $\int_0^1 U'v'\,dx$.

The equation (6.10) expresses the fact that the residual error $-u'' - f$ of the exact solution is orthogonal to *all* test functions v and (6.11) is a reformulation in weak form. Similarly, (6.12) is a way of forcing in weak form the residual error of the finite element solution U to be orthogonal to the finite dimensional set of test functions v in V_h.

6.2.3. The discrete system of equations

Using the basis of hat functions $\{\varphi_j\}_{j=1}^M$, we have

$$U(x) = \sum_{j=1}^M \xi_j \varphi_j(x)$$

and determine the nodal values $xi_j = U(x_j)$ using the Galerkin orthogonality (6.12). Substituting, we get

$$\sum_{j=1}^{M} \xi_j \int_0^1 \varphi_j' v' \, dx = \int_0^1 f v \, dx, \tag{6.13}$$

for all $v \in V_h$. It suffices to check (6.13) for the basis functions $\{\varphi_i\}_{i=1}^{M}$, which gives the $M \times M$ linear system of equations

$$\sum_{j=1}^{M} \xi_j \int_0^1 \varphi_j' \varphi_i' \, dx = \int_0^1 f \varphi_i \, dx, \quad i = 1, \dots, M, \tag{6.14}$$

for the unknown coefficients $\{\xi_j\}$. We let $\xi = (\xi_j)$ denote the vector of unknown coefficients and define the $M \times M$ *stiffness matrix* $A = (a_{ij})$ with coefficients

$$a_{ij} = \int_0^1 \varphi_j' \varphi_i' \, dx,$$

and the *load vector* $b = (b_i)$ with

$$b_i = \int_0^1 f \varphi_i \, dx.$$

These names originate from early applications of the finite element method in structural mechanics. Using this notation, (6.14) is equivalent to the linear system

$$A\xi = b. \tag{6.15}$$

In order to solve for the coefficients of U, we first have to compute the stiffness matrix A and load vector b. For the stiffness matrix, we note that a_{ij} is zero unless $i = j - 1$, $i = j$, or $i = j + 1$ because otherwise either $\varphi_i(x)$ or $\varphi_j(x)$ is zero on each sub-interval occurring in the integration. We illustrate this in Fig. 6.9. We compute a_{ii} first. Using the definition of the φ_i,

$$\varphi_i(x) = \begin{cases} (x - x_{i-1})/h_i, & x_{i-1} \le x \le x_i, \\ (x_{i+1} - x)/h_{i+1}, & x_i \le x \le x_{i+1}, \end{cases}$$

and $\varphi_i(x) = 0$ elsewhere, the integration breaks down into two integrals:

$$a_{ii} = \int_{x_{i-1}}^{x_i} \left(\frac{1}{h_i}\right)^2 dx + \int_{x_i}^{x_{i+1}} \left(\frac{-1}{h_i}\right)^2 dx = \frac{1}{h_i} + \frac{1}{h_{i+1}}$$

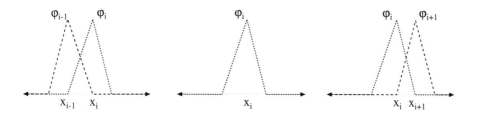

Figure 6.9: Three possibilities to obtain a non-zero coefficient in the stiffness matrix.

since $\varphi_i' = 1/h_i$ on (x_{i-1}, x_i) and $\varphi_i' = -1/h_{i+1}$ on (x_i, x_{i+1}), and φ_i is zero on the rest of the sub-intervals. Similarly,

$$a_{i\,i+1} = \int_{x_i}^{x_{i+1}} \frac{-1}{h_{i+1}} \frac{1}{h_{i+1}}\, dx = -\frac{1}{h_{i+1}}.$$

Problem 6.8. Prove that $a_{i-1\,i} = -1/h_i$ for $i = 2, 3, ..., M$.

Problem 6.9. Determine the stiffness matrix A in the case of a uniform mesh with meshsize $h_i = h$ for all i.

We compute the coefficients of b in the same way to get

$$b_i = \int_{x_{i-1}}^{x_i} f(x)\frac{x - x_{i-1}}{h_i}\, dx + \int_{x_i}^{x_{i+1}} f(x)\frac{x_{i+1} - x}{h_{i+1}}\, dx, \quad i = 1, ..., M.$$

Problem 6.10. Verify this formula.

Problem 6.11. Using a uniform mesh with $h = .25$, compute the Galerkin finite element approximation for $f(x) = x$ by hand.

The matrix A is a *sparse* matrix in the sense that most of its entries are zero. In this case, A is a *banded* matrix with non-zero entries occurring only in the diagonal, super-diagonal and sub-diagonal positions. This contrasts sharply with the coefficient matrix in the first example which is "full" in the sense that all of its entries are non-zero. The bandedness of A reflects the fact that the basis functions $\{\varphi_i\}$ for V_h are "nearly" orthogonal, unlike the basis used in the first example. Moreover, A is a *symmetric* matrix since $\int_0^1 \varphi_i'\varphi_j'\, dx = \int_0^1 \varphi_j'\varphi_i'\, dx$. Finally, it is possible to show that A is *positive-definite*, which means that

$\sum_{i,j=1}^{M} \eta_i a_{ij} \eta_j > 0$ unless all $\eta_i = 0$, which implies that A is invertible, so that (6.15) has a unique solution.

We expect to increase the accuracy of the approximate solution by increasing the dimension M. Systems of dimension $100 - 1,000$ in one dimension and up to $100,000$ in higher dimensions are common. Thus, it is important to study the solution of (6.15) in detail and we begin to do this in Chapter 7.

Problem 6.12. Prove that A is positive-definite. Hint: use that $\eta^{\mathrm{T}} A\eta = \int_0^1 (v(x)')^2 \, dx$ where $v(x)$ is the function in V_h defined by $v(x) = \sum_i \eta_i \varphi_i(x)$.

6.2.4. A concrete example

As a concrete example, we consider (6.9) with

$$f(x) = 10(1 - 10(x - .5)^2) \, e^{-5(x-.5)^2}.$$

This is an exceptional case in which the solution $u(x)$ can be computed exactly:

$$u(x) = e^{-5(x-.5)^2} - e^{-5/4},$$

and we may easily compare the approximation with the exact solution.

In this example, f varies quite a bit over the interval. We plot it in Fig. 6.10. Recalling the discussions in Chapters 3 and 5, we expect the

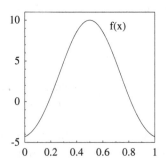

Figure 6.10: Load function f for the first example.

mesh size to vary if we want the computational method to be efficient with a minimal number of mesh points. The following choice of the mesh points minimizes $\|u' - U'\|_{L_2(0,1)}$ while keeping $M = 16$:

$$
\begin{aligned}
h_1 = h_{16} &\approx .2012 \\
h_2 = h_{15} &\approx .0673 \\
h_3 = h_{14} &\approx .0493 \\
h_4 = h_{13} &\approx .0417 \\
h_5 = h_{12} &\approx .0376 \\
h_6 = h_{11} &\approx .0353 \\
h_7 = h_{10} &\approx .0341 \\
h_8 = h_9 &\approx .0335
\end{aligned}
$$

This partition is the result of applying an adaptive algorithm based on information obtained from the data f and the computed approximation U, and does not require any knowledge of the exact solution u. We will present the adaptive algorithm in Chapter 8 and explain how it is possible to get around the apparent need to know the exact solution.

In Fig. 6.11, we plot the exact solution u and the Galerkin finite element approximation U together with their derivatives. Notice that

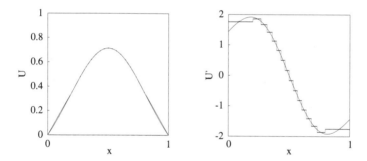

Figure 6.11: The true solution u and the finite element approximation U together with their derivatives.

the adaptive procedure has clustered mesh points towards the middle of the interval. With $M = 16$ we have $\|u' - U'\|_2 \le .25$, while $\|u - U\|_2 \le .01$. If we perform the computation using a mesh with uniform spacing we need to take $M = 24$ to obtain the same accuracy for the derivatives. This disparity increases as the number of elements is increased and may be much more substantial for more complex problems. We will meet many examples below.

6.3.   Galerkin's method with trigonometric polynomials

We noted at the very end of Chapter 3 that for $n = 1, 2, ...$, the trigono-
metric function $\sin(n\pi x)$ solves the boundary value problem $u'' + n^2\pi^2 u = 0$ in $(0, 1)$, $u(0) = u(1) = 0$. In fact, the functions $\sin(n\pi x)$, $n = 1, 2, ...$,
are the solutions to the *eigenvalue problem* of finding nonzero functions
$\varphi(x)$ and scalars λ such that

$$\begin{cases} -\varphi'' = \lambda\varphi & \text{in } (0, 1), \\ \varphi(0) = \varphi(1) = 0. \end{cases} \tag{6.16}$$

Eigenvalue problems play an important role in mathematics, mechanics
and physics and we will study such problems in more detail below. This
gives a different perspective on the trigonometric functions $\sin(n\pi x)$ as
the eigenfunctions of the particular boundary value problem (6.16). In
particular the mutual orthogonality of the functions $\sin(n\pi x)$ on $(0, 1)$
reflect a general orthogonality property of eigenfunctions. The general
idea of the *spectral* Galerkin method is to seek a solution of a given
boundary value problem as a linear combination of certain eigenfunctions. The given boundary value and the eigenvalue problem do not
have to be directly related in the sense that the differential operators involved are the same, but the boundary conditions should match. As an
example we consider the application of the spectral Galerkin method to
the model (6.8) of stationary heat flow with variable heat conductivity

$$\begin{cases} -\big((1+x)u'(x)\big)' = f(x) & \text{for } 0 < x < 1, \\ u(0) = u(1) = 0, \end{cases} \tag{6.17}$$

where $f(x) = 1 + (1 + 3x - x^2)\exp(-x)$, based on the eigenfunctions
$\sin(n\pi x)$, $n = 1, 2, ...$, of (6.16). We then seek to express the solution
$u(x)$ of (6.17) as a sine series, cf. (5.33),

$$u(x) = \sum_{n=1}^{\infty} b_n \sin(n\pi x).$$

In the corresponding discrete analog we seek for some fixed $q > 0$ an
approximate solution $U(x)$ of the form

$$U(x) = \sum_{j=1}^{q} \xi_j \sin(j\pi x), \tag{6.18}$$

using Galerkin's method to determine the coefficients ξ_j. Denoting by $V^{(q)}$ the space of trigonometric polynomials that are linear combinations of the functions $\{\sin(i\pi x)\}_{i=1}^q$, Galerkin's method reads: find $U \in V^{(q)}$ such that

$$- \int_0^1 \big((1+x)U'(x)\big)' v(x)\, dx = \int_0^1 f(x) v(x)\, dx \quad \text{for all } v \in V^{(q)}. \tag{6.19}$$

Because the functions in $V^{(q)}$ are smooth, we have no difficulty defining the residual error $\big((1+x)U'(x)\big)' - f(x)$ of the approximate solution U in $V^{(q)}$. By (6.19), the Galerkin approximation U is the function in $V^{(q)}$ whose residual error is orthogonal to all functions in $V^{(q)}$. Using integration by parts, we can rewrite this as

$$\int_0^1 (1+x)U'(x) v'(x)\, dx = \int_0^1 f(x) v(x)\, dx \quad \text{for all } x \in V^{(q)}, \tag{6.20}$$

which is the usual formulation of a Galerkin method for the boundary value problem (6.17). As before, we substitute the expansion (6.18) for U into (6.20) and choose $v = \sin(i\pi x)$, $i = 1, ..., q$. This gives a linear system of equations

$$A\xi = d$$

where $A = (a_{ij})$ and $d = (d_i)$ have coefficients

$$a_{ij} = \pi^2 ij \int_0^1 (1+x) \cos(j\pi x) \cos(i\pi x)\, dx$$

$$d_i = \int_0^1 \big(1 + (1 + 3x - x^2) \exp(-x)\big) \sin(i\pi x)\, dx. \tag{6.21}$$

Problem 6.13. Prove that $V^{(q)}$ is a subspace of the continuous functions on $[0, 1]$ that satisfy the boundary conditions and show that $\{\sin(i\pi x)\}_{i=1}^q$ is an orthogonal basis for $V^{(q)}$ with respect to the L_2 inner product.

Problem 6.14. Derive (6.20).

Problem 6.15. Verify the formulas for A and d then compute explicit formulas for the coefficients.

In this problem, we are able to compute these integrals exactly using a symbolic manipulation program, though the computations are messy. For a general problem we are usually unable to evaluate the corresponding integrals exactly. For these reasons, it is natural to also consider the use of quadrature to evaluate the integrals giving the coefficients of A and d. We examine three quadrature rules discussed in Chapter 5, the composite rectangle, midpoint, and trapezoidal rules respectively. We partition $[0, 1]$ into $M+1$ intervals with equally spaced nodes $x_l = l/(M+1)$ for $l = 0, ..., M+1$. For example, the composite rectangle rule is

$$a_{ij} \approx \frac{\pi^2 ij}{M+1} \sum_{l=1}^{M+1} (1 + x_l) \cos(j\pi x_l) \cos(i\pi x_l)$$

$$d_i \approx \frac{1}{M+1} \sum_{l=1}^{M+1} \left(1 + (1 + 3x_l - x_l^2) \exp(-x_l)\right) \sin(i\pi x_l).$$

Problem 6.16. Write out the corresponding formulas for schemes that use the midpoint and trapezoidal rules to evaluate the integrals.

From the discussion in Chapter 5, we expect the composite midpoint and trapezoidal rules to be more accurate than the composite rectangle rule on a given partition. In Fig. 6.12, we plot the results obtained by using the four indicated methods. In this case, using the lower order accurate rectangle rule affects the results significantly. We list the coefficients obtained by the four methods below.

q	U	U_r	U_m	U_t
1	0.20404	0.21489	0.20419	0.20375
2	0.017550	0.031850	0.017540	0.017571
3	0.0087476	0.013510	0.0087962	0.0086503
4	0.0022540	0.0095108	0.0022483	0.0022646
5	0.0019092	0.0048023	0.0019384	0.0018508
6	0.00066719	0.0054898	0.00066285	0.00067495
7	0.00069279	0.0026975	0.00071385	0.00065088
8	0.00027389	0.0038028	0.00026998	0.00028064
9	0.00030387	0.0014338	0.00032094	0.00027024

Several questions about the error arise in this computation. We would like to know the accuracy of the Galerkin approximation U and

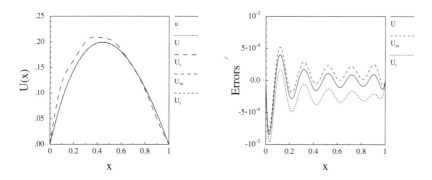

Figure 6.12: On the left, we plot four approximate solutions of (6.17)
computed with $q = 9$. U was computed by evaluating the
integrals in (6.21) exactly. U_r is computed using the composite rectangle rule, U_m is computed using the composite
midpoint rule, and U_t is computed using the composite
trapezoidal rule with $M = 20$ to evaluate the integrals
respectively. On the right, we plot the errors versus x for
U, U_m, and U_t.

whether the approximation converges as we increase q. Moreover, we
would also like to know the effects of choosing different quadrature rules
on the accuracy. Moreover, since $a = (1 + x)$ and $f(x) = 1 + (1 +
3x - x^2)e^{-x}$ vary through $[0, 1]$, it would be better to adapt the mesh
used for the quadrature rules in order to achieve the desired accuracy
in the coefficients (6.21). This also requires knowledge of the effects of
quadrature error on the accuracy of the approximation.

Problem 6.17. Formulate a Galerkin approximation method using trigonometric polynomials for (6.9). Note in this case that the linear system giving
the coefficients of the approximation is trivial to solve because it is diagonal. Why didn't this happen when we discretized (6.17)? Interpret the
formula for the coefficients of the data vector in terms of Fourier series.

Problem 6.18. Use the spectral method to discretize $-u'' + u' = f$ in
$(0, 1)$ with $u(0) = u(1) = 0$.

6.4.   Comments on Galerkin's method

We have considered Galerkin's method for three kinds of approxima-tions: the q-method (6.2) with global polynomials of order q (this method is also often referred to as the p-method), the h-method (6.12) with con-tinuous piecewise linear polynomials on a partition $\mathcal{T}_h$ of mesh size h, and the spectral method (6.20) with global trigonometric polynomials. In rough terms, we can distinguish different Galerkin approximations by the following properties of the basis functions:

- local or global support

- (near) orthogonality or non-orthogonality.

The *support* of a function is the set of points where the function is dif-ferent from zero. A function is said to have *local support* if it is zero outside a set of small size in comparison to the domain, while a function with *global support* is non-zero throughout the domain except at a few points. The h-method uses basis functions with local support, while the basis functions of the q-method in Section 6.1 or the spectral method are polynomials or trigonometric polynomials with global support. The basis functions used for the spectral method are mutually orthogonal, while the basis functions used in Section 6.1 for the global polynomials was not. Using orthogonal basis functions tends to reduce the effects of rounding errors that occur when computing the approximation. For example, in the spectral method in Section 6.3, the matrix determining the approximation is sparse with roughly half the coefficients equal to zero. The basis functions used in the h-method in Section 6.2 are not orthogonal, but because they have local support, they are "nearly" or-thogonal and again the matrix determining the approximation is sparse. Recent years have seen the development of *wavelet methods*, which are Galerkin methods with basis functions that combine L_2 orthogonality with small support. Wavelets are finding increasing applications in im-age processing, for example.

In Galerkin's method, we expect to increase accuracy by increasing the dimension of the trial and test spaces used to define the approxima-tion. In the h-method this is realized by decreasing the mesh size and in the $q-method$ by increasing the degree of the polynomials. More gener-ally, we can use a combination of decreasing the mesh size and increasing the polynomial degree, which is called the (h, q)-method. Finally, in the spectral Galerkin method, we try to improve accuracy by increasing the degree of the trigonometric polynomial approximation.

Galerkin was born in 1871 in Russia and was educated at the St. Petersburg Technological Institute and in the Kharkhov locomotive works. He began doing research in engineering while he was in prison in 1906-7 for his participation in the anti-Tsarist revolutionary movement. He later made a distinguished academic career and had a strong influence on engineering in the Soviet Union. This was due in no small part to the success of Galerkin's method, first introduced in a paper on elasticity published in 1915. From its inception, which was long before the computer age, Galerkin's method was used to obtain high accuracy with minimal computational effort using a few cleverly chosen basis functions. Galerkin's method belongs to the long tradition of variational methods in mathematics, mechanics and physics going back to the work by Euler and Lagrange, and including important work by e.g. Hamilton, Rayleigh, Ritz and Hilbert. The finite element method, first developed in aerospace engineering in the 1950s, may be viewed as the computer-age realization of this variational methodology. In the advanced companion volume we give a more detailed account of the development of the finite element method into the first general technique for solving differential equations numerically, starting with applications to elasticity problems in the 1960s, to the diffusion problems in the 1970s and to fluid flow and wave propagation in the 1980s.

6.5. Important issues

The examples we have presented raise the following questions concerning a finite element approximation U of the solution u of a given differential equation:

- How big is the *discretization error* $u - U$ or $u' - U'$ in some norm?

- How should we choose the mesh to control the discretization error $u - U$ or $u' - U'$ to a given tolerance in some norm using the least amount of work (computer time)?

- How can we compute the discrete solution U efficiently and what is the effect of errors made in computing U?

This book seeks to answer these questions for a variety of problems.

Between my finger and my thumb
The squat pen rests
I'll dig with it. (Heaney)

The *calculus ratiocinator* of Leibniz merely needs to have an engine put into it to become a *machina ratiocinatrix*. The first step in this direction is to proceed from the calculus to a system of *ideal* reasoning machines, and this was taken several years ago by Turing. (Wiener)

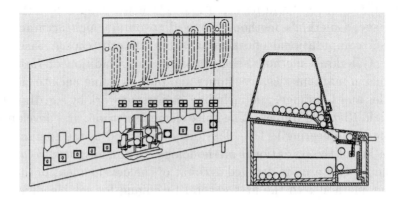

Figure 6.13: Proposal by Leibniz for a "ball-computer" for the multiplication of binary numbers.

7

Solving Linear Algebraic Systems

All thought is a kind of computation. (Hobbes)

In the examples in Chapter 6, the coefficients of the Galerkin approximations were computed by solving systems of linear algebraic equations. In general, discretization of a differential equation yields a large system of equations, whose solution gives the approximation. In fact, a code for the numerical solution of differential equations typically spends the greatest portion of computer time solving systems of linear algebraic equations. In this chapter, we introduce the two basic types of methods for this purpose, namely *direct methods* based on *Gaussian elimination* that produces the solution after a finite number of arithmetic operations and *iterative methods* based on generating a sequence of approximate solutions that (hopefully) converges to the true solution. We consider only the most basic solution methods in this book. We discuss modern efficient methods in the advanced companion volume, building on the ideas introduced here.

We are interested in solving a linear system

$$A\xi = b$$

where A is a $n \times n$ matrix and b is a n-vector. Recall that such a system has a unique solution if the matrix is invertible, which means there is a unique matrix A^{-1} with the property $A^{-1}A = AA^{-1} = I$, where I is the $n \times n$ identity matrix. In this case, it is common in linear algebra to write the solution as $\xi = A^{-1}b$ and pursue the matter no further. In practice, however, we encounter a couple of difficulties in solving a system. The first we already encountered in Section 6.1, when we tried to solve for the high order Galerkin approximation of a differential equation. In this problem, the matrix is close to being non-invertible, or is *ill conditioned*, and this combined with the finite precision of a computer caused the resulting numerical solution to be highly inaccurate. Therefore, we have to study the effect of the ill conditioning of A on the solution and estimate the error of the numerical solution. The second practical problem we encounter is the cost of obtaining the solution of the system. We study different methods of computing the solution and analyze their efficiency. We shall see that computing A^{-1} is never the most efficient way to compute the solution.

We begin by deriving the linear system of equations describing the equilibrium position of a system of masses connected by elastic springs. This system falls into the class of symmetric, positive-definite systems, which plays a basic role in mathematical modeling. Such systems can be formulated as equivalent quadratic minimization problems. Mass-spring systems have a finite number of degrees of freedom and are called *discrete systems* as opposed to the *continuous systems* typically modeled by a differential equation. The processes of passing from a discrete to a continuous system by refining the scale of the discrete system, or *passing to the continuum limit*, and from a continuous to a discrete system by *discretization*, are fundamental in modeling in both mathematics and physics. In particular, later in the book we consider continuous analogs of the discrete mass-spring systems discussed here.

After this experience in deriving a linear system, we focus on methods for computing solutions, beginning with direct methods both for general systems and some special systems arising from discretization of differential equations. Then we continue with iterative methods, first deriving an iterative method for symmetric, positive-definite systems based on finding the minimum of the equivalent quadratic minimization problem by computing a sequence of values that successively decrease the function to be minimized. We analyze the convergence rate of the iterative method in terms of the ratio of largest to smallest eigenvalue of the system. After this, we present a general framework for the construction and analysis of iterative methods, and then explain how the first iterative method fits into this setting.

Finally, we conclude with a section describing a method for estimating the error of the computed solution of a system of equations based on an *a posteriori* error analysis (which means an analysis that is performed after the computation is completed).

We emphasize that this chapter serves only as an elementary introduction to the subject of solving linear systems of equations and covers only the least sophisticated examples of different techniques. We also consider only the case of scalar computers. Parallel computers require a different approach to implementation and analysis of costs. In the advanced companion volume, we discuss in detail so-called multigrid methods, which are very efficient iterative methods based on using the simple iterative methods presented in this chapter on a hierarchy of meshes. We also refer the reader to Atkinson ([2]), Isaacson and Keller ([9]), and Golub and Van Loan ([7]).

7.1. Stationary systems of masses and springs

In this section, we derive a model for stationary mass-spring systems. We start by considering a body of mass m suspended vertically by an elastic spring attached to a fixed support; see Fig. 7.1. We seek the elongation,

Figure 7.1: A system of one mass and one spring.

or displacement, ξ of the spring under the weight of the mass measured from the reference configuration with zero spring force. *Hooke's law* states that the spring force σ is proportional to the displacement: $\sigma =$

$k\xi$, where the constant of proportionality $k > 0$ is called the *spring constant*. Hooke's law is a linear approximation of what happens in a real spring and is generally valid if the spring is not stretched too much. The elongation ξ is determined by the fact that at equilibrium the weight of the body is precisely balanced by the spring force:

$$k\xi = m, \tag{7.1}$$

where we assume the gravitational constant is one. This is the simplest linear system of equations. Solving for the unknown ξ, we find $\xi = m/k$.

We may also derive the equation (7.1) from a basic principle of mechanics stating that at equilibrium, the true elongation ξ minimizes the *total energy*,

$$F(\eta) = \frac{k}{2}\eta^2 - m\eta, \tag{7.2}$$

over all possible elongations η. The total energy $F(\eta)$ has contributions from two sources. The first is the internal elastic energy $k\eta^2/2$ stored in the spring when it is stretched a distance η. To obtain this expression for the internal energy, we notice that the work ΔW needed to change the length of the spring from η to $\eta + \Delta\eta$ is equal to $k\eta \times \Delta\eta$ by the definition of work as "force$\times$distance". In the limit of small change, this yields the differential equation $W'(\eta) = k\eta$ with solution $W(\eta) = k\eta^2/2$ when $W(0) = 0$. The second part of the total energy is the loss of potential energy $-m\eta$ when the mass is lowered in height a distance η.

To see the connection with the equilibrium equation $k\xi = m$, we compute the minimum of $F(\eta)$ using a technique that generalizes naturally to systems of masses and springs. We start by observing that since ξ minimizes $F(\eta)$, so $F(\xi) \le F(\eta)$ for all $\eta \in \mathbb{R}$, then for any $\zeta \in \mathbb{R}$, $F(\xi) \le F(\xi+\epsilon\zeta)$ for all $\epsilon \in \mathbb{R}$. Introducing the function $g(\epsilon) = F(\xi+\epsilon\zeta)$ with ξ and ζ fixed, we write the last inequality in the equivalent form $g(0) \le g(\epsilon)$ for all $\epsilon \in \mathbb{R}$. Thus, the function $g(\epsilon)$ is minimized for $\epsilon = 0$, which implies that the derivative $g'(\epsilon)$ must vanish for $\epsilon = 0$ if it exists. But

$$g(\epsilon) = \frac{k}{2}\xi^2 + k\epsilon\xi\zeta + \frac{k}{2}\epsilon^2\zeta^2 - m\xi - \epsilon m\zeta,$$

so that

$$g'(0) = k\xi\zeta - m\zeta,$$

which implies that $k\xi - m = 0$, since ζ is arbitrary. Therefore, the minimizer ξ is the solution $\xi = m/k$ of the equation $k\xi = m$. Conversely, if ξ solves $k\xi = m$, then $F'(\xi) = 0$ and ξ must be the minimizer since F is convex because $F'' > 0$. We now have proved that the condition for equilibrium of the mass-spring system may be expressed in two equivalent ways: as a linear equation $k\xi = m$, and as minimization of the total energy, $F(\xi) \leq F(\eta)$ for all η. This illustrates the fundamental connection between two basic physical principles: equilibrium of forces and energy minimization. We shall see that this useful principle has a wide range of application.

As the next example, we apply the energy minimization principle to a stationary system of two masses m_1 and m_2 connected to each other and to fixed supports by three springs with spring constants k_1, k_2, and k_3; see Fig. 7.2. We seek the vertical displacements ξ_1 and ξ_2 of the

Figure 7.2: A system of two masses and three springs.

masses from the reference position with zero spring forces. The true displacement $\xi = (\xi_1, \xi_2)^\mathsf{T}$ minimizes the total energy

$$F(\eta) = \frac{k_1}{2}\eta_1^2 + \frac{k_2}{2}(\eta_2 - \eta_1)^2 + \frac{k_3}{2}\eta_2^2 - m_1\eta_1 - m_2\eta_2,$$

which is the sum of the internal elastic energies and the load potential as above. For example, the elastic energy of the middle spring is determined

by its elongation $\eta_2 - \eta_1$. In matrix form, the total energy may be
expressed as

$$F(\eta) = \frac{1}{2}(DB\eta, B\eta) - (b, \eta),$$

where $\eta = (\eta_1, \eta_2)^\top$, $b = (m_1, m_2)^\top$, $(\cdot, \cdot) = (\cdot, \cdot)_2$, and

$$B = \begin{pmatrix} 1 & 0 \\ -1 & 1 \\ 0 & 1 \end{pmatrix}, \quad \text{and } D = \begin{pmatrix} k_1 & 0 & 0 \\ 0 & k_2 & 0 \\ 0 & 0 & k_3 \end{pmatrix}.$$

Problem 7.1. Verify this.

If $\xi \in \mathbb{R}^2$ is the minimizer, then for any $\zeta \in \mathbb{R}^2$ the derivative of
$F(\xi + \epsilon\zeta)$ with respect to ϵ must vanish for $\epsilon = 0$. Computing, we get
$(DB\xi, B\zeta) = (b, \zeta)$ or $(B^\top DB\xi, \zeta) = (b, \zeta)$. Since ζ is arbitrary, we
obtain the following linear system for the displacement ξ

$$A\xi = b \quad \text{with } A = B^\top DB, \tag{7.3}$$

where

$$A = \begin{pmatrix} k_1 + k_2 & -k_2 \\ -k_2 & k_2 + k_3 \end{pmatrix}.$$

Problem 7.2. Provide the details of these computations.

We may rewrite the system (7.3) as

$$B^\top \sigma = b \quad \text{with } \sigma = DB\xi \tag{7.4}$$

where σ represents the vector of spring forces. The relation $B^\top \sigma = b$
is the *equilibrium equation* and the relation $\sigma = DB\xi$ expresses Hooke's
law relating the displacements to the spring forces.

Problem 7.3. Verify that (7.4) correctly expresses the equilibrium of
forces and the relation between displacements and spring forces.

The system matrix $A = B^\top DB$ in (7.3) is symmetric since $(B^\top DB)^\top$
$= B^\top D^\top (B^\top)^\top = B^\top DB$ and positive semi-definite since for any vector
$v \in \mathbb{R}^2$, $(B^\top DBv, v) = (DBv, Bv) = (Dw, w) \geq 0$, where $w = Bv$. Further, if $(B^\top DBv, v) = (Dw, w) = 0$, then $w = Bv = 0$ and consequently
$v = 0$, so that A is in fact positive-definite.

Using this approach, we can easily derive the linear system modeling the equilibrium position of an arbitrary arrangement of masses and springs, as long as we assume the displacements to be small so that the change of geometry of the systems from the unloaded reference position due to the weight of the masses is negligible. In every case, we obtain a symmetric positive-definite linear system of the form $B^T DB\xi = b$. We may set up this system for a specific configuration of masses and springs, either by using equilibrium of forces and Hooke's law, or minimization of the total energy. Usually the second alternative is preferable. We illustrate this in the following problem, which we urge the reader to solve before proceeding.

Problem 7.4. Consider the configuration of 9 masses and 16 springs shown in the reference position in Fig. 7.3. (a) Assume that the displacements of

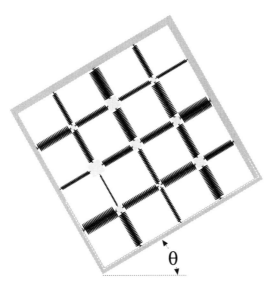

Figure 7.3: A system of nine masses in the reference position.

the masses are small, so that the elongation of a spring may be set equal to the difference of the displacements of the endpoints of the spring in the direction of the spring in its reference position. Express the total energy in terms of the displacements from the reference position and then determine the corresponding linear system of equations satisfied by the minimizer of the total energy. Hint: the internal energy is a half of the sum of the squares of the elongations of the springs times the corresponding spring

constants. It is convenient to choose the coordinate system to be aligned
with the frame.

(b) Derive the corresponding system of equations in the general case, considering also large displacements.

7.2. Direct methods

We begin by noting that some linear systems $Ax = b$, with A an $n \times n$-matrix (where we change notation from ξ to x for convenience), are easier to solve than others. For example, if $A = (a_{ij})$ is *diagonal*, which means that $a_{ij} = 0$ if $i \neq j$, then the system is solved in n operations: $x_i = b_i/a_{ii}$, $i = 1, ..., n$. Further, if the matrix is *upper triangular*, which means that $a_{ij} = 0$ if $i > j$, or *lower triangular*, which means that $a_{ij} = 0$ if $i < j$, then the system can be solved by *backward substitution* or *forward substitution* respectively; see Fig. 7.4 for an illustration of these different types. For example, if A is upper triangular, the "pseudo-

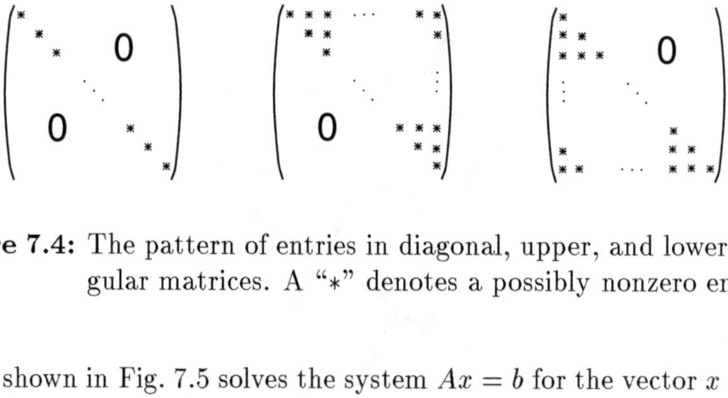

Figure 7.4: The pattern of entries in diagonal, upper, and lower triangular matrices. A "$*$" denotes a possibly nonzero entry.

code" shown in Fig. 7.5 solves the system $Ax = b$ for the vector $x = (x_i)$ given the vector $b = (b_i)$ (assuming that $a_{kk} \neq 0$):

Problem 7.5. Using a similar format, write down algorithms to solve a diagonal system and then a lower triangular system using forward substitution. Determine the number of arithmetic operations needed to compute the solution.

In all three cases, the systems have a unique solution as long as the diagonal entries of A are nonzero.

Direct methods are based on Gaussian elimination, which in turn is based on the observation that the solution of a linear system is not changed under the following *elementary row operations*:

$$x_n = b_n/a_{nn}$$

for $k = n\text{-}1, n\text{-}2, ..., 1,$ do

    sum = 0

    for $j = k\text{+}1, ..., n,$ do

        sum = sum + $a_{kj} \cdot x_j$

    $x_k = (b_k - \text{sum})/a_{kk}$

Figure 7.5: An algorithm for solving an upper triangular system by back substitution.

- interchanging two equations

- adding a multiple of one equation to another

- multiplying an equation by a nonzero constant.

The idea behind Gaussian elimination is to transform using these operations a given system into an upper triangular system, which is solved by back substitution. For example, to solve the system

$$x_1 + x_2 + x_3 = 1$$
$$x_2 + 2x_3 = 1$$
$$2x_1 + x_2 + 3x_3 = 1,$$

we first subtract 2 times the first equation from the third to get the equivalent system,

$$x_1 + x_2 + x_3 = 1$$
$$x_2 + 2x_3 = 1$$
$$-x_2 + x_3 = -1.$$

We define the *multiplier* to be the factor 2. Next, we subtract -1 times the second row from the third to get

$$x_1 + x_2 + x_3 = 1$$
$$x_2 + 2x_3 = 1$$
$$3x_3 = 0.$$

In this case, the multiplier is -1. The system is now upper triangular and using back substitution, we obtain $x_3 = 0$, $x_2 = 1$, and $x_1 = 0$. Gaussian elimination can be coded in a straightforward way using matrix notation.

7.2.1. Matrix factorization

There is another way to view Gaussian elimination that is useful for the purposes of programming and handling special cases. Namely, Gaussian elimination is equivalent to computing a *factorization* of the coefficient matrix, $A = LU$, where L is a lower triangular and U an upper triangular $n \times n$ matrix. Given such a factorization of A, solving the system $Ax = b$ is straightforward. We first set $y = Ux$, then solve $Ly = b$ by forward substitution and finally solve $Ux = y$ by backward substitution.

To see that Gaussian elimination gives an LU factorization of A, consider the example above. We performed row operations that brought the system into upper triangular form. If we view these operations as row operations on the matrix A, we get the sequence

$$\begin{pmatrix} 1 & 1 & 1 \\ 0 & 1 & 2 \\ 2 & 1 & 3 \end{pmatrix} \to \begin{pmatrix} 1 & 1 & 1 \\ 0 & 1 & 2 \\ 0 & -1 & 1 \end{pmatrix} \to \begin{pmatrix} 1 & 1 & 2 \\ 0 & 1 & 2 \\ 0 & 0 & 3 \end{pmatrix},$$

which is an upper triangular matrix. This is the "U" in the LU decomposition.

The matrix L is determined by the observation that the row operations can be performed by multiplying A on the left by a sequence of special matrices called *Gauss transformations*. These are lower triangular matrices that have at most one nonzero entry in the off-diagonal positions and 1s down the diagonal. We show a Gauss transformation in Fig. 7.6. Multiplying A on the left by the matrix in Fig. 7.6 has the effect of adding α_{ij} times row j of A to row i of A. Note that the inverse of this matrix is obtained changing α_{ij} to $-\alpha_{ij}$; we will use this below.

Problem 7.6. Prove the last claim.

To perform the first row operation on A above, we multiply A on the left by

$$L_1 = \begin{pmatrix} 1 & 0 & 0 \\ 0 & 1 & 0 \\ -2 & 0 & 1 \end{pmatrix},$$

$$\begin{pmatrix}
1 & 0 & & & & \cdots & & & 0 \\
0 & 1 & 0 & & & & & & 0 \\
 & \ddots & 1 & \ddots & & 0 & & & \\
\vdots & & & & & & & & \vdots \\
 & 0 & 0 & 0 & \ddots & 0 & & & \\
 & 0 & \alpha_{ij} & 0 & \ddots & 1 & \ddots & & \\
 & 0 & 0 & 0 & & & \ddots & & \\
 & & & & & & 0 & 1 & 0 \\
0 & & \cdots & & & & & 0 & 1
\end{pmatrix}$$

Figure 7.6: A Gauss transformation.

to get

$$L_1 A = \begin{pmatrix} 1 & 1 & 1 \\ 0 & 1 & 2 \\ 0 & -1 & -1 \end{pmatrix}.$$

The effect of pre-multiplication by L_1 is to add $-2\times$ row 1 of A to row 3. Note that L_1 is lower triangular and has ones on the diagonal.

Next we multiply $L_1 A$ on the left by

$$L_2 = \begin{pmatrix} 1 & 0 & 0 \\ 0 & 1 & 0 \\ 0 & 1 & 1 \end{pmatrix},$$

and get

$$L_2 L_1 A = \begin{pmatrix} 1 & 1 & 1 \\ 0 & 1 & 2 \\ 0 & 0 & 3 \end{pmatrix} = U.$$

L_2 is also lower triangular with ones on the diagonal. It follows that $A = L_1^{-1} L_2^{-1} U$ or $A = LU$, where

$$L = L_1^{-1} L_2^{-1} = \begin{pmatrix} 1 & 0 & 0 \\ 0 & 1 & 0 \\ 2 & -1 & 1 \end{pmatrix}.$$

It is easy to see that L is also lower triangular with 1's on the diagonal with the multipliers (with sign change) occurring at the corresponding positions. We thus get the factorization

$$A = LU = \begin{pmatrix} 1 & 0 & 0 \\ 0 & 1 & 0 \\ 2 & -1 & 1 \end{pmatrix} \begin{pmatrix} 1 & 1 & 1 \\ 0 & 1 & 2 \\ 0 & 0 & 3 \end{pmatrix}.$$

Note that the entries in L below the diagonal are exactly the multipliers used to perform Gaussian elimination on A.

A general linear system can be solved in exactly the same fashion by Gaussian elimination using a sequence of Gauss transformations to obtain a factorization $A = LU$.

Problem 7.7. Verify that the product $L = L_1^{-1} L_2^{-1}$ has the stated form.

Problem 7.8. Show that the product of two Gauss transformations is a lower triangular matrix with ones on the diagonal and the inverse of a Gauss transformation is a Gauss transformation.

Problem 7.9. Solve the system

$$
\begin{aligned}
x_1 - x_2 - 3x_3 &= 3 \\
-x_1 + 2x_2 + 4x_3 &= -5 \\
x_1 + x_2 &= -2
\end{aligned}
$$

by computing an LU factorization of the coefficient matrix and using forward/backward substitution.

An LU factorization can be performed *in situ* using the storage space allotted to the matrix A. The fragment of code shown in Fig. 7.7 computes the LU factorization of A, storing U in the upper triangular part of A and storing the entries in L below the diagonal in the part of A below the diagonal. We illustrate the storage of L and U in Fig. 7.8.

Problem 7.10. On some computers, dividing two numbers is up to ten times more expensive than computing the reciprocal of the denominator and multiplying the result with the numerator. Alter this code to avoid divisions. Note: the reciprocal of the diagonal element a_{kk} has to be computed just once.

Problem 7.11. Write some pseudo-code that uses the matrix generated by the code in Fig. 7.7 to solve the linear system $Ax = b$ using forward/backward substitution. Hint: the only missing entries of L are the 1s on the diagonal.

```
for k = 1, ..., n-1, do                    (step through rows)
  for j = k+1, ..., n, do                  (eliminate entries
                                             below diagonal entry)
    a_jk = a_jk/a_kk                       (store the entry of L)
    for m = k+1, ..., n, do                (correct entries
                                             down the row)
      a_jm = a_jm - a_jk × a_km            (store the entry of U)
```

Figure 7.7: An algorithm to compute the LU factorization of A that stores the entries of L and U in the storage space of A.

$$\begin{pmatrix} u_{11} & u_{12} & \cdots & & u_{1n} \\ l_{21} & u_{22} & & & \\ \vdots & & \ddots & & \vdots \\ & & & & \\ l_{n1} & \cdots & & l_{nn-1} & u_{nn} \end{pmatrix}$$

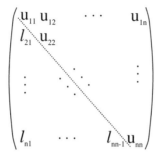

Figure 7.8: The matrix output from the algorithm in Fig. 7.7. L and U are stored in the space allotted to A.

7.2.2. Measuring the cost

The cost of solving a linear system using a direct method is measured in terms of computer time. In practice, the amount of computer time is proportional to the number of arithmetic and storage operations the computer uses to compute the solution. It is traditional (on a sequential computer) to simplify the cost calculation by equating storing a value, addition, and subtraction and equating multiplication and division when counting operations. Moreover, since multiplication (i.e. multiplications and divisions) generally cost much more than addition on older computers, it is also common to simply count the number of multiplications (=multiplications+divisions).

By this measure, the cost of computing the LU decomposition of an $n \times n$ matrix is $n^3 - n/3 = O(n^3/3)$. We introduce some new notation here, the big "O". The actual count is $n^3/3 - n/3$, however when n is large, the lower order term $-n/3$ becomes less significant. In fact,

$$\lim_{n \to \infty} \frac{n^3/3 - n/3}{n^3/3} = 1, \tag{7.5}$$

and this is the definition of the big "O". (Sometimes the big "O" notation means that the limit of the ratio of the two relevant quantities is any constant). With this notation, the operations count of the LU decomposition is just $O(n^3)$.

The cost of the forward and backward substitutions is much smaller:

Problem 7.12. Show that the cost of a backward substitution using an upper triangular matrix of dimension $n \times n$ is $O(n^2/2)$.

Problem 7.13. Determine the cost of multiplying a $n \times n$ matrix with another.

Problem 7.14. One way to compute the inverse of a matrix is based on viewing the equation $AA^{-1} = I$ as a set of linear equations for the columns of A^{-1}. If $a^{(j)}$ denotes the jth column of A^{-1}, then it satisfies the linear system

$$Aa^{(j)} = e_j$$

where e_j is the standard basis vector of $\mathbb{R}^n$ with a one in the jth position. Use this idea to write a pseudo-code for computing the inverse of a matrix using LU factorization and forward/backward substitution. Note that it suffices to compute the LU factorization only once. Show that the cost of computing the inverse in this fashion is $O(4n^3/3)$.

The general LU decomposition is used when the coefficient matrix of the system is *full*. A full matrix has mostly non-zero entries. An example is the matrix for the coefficients of the Galerkin polynomial approximation of the first order differential equation in Section 6.1 in Chapter 6.

Problem 7.15. Write a code that computes the Galerkin polynomial approximation for the first example in Section 6.1 using LU factorization and forward/backward substitution routines.

There are special classes of matrices for which the general LU decomposition is not appropriate because most of the elimination operations are wasted. For example, there is no point in computing the LU decomposition of a matrix that is already in upper or lower triangular form since the elimination would involve reducing coefficients that are already zero. Similarly, we do not need to eliminate most of the entries in the tridiagonal matrix that results from discretizing the two-point boundary value problem in Section 6.2 using the Galerkin finite element method. We discuss special forms of the LU factorization in Section 7.3.

7.2.3. Pivoting

During Gaussian elimination, it sometimes happens that the coefficient of a variable in the "diagonal position" becomes zero as a result of previous eliminations. When this happens of course, it is not possible to use that equation to eliminate the corresponding entries in the same column lying below the diagonal position. If the matrix is invertible, it is possible to find a non-zero coefficient in the same column and below the diagonal position, and by switching the two rows, the Gaussian elimination can proceed. This is called *zero pivoting*, or just *pivoting*.

Problem 7.16. Solve the system

$$x_1 + x_2 + x_3 = 2$$
$$x_1 + x_2 + 3x_3 = 5$$
$$-x_1 - 2x_3 = -1.$$

This requires pivoting.

Adding pivoting to the LU decomposition algorithm is straightforward. Before beginning the elimination using the current diagonal entry, we check to see if that entry is non-zero. If it is zero, we search the entries below in the same column for the first non-zero value, then interchange the row corresponding to that non-zero entry with the row corresponding to the current diagonal entry which is zero. Because the row interchanges involve rows in the "un-factored" part of A, the form of L and U are not affected. We illustrate this in Fig. 7.9.

To obtain the correct solution of the linear system $Ax = b$, we have to mirror all pivots performed on A in the data b. This is easy to do with the following trick. We define the vector of integers $p = (1\ 2\ \dots\ n)^\mathsf{T}$.

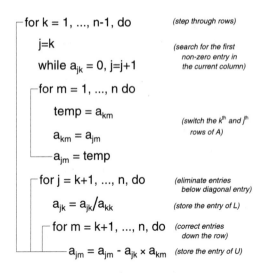

Figure 7.9: An algorithm to compute the LU factorization of A that used pivoting to avoid zero-valued diagonal entries.

This vector is passed to the LU factorization routine and whenever two rows of A are interchanged, we interchange the corresponding entries in p. After getting the altered p vector back, we pass it to the forward/backward routine. Here, we address the vector b indirectly using the vector p, i.e., we use the vector with entries $(b_{p_i})_{i=1}^{n}$, which has the effect of interchanging the rows in b in the correct fashion.

Problem 7.17. Alter the LU decomposition and forward/backward routines to solve a linear system with pivoting.

There are additional reasons to pivot in practice. As we have noted, the computation of the LU decomposition can be sensitive to errors originating from the finite precision of the computer if the matrix A is close to being non-invertible. We discuss this further below. We mention here however that a special kind of pivoting, called *partial pivoting* can be used to reduce this sensitivity. The strategy behind partial pivoting is to search the entries in the same column and below the current diagonal entry for the largest in absolute value. The row corresponding to the largest entry in magnitude is interchanged with the row corresponding to the current entry at the diagonal. The use of partial pivoting generally gives more accurate results than factorization without partial pivoting. One reason is that partial pivoting insures that the multipliers in the elimination process are kept as small as possible and consequently the errors in each entry are magnified by as little as possible during the course of the Gaussian elimination. We illustrate this with an example. Suppose that we solve

$$.000100x_1 + 1.00x_2 = 1.00$$
$$1.00x_1 + 1.00x_2 = 2.00$$

on a computer that holds three digits. Without pivoting, we get

$$.000100x_1 + 1.00x_2 = 1.00$$
$$-10000x_2 = -10000$$

which implies that $x_2 = 1$ and $x_1 = 0$. Note the large multiplier that is required for the elimination. Since the true answer is $x_1 = 1.0001$ and $x_2 = .9999$, the computed result has an error of 100% in x_1. If we switch the two rows before eliminating, which corresponds exactly to the partial pivoting strategy, we get

$$1.00x_1 + 1.00x_2 = 2.00$$
$$1.00x_2 = 1.00$$

which gives $x_1 = x_2 = 1.00$ as a result.

Problem 7.18. Verify the computations in this example.

Problem 7.19. Modify the code in Problem 7.17 to use partial pivoting.

Problem 7.20. Consider Problem 7.15. Compute the Galerkin approximation for $q = 1, 2, ..., 9$ using LU factorization and forward/backward substitution with and without partial pivoting.

Remark 7.2.1. There is a collection of high quality software for solving linear systems called LINPACK that is in the public domain and can be obtained through NETLIB. NETLIB is an online library of numerous public domain numerical analysis packages run by a consortium of industry, government and university groups. Some of these are well tested, production quality codes while others are in the development stage.

Sending an e-mail to *netlib@research.att.com* with the message *send index* causes NETLIB to return instructions and descriptions of the various libraries in the main group. More information about a specific package is obtained by sending an e-mail with the message *send index for linpack* for example. The LINPACK index describes the subroutines in that library and how to obtain the routines by e-mail. The specific calling instructions for a given routine are in the commented section at the beginning of the listing. Warning: some of the libraries are very big.

7.3. Direct methods for special systems

It is often the case that the matrices arising from the Galerkin finite element method applied to a differential equation have special properties that can be useful during the solution of the associated algebraic equations. For example, the stiffness matrix for the Galerkin finite element approximation of the two-point boundary value problem in Section 6.2 in Chapter 6 is symmetric, positive-definite, and tridiagonal. In this section, we examine a couple of different classes of problems that occur frequently.

7.3.1. Symmetric, positive-definite systems

As we mentioned, symmetric, positive-definite matrices are often encountered when discretizing differential equations (especially if the spatial part of the differential equation is of the type called elliptic). If A is symmetric and positive-definite, then it can be factored as $A = BB^\mathsf{T}$ where B is a lower triangular matrix with positive diagonal entries. This factorization can be computed from the LU decomposition of A, but there is a *compact method* of factoring A that requires only $O(n^3/6)$ multiplications called *Cholesky's method.*:

$$b_{11} = \sqrt{a_{11}}$$

$$b_{i1} = \frac{a_{i1}}{b_{11}}, \quad 2 \le i \le n,$$

$$\begin{cases} b_{jj} = \left(a_{jj} - \sum_{k=1}^{j-1} b_{jk}^2\right)^{1/2}, \\ b_{ij} = \left(a_{ij} - \sum_{k=1}^{j-1} b_{ik}b_{jk}\right)/b_{jj}, \end{cases} \quad 2 \le j \le n,\, j+1 \le i \le n$$

Problem 7.21. Count the cost of this algorithm.

This is called a compact method because it is derived by assuming that the factorization exists and then computing the coefficients of B directly from the equations obtained by matching coefficients in $BB^{\mathsf{T}} = A$. For example, if we compute the coefficient in the first row and column of BB^{T} we get b_{11}^2, which therefore must equal a_{11}. It is possible to do this because A is positive-definite and symmetric, which implies among other things that the diagonal entries of A remain positive throughout the factorization process and pivoting is not required when computing an LU decomposition.

Alternatively, the square roots in this formula can be avoided by computing a factorization $A = CDC^{\mathsf{T}}$ where C is a lower triangular matrix with ones on the diagonal and D is a diagonal matrix with positive diagonal coefficients.

Problem 7.22. Compute the Cholesky factorization of

$$\begin{pmatrix} 4 & 2 & 1 \\ 2 & 3 & 0 \\ 1 & 0 & 2 \end{pmatrix}$$

Problem 7.23. Compute the Cholesky factorization of the matrix in the system in Problem 7.4.

7.3.2. Banded systems

Banded systems are matrices with non-zero coefficients only in some number of diagonals centered around the main diagonal. In other words, $a_{ij} = 0$ for $j \leq i - d_l$ and $j \geq i + d_u$, $1 \leq i, j \leq n$, where d_l is the *lower bandwidth*, d_u is the *upper bandwidth*, and $d = d_u + d_l - 1$ is called the *bandwidth*. We illustrate this in Fig. 7.10. The stiffness matrix computed for the two-point boundary value problem in Section 6.2 is an example of a tridiagonal matrix, which is a matrix with lower bandwidth 2, upper bandwidth 2, and bandwidth 3.

When performing the Gaussian elimination used to compute the LU decomposition, we see that the entries of A that are already zero do not have to be reduced further. If there are only relatively few diagonals with non-zero entries, then the potential saving is great. Moreover, there is no need to store the zero-valued entries of A. It is straightforward to adapt

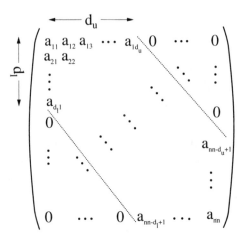

Figure 7.10: The notation for a banded matrix.

the LU factorization and forward/backward substitution routines to a banded pattern, once a storage scheme has been devised. For example, we can store a tridiagonal matrix as a $3 \times n$ matrix:

$$\begin{pmatrix} a_{21} & a_{31} & 0 & & \cdots & & 0 \\ a_{12} & a_{22} & a_{32} & 0 & \cdots & & 0 \\ 0 & a_{13} & a_{23} & a_{33} & 0 & \cdots & \vdots \\ & \ddots & \ddots & \ddots & \ddots & \ddots & 0 \\ \vdots & & & 0 & a_{1n-1} & a_{2n-1} & a_{3n-1} \\ 0 & & \cdots & & 0 & a_{1n} & a_{2n} \end{pmatrix}.$$

The routine displayed in Fig. 7.11 computes the LU factorization, while the routine in Fig. 7.12 performs the forward/backward substitution.

Problem 7.24. Show that the operations count for solving a tridiagonal system in this way is $O(5n)$.

The cost of this routine grows linearly with the dimension, rather than at a cubic rate as in the full case. Moreover, we use only the equivalent of six vectors of dimension n for storage. A more efficient version, derived as a compact method, uses even less.

$$\text{for } k = 2, ..., n, \text{ do}$$
$$a_{1k} = a_{1k}/a_{2k-1}$$
$$a_{2k} = a_{2k} - a_{1k} \times a_{3k-1}$$

Figure 7.11: A routine for computing the LU factorization of a tridiagonal system.

$$y_1 = b_1$$
$$\text{for } k = 2, ..., n, \text{ do}$$
$$y_k = b_k - a_{1k} \times y_{k-1}$$

$$x_n = y_n/a_{2n}$$
$$\text{for } k = n-1, ..., 1, \text{ do}$$
$$x_k = \left(y_k - a_{3k} \times x_{k+1}\right)/a_{2k}$$

Figure 7.12: Using forward and backward substitution to solve a tridiagonal system given the LU factorization.

Problem 7.25. Find an algorithm to solve a tridiagonal system that stores only four vectors of dimension n.

This algorithm assumes that no pivoting is required to factor A. Pivoting during the factorization of a banded matrix raises the difficulty that the bandwidth becomes larger. This is easy to see in a tridiagonal matrix, in which case we have to store an extra vector to hold the extra elements above the diagonal that result if two adjacent rows are switched.

Problem 7.26. Convince yourself of this.

In the case of the two-point boundary value problem in Chapter 6, A is symmetric and positive-definite as well as tridiagonal. As we mentioned above, this guarantees that pivoting is not required when factoring A.

Problem 7.27. A factorization of a tridiagonal solver can be derived as a compact method. Assume that A can be factored as

$$
A = \begin{pmatrix}
\alpha_1 & 0 & & \cdots & & 0 \\
\beta_2 & \alpha_2 & 0 & & & \\
0 & \beta_3 & \alpha_3 & & & \\
\vdots & & & \ddots & & 0 \\
0 & \cdots & & 0 & \beta_n & \alpha_n
\end{pmatrix}
\begin{pmatrix}
1 & \gamma_1 & 0 & \cdots & & 0 \\
0 & 1 & \gamma_2 & 0 & & \\
\vdots & & & \ddots & & \ddots \\
& & & & 1 & \gamma_{n-1} \\
0 & \cdots & & & 0 & 1
\end{pmatrix}
$$

Multiply out the factors and equate the coefficients to get equations for α, β, and γ. Derive some code based on these formulas.

Problem 7.28. Write some code to solve the tridiagonal system resulting from the Galerkin finite element discretization of the two-point boundary value problem in Section 6.2. Using $M = 50$, compare the time it takes to solve the system with this tridiagonal solver to the time using a full LU decomposition routine.

As for a tridiagonal matrix, it is straightforward to program special LU factorization and forward/backward substitution routines for a matrix with bandwidth d. The operations count is $O(nd^2/2)$ and the storage requirement is a matrix of dimension $d \times n$ if no pivoting is required. If d is much less than n, the savings in a special approach are considerable.

Problem 7.29. Show that the operations count of a banded solver for a $n \times n$ matrix with bandwidth d is $O(nd^2/2)$.

Problem 7.30. Write code to solve a linear system with bandwidth five centered around the main diagonal. What is the operations count for your code?

While it is true that if A is banded, then L and U are also banded, it is also true that in general L and U have non-zero entries in positions where A is zero. This is called *fill-in*. In particular, the stiffness matrix for a boundary value problem in several variables is banded and moreover most of the sub-diagonals in the band have zero coefficients. However, L and U do not have this property and we may as well treat A as if all the diagonals in the band have non-zero entries.

Banded matrices are one example of the class of *sparse* matrices. Recall that a sparse matrix is a matrix with mostly zero entries. As for banded matrices, it is possible to take advantage of sparsity to reduce the cost of factoring A in terms of time and storage. However, it is more difficult to do this than for banded matrices if the sparsity pattern puts non-zero entries at any location in the matrix. One approach to this problem is based on rearranging the equations and variables, or equivalently rearranging the rows and columns to form a banded system.

Remark 7.3.1. In reference to Remark 7.2.1, LINPACK also contains routines to handle banded matrices.

7.4. Iterative methods

Instead of solving $Ax = b$ directly, we now consider iterative solution methods based on computing a sequence of approximations $x^{(k)}$, $k = 1, 2, ...$, such that

$$\lim_{k \to \infty} x^{(k)} = x \quad \text{or} \quad \lim_{k \to \infty} \|x^{(k)} - x\| = 0,$$

for some norm $\| \cdot \|$ (recall that we discussed norms in Chapter 4).

Note that the finite precision of a computer has a different effect on an iterative method than it has on a direct method. A theoretically convergent sequence can not reach its limit in general on a computer using a finite number of digits. In fact, at the point at which the change from one iterate to the next occurs outside the range of digits held by the computer, the sequence simply stops changing. Practically speaking, there is no point computing iterations past this point, even if the limit has not been reached. On the other hand, it is often sufficient to have less accuracy than the limit of machine precision, and thus it is important to be able to estimate the accuracy of the current iterate.

7.4.1. Minimization algorithms

We first construct iterative methods for a linear system $Ax = b$ where A is symmetric and positive-definite. In this case, the solution x can be characterized equivalently as the solution of the quadratic minimization problem: find $x \in \mathbb{R}^n$ such that

$$F(x) \leq F(y) \quad \text{for all } y \in R^n, \tag{7.6}$$

where

$$F(y) = \frac{1}{2}(Ay, y) - (b, y),$$

with $(\cdot, \cdot)$ denoting the usual Euclidean scalar product.

Problem 7.31. Prove that the solution of (7.6) is also the solution of $Ax = b$.

We construct an iterative method for the solution of the minimization problem (7.6) based on the following simple idea: given an approximation $x^{(k)}$, compute a new approximation $x^{(k+1)}$ such that $F(x^{(k+1)}) < F(x^{(k)})$. On one hand, since F is a quadratic function, there must be a "downhill" direction from the current position, unless we are at the minimum. On the other hand, we hope that computing the iterates so that their function values are strictly decreasing, will force the sequence to converge to the minimum point x. Such an iterative method is called a *minimization method*.

Writing $x^{(k+1)} = x^{(k)} + \alpha_k d^{(k)}$, where $d^{(k)}$ is a *search direction* and α_k is a *step length*, by direct computation we get

$$F(x^{(k+1)}) = F(x^{(k)}) + \alpha_k (Ax^{(k)} - b, d^{(k)}) + \frac{\alpha_k^2}{2}(Ad^{(k)}, d^{(k)}),$$

where we used the symmetry of A to write $(Ax^{(k)}, d^{(k)}) = (x^{(k)}, Ad^{(k)})$. If the step length is so small that the second order term in α_k can be neglected, then the direction $d^{(k)}$ in which F decreases most rapidly, or the direction of *steepest descent*, is

$$d^{(k)} = -(Ax^{(k)} - b) = -r^{(k)},$$

which is the opposite direction to the residual error $r^{(k)} = Ax^{(k)} - b$. This suggests using an iterative method of the form

$$x^{(k+1)} = x^{(k)} - \alpha_k r^{(k)}. \tag{7.7}$$

A minimization method with this choice of search direction is called a *steepest descent method*. The direction of steepest descent is perpendicular to the *level curve* of F through $x^{(k)}$, which is the curve in the graph of F generated by the points where F has the same value as at $x^{(k)}$. We illustrate this in Fig. 7.13.

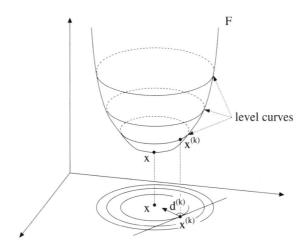

Figure 7.13: The direction of steepest descent of F at a point is perpendicular to the level curve of F through the point.

Problem 7.32. Prove that the direction of steepest descent at a point is perpendicular to the level curve of F through the same point.

Problem 7.33. Show that $r^{(k)} = Ax^{(k)} - b = \nabla F(x^{(k)})$, where ∇F is the gradient of F, that is the vector formed by the derivatives of $F(\eta_1, ..., \eta_n)$ with respect to the variables η_i.

It remains to choose the step lengths α_k. Staying with the underlying principle, we choose α_k to give the minimum value of F in the direction of $d^{(k)}$ starting from $x^{(k)}$. Differentiating $F(x^{(k)} + \alpha_k r^{(k)})$ with respect to α_k and setting the derivative zero gives

$$\alpha_k = -\frac{\left(r^{(k)}, d^{(k)}\right)}{\left(d^{(k)}, Ad^{(k)}\right)}. \tag{7.8}$$

Problem 7.34. Prove this formula.

As a simple illustration, we consider the case

$$A = \begin{pmatrix} \lambda_1 & 0 \\ 0 & \lambda_2 \end{pmatrix}, \quad 0 < \lambda_1 < \lambda_2, \tag{7.9}$$

and $b = 0$, corresponding to the minimization problem

$$\min_{y \in \mathbb{R}^n} \frac{1}{2} \left(\lambda_1 y_1^2 + \lambda_2 y_2^2 \right),$$

with solution $x = 0$.

Problem 7.35. Prove that the level curves of F in this case are ellipses with major and minor axes proportional to $1/\sqrt{\lambda_1}$ and $1/\sqrt{\lambda_2}$, respectively.

Applying (7.7) to this problem, we iterate according to

$$x^{(k+1)} = x^{(k)} - \alpha_k A x^{(k)},$$

using for simplicity a constant step length with $\alpha_k = \alpha$ instead of (7.8). In Fig. 7.14, we plot the iterations computed with $\lambda_1 = 1$, $\lambda_2 = 9$, and $x^{(0)} = (9, 1)^\top$. The convergence in this case is quite slow. The reason is that if $\lambda_2 \gg \lambda_1$, then the search direction $-(\lambda_1 x_1^{(k)}, \lambda_2 x_2^{(k)})^\top$ and the direction $-(x_1^{(k)}, x_2^{(k)})^\top$ to the solution at the origin, are very different. As a result the iterates swing back and forth across the long, narrow "valley".

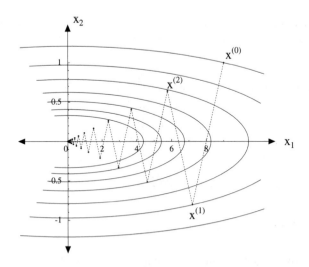

Figure 7.14: A sequence generated by the steepest descent method for (7.9) plotted together with some level curves of F.

Problem 7.36. Perform this computation.

It turns out that the rate at which the steepest descent method converges in general depends on the *condition number* $\kappa(A) = \lambda_n/\lambda_1$ of A, where $\lambda_1 \leq \lambda_2 \leq \ldots \leq \lambda_n$ are the eigenvalues of A (counted with multiplicity). In other words, the condition number of a symmetric positive definite matrix is the ratio of the largest eigenvalue to the smallest eigenvalue.

Remark 7.4.1. The general definition of the condition number of a matrix A in terms of a norm $\|\cdot\|$ is $\kappa(A) = \|A\|\|A^{-1}\|$. In the $\|\cdot\|_2$ norm, the two definitions are equivalent on symmetric matrices. Using any definition, a matrix is said to be *ill-conditioned* if the $\log(\kappa(A))$ is of the order of the number of digits used in the computer. As we said, we can expect to have difficulty solving an ill-conditioned system; which in terms of direct methods means large errors due to rounding errors and in terms of iterative methods means slow convergence.

We now analyze the steepest descent method for $Ax = b$ in the case of a constant step length α, where we iterate according to

$$x^{(k+1)} = x^{(k+1)} - \alpha(Ax^{(k)} - b).$$

Since the exact solution x satisfies $x = x - \alpha(Ax - b)$, we get the following equation for the error $e^{(k)} = x - x^{(k)}$:

$$e^{(k+1)} = (I - \alpha A)e^{(k)}.$$

The iterative method converges if the error tend to zero. Taking norms, we get

$$\|e^{(k+1)}\| \leq \mu \|e^{(k)}\| \tag{7.10}$$

where we use the spectral estimate (4.3) to write

$$\mu = \|I - \alpha A\| = \max_j |1 - \alpha\lambda_j|,$$

since the eigenvalues of the matrix $I - \alpha A$ are $1 - \alpha\lambda_j$, $j = 1, \ldots, n$. Iterating this estimate we get

$$\|e^{(k+1)}\| \leq \mu^k \|e^{(0)}\|, \tag{7.11}$$

where $e^{(0)}$ is the initial error.

To understand when (7.10), or (7.11), guarantees convergence, consider the scalar sequence $\{\lambda^k\}$ for $k \geq 0$. If $|\lambda| < 1$, then $\lambda^k \to 0$; if $\lambda = 1$, then the sequence is always 1; if $\lambda = -1$, the sequence alternates between 1 and -1 and does not converge; and if $|\lambda| > 1$, then the sequence diverges. Therefore if we want the iteration to converge for any initial value, then we must choose α so that $\mu < 1$. Since the λ_j are positive by assumption, $1 - \alpha\lambda_j < 1$ automatically, and we can guarantee that $1 - \alpha\lambda_j > -1$ if α satisfies $\alpha < 2/\lambda_n$. Choosing $\alpha = 1/\lambda_n$, which is not so far from optimal, we get

$$\mu = 1 - 1/\kappa(A).$$

Problem 7.37. Show that the choice $\alpha = 2(\lambda_1 + \lambda_n)^{-1}$ is optimal, and gives $\mu = 1 - 2/(\kappa(A) + 1)$.

If $\kappa(A)$ is large, then the convergence can be slow because then the reduction factor $1 - 1/\kappa(A)$ is close to one. More precisely, the number of steps required to lower the error by a given amount is proportional to the condition number; see the following problem.

Problem 7.38. Prove that $(1 - 1/\kappa(A))^k \leq \epsilon$ if $k \geq \kappa(A)\log(1/\epsilon)$.

When an iteration converges in this fashion, i.e. the error decreases (more or less) by a given factor in each iteration, then we say that the iteration converges *linearly*. We define the *rate of convergence* to be $-\log(\mu)$. The motivation is that the number of iterations are required to reduce the error by a factor of 10^{-m} is approximately $-m\log(\mu)$. Note that a faster rate of convergence means a smaller value of μ.

This is an *a priori* estimate of the error reduction per iteration, since we estimate the error before the computation. Such an analysis must account for the slowest possible rate of convergence because it holds for all initial vectors.

Consider the system $Ax = 0$ with

$$A = \begin{pmatrix} \lambda_1 & 0 & 0 \\ 0 & \lambda_2 & 0 \\ 0 & 0 & \lambda_3 \end{pmatrix}, \tag{7.12}$$

where $0 < \lambda_1 < \lambda_2 < \lambda_3$. For an initial guess $x^{(0)} = (x_1^0, x_2^0, x_3^0)^\top$, the steepest descent method with $\alpha = 1/\lambda_3$ gives the sequence

$$x^{(k)} = \left(\left(1 - \frac{\lambda_1}{\lambda_3}\right)^k x_1^0, \left(1 - \frac{\lambda_2}{\lambda_3}\right)^k x_2^0, 0 \right), \quad k = 1, 2, \ldots,$$

and,

$$\|e^{(k)}\| = \sqrt{\left(1 - \frac{\lambda_1}{\lambda_3}\right)^{2k} \left(x_1^0\right)^2 + \left(1 - \frac{\lambda_2}{\lambda_3}\right)^{2k} \left(x_2^0\right)^2}, \quad k = 1, 2, \ldots$$

Problem 7.39. Verify these last two formulas.

Thus for a general initial guess, the size of the error is given by the root mean square average of the corresponding iterate and the rate that the errors decrease is the root mean square average of the rates of decrease of the components. Therefore, depending on the initial vector, initially the iterates will generally converge more quickly than the rate of decrease of the first, i.e. slowest, component. In other words, more quickly than the rate predicted by (7.10), which bounds the rate of decrease of the errors by the rate of decrease in the slowest component. However, as the iteration proceeds, the second component eventually becomes much smaller than the first component (as long as $x_1^0 \neq 0$) and we can neglect that term in the expression for the error, i.e.

$$\|e^{(k)}\| \approx \left(1 - \frac{\lambda_1}{\lambda_3}\right)^k |x_1^0| \quad \text{for } k \text{ sufficiently large.} \tag{7.13}$$

In other words, the rate of convergence of the error for almost all initial vectors eventually becomes dominated by the rate of convergence of the slowest component. It is straightforward to show that the number of iterations that we have to wait for this approximation to be valid is determined by the relative sizes of the first and second components of $x^{(0)}$.

Problem 7.40. Compute the iteration corresponding to $\lambda_1 = 1$, $\lambda_2 = 2$, $\lambda_3 = 3$, and $x^{(0)} = (1, 1, 1)^\top$. Make a plot of the ratios of successive errors versus the iteration number. Do the ratios converge to the ratio predicted by the error analysis?

Problem 7.41. Show that the number of iterations it takes for the second component of $x^{(k)}$ to be less than 10% of the first component is proportional to $\log(|x_2^0|/10|x_1^0|)$.

Problem 7.42. Prove that the estimate (7.13) generalizes to any symmetric positive-definite matrix A, diagonal or not. Hint: use the fact that there is a set of eigenvectors of A that form an orthonormal basis for $\mathbb{R}^n$ and write the initial vector in terms of this basis. Compute a formula for the iterates and then the error.

This simple error analysis does not apply to the unmodified steepest descent method with varying α_k. However, it is generally true that the rate of convergence depends on the condition number of A, with a larger condition number meaning slower convergence. If we again consider the 2×2 example (7.9) with $\lambda_1 = 1$ and $\lambda_2 = 9$, then the estimate (7.10) for the simplified method suggests that the error should decrease by a factor of $1 - \lambda_1/\lambda_2 \approx .89$ in each iteration. The sequence generated by $x^{(0)} = (9,1)^\mathsf{T}$ decreases by exactly .8 in each iteration. The simplified analysis overpredicts the rate of convergence for this particular sequence, though not by a lot. By way of comparison, if we choose $x^{(0)} = (1,1)^\mathsf{T}$, we find that the ratio of successive iterations alternates between $\approx .126$ and $\approx .628$, because α_k oscillates in value, and the sequence converges much more quickly than predicted. On the other hand, there are initial guesses leading to sequences that converge at the predicted rate.

Problem 7.43. (a) Compute the steepest descent iterations for (7.9) corresponding to $x^{(0)} = (9,1)^\mathsf{T}$ and $x^{(0)} = (1,1)^\mathsf{T}$, and compare the rates of convergence. Try to make a plot like Fig. 7.14 for each. Try to explain the different rates of convergence.

Problem 7.44. Find an initial guess which produces a sequence that decreases at the rate predicted by the simplified error analysis.

The stiffness matrix A of the two-point boundary value problem in Section 6.2 in Chapter 6 is symmetric and positive-definite, and its condition number $\kappa(A) \propto h^{-2}$. Therefore the convergence of the steepest descent method is very slow if the number of mesh points is large. We return to this question Chapters 15 and 20.

7.4.2. A general framework for iterative methods

We now briefly discuss iterative methods for a general, linear system $Ax = b$, following the classical presentation of iterative methods in Isaacson and Keller ([9]). Recall that some matrices, like diagonal and triangular matrices, are relatively easy and cheap to invert, and Gaussian elimination can be viewed as a method of factoring A into such matrices. One way to view an iterative method is an attempt to approximate A^{-1} by the inverse of a part of A that is easier to invert. This is called an approximate inverse of A, and we use this to produce an approximate solution to the linear system. Since we don't invert the matrix A, we try to improve the approximate solution by repeating the partial inversion over and over. With this viewpoint, we start by *splitting* A into two parts:

$$A = N - P,$$

where the part N is chosen so that the system $Ny = c$ for some given c is relatively inexpensive to solve. Noting that the true solution x satisfies $Nx = Px + b$, we compute $x^{(k+1)}$ from $x^{(k)}$ by solving

$$Nx^{(k+1)} = Px^{(k)} + b \quad \text{for } k = 1, 2, ..., \tag{7.14}$$

where $x^{(0)}$ is an initial guess. For example, we may choose N to be the diagonal of A:

$$N_{ij} = \begin{cases} a_{ij}, & i = j, \\ 0, & i \neq j, \end{cases}$$

or triangular:

$$N_{ij} = \begin{cases} a_{ij}, & i \geq j, \\ 0, & i < j. \end{cases}$$

In both cases, solving the system $Nx^{(k+1)} = Px^{(k)} + b$ is cheap compared to doing a complete Gaussian elimination on A. so we could afford to do it many times.

As an example, suppose that

$$A = \begin{pmatrix} 4 & 1 & 0 \\ 2 & 5 & 1 \\ -1 & 2 & 4 \end{pmatrix} \quad \text{and} \quad b = \begin{pmatrix} 1 \\ 0 \\ 3 \end{pmatrix}, \tag{7.15}$$

and we choose

$$N = \begin{pmatrix} 4 & 0 & 0 \\ 0 & 5 & 0 \\ 0 & 0 & 4 \end{pmatrix} \quad \text{and} \quad P = \begin{pmatrix} 0 & -1 & 0 \\ -2 & 0 & -1 \\ 1 & -2 & 0 \end{pmatrix},$$

in which case the equation $Nx^{(k+1)} = Px^{(k)} + b$ reads

$$4x_1^{k+1} = -x_2^k + 1$$
$$5x_2^{k+1} = -2x_1^k - x_3^k$$
$$4x_3^{k+1} = x_1^k - 2x_2^k + 3.$$

Being a diagonal system it is easily solved, and choosing an initial guess and computing, we get

$$x^{(0)} = \begin{pmatrix} 1 \\ 1 \\ 1 \end{pmatrix}, \ x^{(1)} = \begin{pmatrix} 0 \\ -.6 \\ .5 \end{pmatrix}, \ x^{(2)} = \begin{pmatrix} .4 \\ -.1 \\ 1.05 \end{pmatrix}, \ x^{(3)} = \begin{pmatrix} .275 \\ -.37 \\ .9 \end{pmatrix},$$

$$x^{(4)} = \begin{pmatrix} .3425 \\ -.29 \\ 1.00375 \end{pmatrix}, \ \cdots \ x^{(15)} = \begin{pmatrix} .333330098 \\ -.333330695 \\ .999992952 \end{pmatrix}, \ \cdots$$

The iteration appears to converge to the true solution $(1/3, -1/3, 1)^{\mathsf{T}}$.

In general, we could choose $N = N_k$ and $P = P_k$ to vary with each iteration.

Problem 7.45.   Prove that the method of steepest descent corresponds to choosing

$$N = N_k = \frac{1}{\alpha_k} I, \text{ and } P = P_k = \frac{1}{\alpha_k} I - A,$$

with suitable α_k.

To analyze the convergence of (7.14), we subtract (7.14) from the equation $Nx = Px + b$ satisfied by the true solution to get an equation for the error $e^{(k)} = x - x^{(k)}$:

$$e^{(k+1)} = Me^{(k)},$$

where $M = N^{-1}P$ is the *iteration matrix*. Iterating on k gives

$$e^{(k+1)} = M^{k+1}e^{(0)}. \tag{7.16}$$

Rephrasing the question of convergence, we are interested in whether $e^{(k)} \to 0$ as $k \to \infty$. By analogy to the scalar case discussed above, if M is "small", then the errors $e^{(k)}$ should tend to zero. Note that the issue of convergence is independent of the data b.

If $e^{(0)}$ happens to be an eigenvector of M, then it follows from (7.16)

$$\|e^{(k+1)}\| = |\lambda|^{k+1}\|e^{(0)}\|,$$

and we conclude that if the method converges then we must have $|\lambda| < 1$ (or $\lambda = 1$). Conversely, one can show that if $|\lambda| < 1$ for all eigenvalues of M, then the method (7.14) indeed does converge:

Theorem 7.1. *An iterative method converges for all initial vectors if and only if every eigenvalue of the associated iteration matrix is less than one in magnitude.*

This theorem is often expressed using the *spectral radius* $\rho(M)$ of M, which is the maximum of the magnitudes of the eigenvalues of A. An iterative method converges for all initial vectors if and only if $\rho(M) < 1$. In general, the asymptotic limit of the ratio of successive errors computed in $\| \|_\infty$ is close to $\rho(M)$ as the number of iterations goes to infinity. We define the *rate of convergence* to be $R_M = -\log(\rho(M))$. The number of iterations required to reduce the error by a factor of 10^m is approximately m/R_M.

Problem 7.46. Prove this claim.

Practically speaking, "asymptotic" means that the ratio can vary as the iteration proceeds, especially in the beginning. In previous examples, we saw that this kind of a priori error result can underestimate the rate of convergence even in the special case when the matrix is symmetric and positive-definite (and therefore has an orthonormal basis of eigenvectors) and the iterative method uses the steepest descent direction. The general case now considered is more complicated, because interactions may occur in direction as well as magnitude, and a spectral radius estimate may overestimate the rate of convergence initially. As an example, consider the non-symmetric (even non-normal) matrix

$$A = \begin{pmatrix} 2 & -100 \\ 0 & 4 \end{pmatrix} \tag{7.17}$$

choosing

$$N = \begin{pmatrix} 10 & 0 \\ 0 & 10 \end{pmatrix} \text{ and } P = \begin{pmatrix} 8 & 100 \\ 0 & 6 \end{pmatrix} \text{ gives } M = \begin{pmatrix} .9 & 10 \\ 0 & .8 \end{pmatrix}.$$

In this case, $\rho(M) = .9$ and we expect the iteration to converge. Indeed it does converge, but the errors become quite large before they start to approach zero. We plot the iterations starting from $x^{(0)} = (1,1)^\top$ in Fig. 7.15.

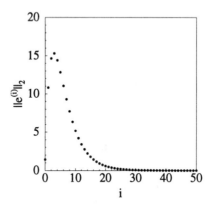

Figure 7.15: The results of an iterative method computed using a non-normal matrix.

Problem 7.47. Compute the eigenvalues and eigenvectors of the matrix A in (7.17) and show that A is not normal.

Problem 7.48. Prove that the matrix $\begin{pmatrix} 1 & -1 \\ 1 & 1 \end{pmatrix}$ is normal.

The goal is obviously to choose an iterative method so that the spectral radius of the iteration matrix is small. Unfortunately, computing $\rho(M)$ in the general case is much more expensive than solving the original linear system and is impractical in general. We recall from Chapter 4 that $|\lambda| \le \|A\|$ holds for any norm and any eigenvalue λ of A. The following theorem indicates a practical way to check for convergence.

Theorem 7.2. *Assume that $\|N^{-1}P\| \le \mu$ for some constant $\mu < 1$ and matrix norm $\|\cdot\|$. Then the iteration converges and $\|e^{(k)}\| \le \mu^k \|e^{(0)}\|$ for $k \ge 0$.*

Problem 7.49. Prove Theorem 7.2. Hint: take norms in (7.16) and use the assumptions and results from Chapter 4.

This theorem is also an a priori convergence result and suffers from the same deficiency as the analysis of the simplified steepest descent method presented above. In fact, choosing an easily computable matrix norm, like $\| \ \|_\infty$, generally leads to an even more inaccurate estimate of the convergence rate than would be obtained by using the spectral radius. In the worst case, it is entirely possible that $\rho(M) < 1 < \|M\|$ for the chosen norm, and hence the iterative method converges even though the theorem does not apply. The amount of "slack" in the bound in Theorem 7.2 depends on how much larger $\|A\|_\infty$ is than $\rho(A)$.

For the 3×3 example (7.15), we compute $\|N^{-1}P\|_\infty = 3/4 = \lambda$ and therefore we know the sequence converges. The theorem predicts that the error will get reduced by a factor of $3/4$ every iteration. If we examine the error of each iterate along with the ratios of successive errors after the first iteration:

i	$\|e^{(i)}\|_\infty$	$\|e^{(i)}\|_\infty / \|e^{(i-1)}\|_\infty$
0	1.333	
1	.5	.375
2	.233	.467
3	.1	.429
4	.0433	.433
5	.0194	.447
6	.00821	.424
7	.00383	.466
8	.00159	.414
9	.000772	.487

we find that after the first few iterations, the errors get reduced by a factor in the range of .4–.5 each iteration and not the factor $3/4$ predicted above. The ratio of $e^{(40)}/e^{(39)}$ is approximately .469. If we compute the eigenvalues of M, we find that $\rho(M) \approx .476$ which is close to the ratio of successive errors. To decrease the initial error by a factor of 10^{-4} using the predicted decrease of .75 per iteration, we would compute 33 iterations, while only 13 iterations are actually needed.

Problem 7.50. Compute 10 iterations using the A and b in (7.15) and the initial guess $x^{(0)} = (-1, \ 1, \ -1)^\top$. Compute the errors and the ratios of successive errors and compare to the results above.

Problem 7.51. Repeat Problem 7.50 using

$$A = \begin{pmatrix} 4 & 1 & 100 \\ 2 & 5 & 1 \\ -1 & 2 & 4 \end{pmatrix} \quad \text{and} \quad b = \begin{pmatrix} 1 \\ 0 \\ 3 \end{pmatrix}.$$

Does Theorem 7.2 apply to this matrix?

We get different methods, and different rates of convergence, by choosing different N and P. The method used in the example above is called the *Jacobi* method. In general, this consists of choosing N to be the "diagonal part" of A and P to be the negative of the "off-diagonal" part of A. This gives the set of equations

$$x_i^{k+1} = -\frac{1}{a_{ii}}\left(\sum_{j \neq i} a_{ij}x_j^k - b_i\right), \quad i = 1, ..., n.$$

To derive a more sophisticated method, we write out these equations in Fig. 7.16. The idea behind the *Gauss-Seidel* method is to use the new

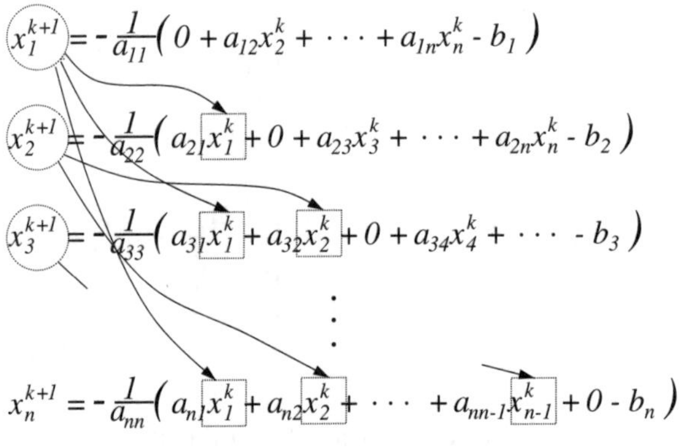

Figure 7.16: The Gauss-Seidel method substitutes new values of the iteration as they become available.

values of the approximation in these equations as they become known. The substitutions are drawn in Fig. 7.16. Presumably, the new values are more accurate than the old values, hence we might guess that this iteration will converge more quickly. The equations can be written

$$x_i^{k+1} = \frac{1}{a_{ii}}\left(-\sum_{j=1}^{i-1} a_{ij}x_j^{k+1} - \sum_{j=i+1}^{n} a_{ij}x_j^k + b_i\right).$$

If we decompose A into the sum of its lower triangular L, diagonal D, and upper triangular U parts, $A = L + D + U$, then the equations can be written $Dx^{(k+1)} = -Lx^{(k+1)} - Ux^{(k)} + b$ or

$$(D + L)x^{(k+1)} = -Ux^{(k)} + b.$$

Therefore, $N = D + L$ and $P = -U$. The iteration matrix is $M_{GS} = N^{-1}P = -(D + L)^{-1}U$.

Problem 7.52. Show that for the Jacobi iteration, $N = D$ and $P = -(L + U)$ and the iteration matrix is $M_J = -D^{-1}(L + U)$

Problem 7.53. (a) Solve (7.15) using the Gauss-Seidel method and compare the convergence with that of the Jacobi method. Also compare $\rho(M)$ for the two methods. (b) Do the same for the system in Problem 7.51.

Problem 7.54. (Isaacson and Keller ([9])) Analyze the convergence of the Jacobi and Gauss-Seidel methods for the matrix

$$A = \begin{pmatrix} 1 & \rho \\ \rho & 1 \end{pmatrix}$$

in terms of the parameter ρ.

In general it is difficult to compare the convergence of the Jacobi method with that of the Gauss-Seidel method. There are matrices for which the Jacobi method converges and the Gauss-Seidel method fails and vice versa. There are two special classes of matrices for which convergence can be established without further computation. A matrix A is *diagonally dominant* if

$$|a_{ii}| > \sum_{\substack{j=1 \\ j \neq i}}^{n} |a_{ij}|, \quad i = 1, ..., n.$$

If A is diagonally dominant then the Jacobi method converges.

Problem 7.55. Prove this claim.

A diagonally dominant matrix often occurs when a parabolic problem is discretized. We have already seen the other case, if A is symmetric and positive-definite then the Gauss-Seidel method converges. This is quite hard to prove, see Isaacson and Keller ([9]) for details.

7.5. Estimating the error of the solution

> How much does it cost to ride this train,
> Conductor, won't you tell me 'fore I go insane.
> I don't want a Greyhound or a fast jet plane.
> How much does it cost to ride this train? (D. Singleton)

The issue of estimating the error of the numerical solution of a linear system $Ax = b$ arises both in Gaussian elimination, because of the cumulative effects of round-off errors, and when using iterative methods, where we need a stopping criterion. Therefore it is important to be able to estimate the error in some norm with a fair degree of accuracy.

We discussed this problem in the context of iterative methods in the last section when we analyzed the convergence of iterative methods and Theorem 7.2 gives an *a priori* estimate for the convergence rate. It is an a priori estimate because the error is bounded before the computation begins. Unfortunately, as we saw, the estimate may not be very accurate on a particular computation, and it also requires the size of the initial error. In this section, we describe a technique of *a posteriori* error estimation that uses the approximation after it is computed to give an estimate of the error of that particular approximation.

We assume that x_c is a numerical solution of the system $Ax = b$ with exact solution x, and we want to estimate the error $\|x - x_c\|$ in some norm $\| \cdot \|$. We should point out that we are actually comparing the approximate solution $\tilde{x}_c$ of $\tilde{A}\tilde{x} = \tilde{b}$ to the true solution $\tilde{x}$, where $\tilde{A}$ and $\tilde{b}$ are the finite precision computer representations of the true A and b respectively. The best we can hope to do is compute $\tilde{x}$ accurately. To construct a complete picture, it would be necessary to examine the effects of small errors in A and b on the solution x. To simplify things, we ignore this part of the analysis and drop the ˜ . In a typical use of an iterative method, this turns out to be reasonable. It is apparently less reasonable in the analysis of a direct method, since the errors arising in direct methods are due to the finite precision. However, the initial error caused by storing A and b on a computer with a finite number of digits occurs only once, while the errors in the arithmetic operations involved in Gaussian elimination occur many times, so even in that case it is not an unreasonable simplification.

We start by considering the *residual error*

$$r = Ax_c - b,$$

which measures how well x_c solves the exact equation. Of course, the residual error of the exact solution x is zero but the residual error of x_c is not zero unless $x_c = x$ by some miracle. We now seek to estimate the unknown error $e = x - x_c$ in terms of the computable residual error r.

By subtracting $Ax - b = 0$ from $Ax_c - b = r$, we get an equation relating the error to the residual error:

$$Ae = -r. \qquad (7.18)$$

This is an equation of the same from as the original equation and by solving it numerically by the same method used to compute x_c, we get an approximation of the error e. This simple idea will be used in a more sophisticated form below in the context of a posteriori error estimates for Galerkin methods.

We now illustrate this technique on the linear system arising in the Galerkin finite element discretization of the two-point boundary value problem in Section 6.2. We generate a problem with a known solution so that we can compute the error and test the accuracy of the error estimate. We choose the true solution vector x with components $x_i = \sin(\pi i h)$, where $h = 1/(M + 1)$, corresponding to the function $\sin(\pi x)$ and then compute the data by $b = Ax$, where A is the stiffness matrix. We use the Jacobi method, suitably modified to take advantage of the fact that A is tridiagonal, to solve the linear system. We use $\| \ \| = \| \ \|_2$ to measure the error.

Problem 7.56. Derive an algorithm that uses the Jacobi method to solve a tridiagonal system. Use as few operations and as little storage as possible.

We compute the Jacobi iteration until the residual error becomes smaller than a given *residual tolerance* RESTOL. In other words, we compute the residual $r^{(k)} = Ax^{(k)} - b$ after each iteration and stop the process when $\|r^{(k)}\| \leq$ RESTOL. We present computations using the stiffness matrix generated by a uniform discretization with $M = 50$ elements yielding a finite element approximation with an error of .0056 in the l_2 norm. We choose the value of RESTOL so that the error in the computation of the coefficients of the finite element approximation is about 1% of the error of the approximation itself. This is reasonable since computing the coefficients of the approximation more accurately would not significantly increase the overall accuracy of the approximation. After the computation of $x^{(k)}$ is complete, we use the Jacobi method to approximate the solution of (7.18) and compute the estimate of the error.

Using the initial vector $x^{(0)}$ with all entries equal to one, we compute 6063 Jacobi iterations to achieve $\|r\| < $ RESTOL $= .0005$. The actual error of $x^{(6063)}$, computed using the exact solution, is approximately .0000506233. We solve (7.18) using the Jacobi method for 6063 iterations, reporting the value of the error estimate every 400 iterations:

iteration	estimate of error
1	0.00049862
401	0.00026027
801	0.00014873
1201	0.000096531
1601	0.000072106
2001	0.000060676
2401	0.000055328
2801	0.000052825
3201	0.000051653
3601	0.000051105
4001	0.000050849
4401	0.000050729
4801	0.000050673
5201	0.000050646
5601	0.000050634
6001	0.000050628

We see that the error estimate is quite accurate after 6001 iterations and sufficiently accurate for most purposes after 2000 iterations. In general, we do not require as much accuracy in the error estimate as we do in the solution of the system, so the estimation of the accuracy of the approximate solution is cheaper than the computation of the solution.

Since we estimate the error of the computed solution of the linear system, we can stop the Jacobi iteration once the error in the coefficients of the finite element approximation is sufficiently small so that we are sure the accuracy of the approximation will not be affected. This is a reasonable strategy given an estimate of the error. If we do not estimate the error, then the best strategy to guarantee that the approximation accuracy is not affected by the solution error is to compute the Jacobi iteration until the residual error is on the order of roughly 10^{-p}, where p is the number of digits that the computer uses. Certainly, there is not much point to computing further Jacobi iterations after this. If we assume that the computations are made in single precision, then $p \approx 8$. It takes a total of 11672 Jacobi iterations to achieve this level of residual error using the same initial guess as above. In fact, estimating the error and computing the coefficients of the approximation to a reasonable level of accuracy costs significantly less than this crude approach.

Problem 7.57. Repeat this example using the Gauss-Seidel method.

This approach can also be used to estimate the error of a solution computed by a direct method, provided the effects of finite precision are included. The added difficulty is that in general the residual error of a solution of a linear system computed with a direct method is small, even if the solution is inaccurate. Therefore, care has to be taken when computing the residual error because the possibility that subtractive cancellation makes the calculation of the residual error itself inaccurate. *Subtractive cancellation* is the name for the fact that the difference of two numbers that agree to the first i places has i leading zeroes. If only the first p digits of the numbers are accurate then their difference can have at most $p - i$ accurate significant digits. This can have severe consequences on the accuracy of the residual error if Ax_c and b agree to most of the digits used by the computer. One way to avoid this trouble is to compute the approximation in single precision and the residual in double precision (which means compute the product Ax_c in double precision, then subtract b). The actual solution of (7.18) is relatively cheap since the factorization of A has already been performed and only forward/backward substitution needs to be done.

Problem 7.58. Devise an algorithm to estimate the error of the solution of a linear system using single and double precision as suggested. Repeat the example using a tridiagonal solver and your algorithm to estimate the error.

The principle which I have always observed in my studies and which I believe has helped me the most to gain what knowledge I have, has been never to spend beyond a few hours daily in thoughts which occupy the imagination, and a very few hours yearly in those which occupy the understanding, and to give all the rest of my time to the relaxation of the senses and the repose of the mind. (Descartes)

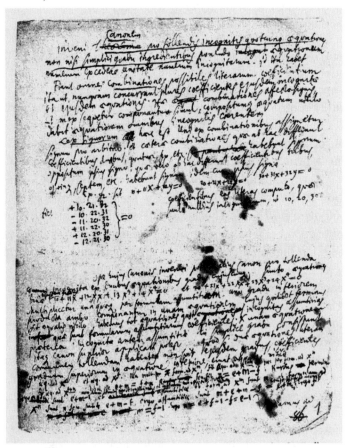

Figure 7.17: Leibniz's 1683 study of algorithms (Canonen) for the solution of linear systems of equations.

Part II

The archetypes

In this part, we explore some basic aspects of Galerkin's method by applying it in three kinds of problems: scalar linear two-point boundary value problems, scalar linear initial value problems, and initial value problems for linear systems of ordinary differential equations. Using these three types of problems, we introduce the three basic types of differential equations: elliptic, parabolic, and hyperbolic problems typically modelling stationary heat conduction, nonstationary heat conduction, and wave propagation respectively. In doing so, we set up a framework that we apply to linear partial differential equations in the third part of this book and to nonlinear systems of differential equations in the companion volume.

8

Two-Point Boundary Value Problems

> In all these cases and even in more complicated ones, our methods are of astonishing and unequaled simplicity. (Leibniz)

> If you take a little trouble, you will attain to a thorough understanding of these truths. For one thing will be illuminated by another, and eyeless night will not rob you from your road till you have looked into the heart of Nature's darkest mysteries. So surely will facts throw light upon facts. (Lucretius)

In this chapter, we continue to study the finite element method introduced for the numerical solution of the two-point boundary value problem (6.9), but now considering the more general problem (6.8) with variable heat conductivity. We originally derived (6.8) as a model of stationary heat conduction. We begin this chapter by re-deriving (6.8) as a model for two problems in elasticity: an elastic string and an elastic bar under load. The two-point boundary value problem (6.8) is an example of an *elliptic problem*, which is the kind of problem that typically arises when modeling stationary phenomena in heat conduction and elasticity.

We describe the structure of the solution of the two-point boundary value problem using both a solution formula in terms of integrals and a formula based on Fourier series. After recalling the Galerkin finite element method, we begin the mathematical study of the error in the finite element solution. Two important questions were raised in Chapter 6: what is the error of an approximation computed on a given mesh and how should we choose a mesh to reach a desired level of accuracy? The answers to both of these questions are based on the error analysis we present in this chapter. The techniques we introduce here are fundamental in the study of the error of the Galerkin approximation of any differential equation.

8.0.1.   A model of an elastic string

We describe the deflection of an elastic string hung horizontally by its ends. An example is a clothes line on which wet clothes are hanging. We assume that the units are chosen so the string occupies the interval $(0, 1)$ and let $u(x)$ denote the vertical displacement of the string at the position x under a load of intensity $f(x)$, where we take the displacement to be zero at the ends. To simplify the model, we assume that the tension in the string is constant and that $u(x)$ and $u'(x)$ stay small. This is reasonable as long as the items we hang on the string are not too heavy. With this assumption, the condition of equilibrium of a segment $(x, x + \Delta x)$ of string can be approximated by the fact that the vertical load $f(x)\Delta x$ on the segment must be carried by the vertical resultant of the forces at the end of the segment. This in turn is proportional to the difference of the slopes u' at x and $x + \Delta x$ with constant of proportionality equal to the tension; see Fig. 8.1. In other words,

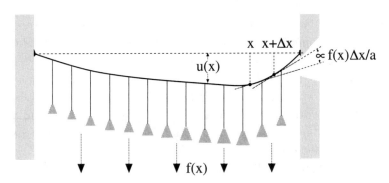

Figure 8.1: An elastic string subject to the load $f(x)$.

$$a(-u'(x + \Delta x) + u'(x)) \approx f(x)\Delta x,$$

where a is the tension. Letting Δx tend to zero yields (6.8) with $a(x) = a$ constant.

An interesting modification of this model includes the effects of vertical springs placed to help to support the string (which may be necessary if the clothes are wet). We assume that there are many springs hanging vertically between the string and a fixed horizontal level. To simplify the model, we consider the *continuum limit* which in this case means imagining that there is a spring at every point x and we assume that the spring at x has spring constant $c(x) \geq 0$. Equivalently, we can imagine a piece of elastic fabric hanging vertically between the string and a fixed horizontal level that has the property that the fabric obeys Hooke's law for a spring (cf. Chapter 7) when stretched in the vertical direction. With $a = 1$, we obtain the model

$$\begin{cases} -u''(x) + c(x)u(x) = f(x) & \text{for } 0 < x < 1, \\ u(0) = 0, u(1) = 0. \end{cases} \tag{8.1}$$

Problem 8.1. Derive (8.1) based on the discussion of elastic springs in Chapter 7.

8.0.2. A model of an elastic bar

We consider a horizontal elastic bar occupying the interval $[0, 1]$ subject to a tangential load with intensity $f(x)$. We let $\sigma(x)$ denote the total *stress* in a cross section of the bar at position x caused by the load and let $u(x)$ denote the corresponding *displacement* of a point in the bar at position x in the positive x direction measured from the reference position with zero load. For an arbitrary sub-interval (x_1, x_2) of $(0, 1)$, equilibrium requires that

$$-\sigma(x_2) + \sigma(x_1) = \int_{x_1}^{x_2} f(x)\, dx.$$

By varying x_1 and x_2, it follows that

$$-\sigma'(x) = f(x) \quad \text{for } 0 < x < 1.$$

Hooke's law in the present context states that the stress σ is proportional to the *strain* u',

$$\sigma = au',$$

where the coefficient of proportionality $a(x) > 0$ is the *modulus of elasticity*. Hooke's law is valid if both displacements and strains are small. Together with the boundary conditions $u(0) = u(1) = 0$ corresponding to zero displacements at the end points, this gives another version of (6.8):

$$\begin{cases} -(a(x)u'(x))' = f(x) & \text{for } 0 < x < 1, \\ u(0) = 0, \ u(1) = 0. \end{cases} \tag{8.2}$$

> Hooke, a contemporary of Newton, was Curator of the Royal Society. The duties of the curator were very onerous. According to his contract at every session of the Society (and they occurred every week except for the summer vacation) Hooke had to demonstrate three or four experiments proving new laws of nature...Towards the end of his life he counted 500 "laws" that he had discovered. Among other things, Hooke discovered the cellular structure of plants, the wave nature of light, Jupiter's red spot and probably also the inverse square law of gravitation. (Arnold)

8.0.3. The structure of the solutions

The two-point boundary value problem (8.2) is the prototypical example of an elliptic problem. We assume that $f(x)$ is continuous in $[0, 1]$ and that $a(x)$ is positive and continuously differentiable on $[0, 1]$. The solution of (8.2) is then required to have two continuous derivatives in $(0, 1)$ and to satisfy the differential equation at every point in $(0, 1)$ and the boundary conditions.

We can express the solution u of (8.2) in terms of data using two integrations:

$$u(x) = \int_0^x \frac{w(y)}{a(y)} \, dy + \alpha_1, \quad w(y) = -\int_0^y f(z) \, dz + \alpha_2,$$

where the constants are chosen so that $u(0) = u(1) = 0$. It follows that u is "twice as differentiable" as the data f, which is known as *elliptic smoothing*. The meaning of elliptic smoothing is obvious if $a = 1$, since then $u'' = -f$.

We also give a solution formula for a solution u of (8.2) in the case $a = 1$ in the form of a Fourier sine series:

$$u(x) = \sum_{j=1}^{\infty} u_j \sin(j\pi x),$$

where the $u_j \in \mathbb{R}$ are the Fourier coefficients of u. We recall from Chapter 5 that a continuous function $v(x)$ with a piecewise continuous derivative and satisfying $v(0) = v(1) = 0$ has a unique Fourier sine series that converges pointwise.

We determine the Fourier coefficients u_j by substituting the series into (8.2) to get

$$\sum_{j=1}^{\infty} j^2 \pi^2 u_j \sin(j\pi x) = f(x). \tag{8.3}$$

Since

$$\int_0^1 \sin(j\pi x) \sin(i\pi x) dx = \begin{cases} 0, & j \neq i, \\ 1/2, & j = i, \end{cases}$$

by multiplying (8.3) by $\sin(ix)$ and integrating over $(0,1)$, we get

$$\frac{1}{2} i^2 \pi^2 u_i = \int_0^1 f\varphi_i \, dx \quad \text{for } i = 1, 2, ...,$$

that is, the Fourier coefficients u_j of u are given by

$$u_j = \frac{1}{j^2 \pi^2} f_j,$$

where

$$f_j = 2 \int_0^1 f(x) \sin(j\pi x) \, dx \tag{8.4}$$

are the Fourier coefficients of $f(x)$. The rate at which Fourier coefficients f_j tend to zero depends on the number of continuous derivatives of an odd periodic extension of f. If $f_j = O(1/j^\mu)$ for some $\mu > 0$ then the Fourier coefficients of u satisfy $u_j = O(1/j^{\mu+2})$ and so the Fourier coefficients of u decay more quickly than those of f, which corresponds to the elliptic smoothing.

Problem 8.2. Compute a Fourier series representation of the solutions of $-u'' = f$ in $(0,1)$ with $u(0) = u(1) = 0$ corresponding to (a) $f = x$, (b) $f = x$ if $0 \leq x < 1/2$ and $f = 1 - x$ if $1/2 \leq x \leq 1$, (c) $f = x(1-x)$. (d) Compare the rate at which the Fourier coefficients of the solutions of (a)-(c) tend to zero as the frequency increases and explain the result.

Problem 8.3. Find some relation between the rate of decay of the Fourier coefficients (8.4) and the smoothness of an odd periodic extension of f. Hint: integrate by parts in (8.4).

In the general case with the coefficient $a(x)$ variable, Fourier's method depends on finding a set of Fourier basis functions that are orthogonal with respect to the weighted L_2 inner product with weight a and also being able to compute the Fourier coefficients of f. Usually, we cannot do these things and have to use numerical approximation.

8.1. The finite element method

Following the route of Chapter 6, we formulate the finite element method for the elliptic model problem (8.2) using Galerkin's method with piecewise linear approximations starting from a variational formulation of (8.2). We begin by assuming that the coefficient $a(x)$ is positive and bounded in the closed interval $[0, 1]$ so that there are positive constants a_1 and a_2 such that $a_1 \leq a(x) \leq a_2$ for $0 \leq x \leq 1$.

8.1.1. The variational formulation

The variational formulation of (8.2) is based on the following observation: multiplying the differential equation $-(au')' = f$ by a test function $v(x)$ satisfying $v(0) = v(1) = 0$ and integrating by parts, we get

$$\int_0^1 fv\, dx = -\int_0^1 (au')'v\, dx = \int_0^1 au'v'\, dx, \qquad (8.5)$$

because of the boundary conditions on v. The variational formulation is to find a function u in an appropriate space of *trial* functions that satisfies (8.5) for all appropriate test functions v. More precisely, if we choose the space of trial and test functions to be the same, and call it V, then the variational formulation reads: find $u \in V$ such that

$$\int_0^1 au'v'\, dx = \int_0^1 fv\, dx \quad \text{for all } v \in V, \qquad (8.6)$$

A natural choice for V turns out to be

$$V = \left\{ v : \int_0^1 v^2\, dx < \infty, \ \int_0^1 (v')^2\, dx < \infty, \ v(0) = v(1) = 0 \right\},$$

or in words, V is the space of functions v defined on $(0, 1)$ satisfying the boundary conditions $v(0) = v(1) = 0$, such that both v and v' are square integrable on $(0, 1)$. We note that if both u and v belong to V and $f \in L_2(0, 1)$, then the integrals occurring in (8.2) are convergent and therefore well defined, since by Cauchy's inequality

$$\int_0^1 |au'v'| \, dx \le \left(\int_0^1 a(u')^2 \, dx \right)^{1/2} \left(\int_0^1 a(v')^2 \, dx \right)^{1/2} < \infty$$

and

$$\int_0^1 |fv| \, dx \le \left(\int_0^1 f^2 \, dx \right)^{1/2} \left(\int_0^1 v^2 \, dx \right)^{1/2} < \infty,$$

where we used the boundedness of the coefficient $a(x)$. Furthermore, it turns out that functions in V are continuous and therefore have well defined boundary values $v(0)$ and $v(1)$, so that the specification $v(0) = v(1) = 0$ makes sense. We prove this in Chapter 21. Another important property of V in the context of Galerkin's method is that the finite element space V_h of continuous piecewise linear functions on $[0, 1]$ satisfying the boundary conditions is a subspace of V.

Problem 8.4. Prove this.

8.1.2. The relation between the two-point boundary value problem and the variational problem

Before getting to the finite element method, we comment on the relationship between the two-point boundary value problem (8.2) and the variational problem (8.6), and in particular, whether these problems have the same solution. A basic difference in the formulation of the two problems is the number of derivatives required, namely the solution of the differential equation is required to have two continuous derivatives while the solution of the variational problem is only required to have a first derivative that is bounded in the $L_2(0, 1)$ norm. In other words, the formulation of the differential equation requires more *regularity* (more derivatives) of its solution than the variational problem. Remember that the choice of requiring less regularity in the variational formulation is deliberate, made in particular so that the continuous piecewise linear approximation U can be substituted into the variational equation.

We have already proved that the solution of (8.2) also is a solution of the variational problem (8.6). This follows because a function that is twice continuously differentiable in $[0, 1]$ is automatically in V. This observation also indicates the main difficulty in showing that the solution of the variational problem (8.6) is also the solution of (8.2), namely to show that the solution of the variational problem is in fact twice continuously differentiable. Once this is proved, it is relatively easy to show that the solution of (8.6) in fact also solves (8.2). To see this, we simply integrate by parts in (8.6) to put two derivatives back on u to get

$$\int_0^1 \left(-(au')' - f\right)v\,dx = 0 \quad \text{for all } v \in V. \tag{8.7}$$

This says that all the weighted averages of $-(au')' - f$ for weights v in V are zero, and from this, we want to conclude that $-(au')' - f \equiv 0$. This is not immediately obvious, because we might think that $-(a(x)u'(x))' - f(x)$ could be positive for some x and negative for others in such a way that there is cancellation in the integral. To see that it is indeed true, assume that $-(a(\xi)u'(\xi))' - f(\xi) > 0$ at some point $0 < \xi < 1$. Because $-(au')' - f$ is continuous, it must actually be positive on some small interval, which we can describe as $[\xi - \delta, \xi + \delta]$. See Fig. 8.2. If we

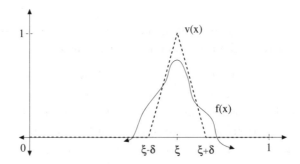

Figure 8.2: Illustration of the variational argument that shows that $-(au')' - f \equiv 0$.

choose v to be the hat function with peak at ξ and touching zero at $\xi - \delta$ and $\xi + \delta$, see Fig. 8.2, then the integrand $\left(-(a(x)u'(x))' - f(x)\right)v(x)$ is non-negative for all x and strictly positive for $\xi - \delta < x < \xi + \delta$. But this means that $\int_0^1 \left(-(au')' - f\right)v\,dx > 0$, which is a contradiction. We get the same result if $-(a(\xi)u'(\xi))' - f(\xi) < 0$ at some point ξ.

When can we expect the solution $u \in V$ of the variational problem (8.6) to have two continuous derivatives? It turns out that this is true under the regularity conditions for $a(x)$ and $f(x)$ demanded in the differential equation formulation. One can understand this by considering that the differential equation relates the second derivative u'' in terms of the first derivative u' and data,

$$u'' = -a^{-1}(a'u' + f).$$

The conclusion is that the variational formulation and the differential equation formulation have the same solution if $f(x)$ is continuous and $a(x)$ is continuously differentiable and positive. But what if $f(x)$ or $a(x)$ are discontinuous for example? Such functions makes sense from a physical point of view; just hang clothes on a part of the clothes line or use a wire made of two different substances joined together so the heat conductivity is discontinuous. In such problems, the meaning of the differential equation formulation is not immediately clear, while the variational formulation is still perfectly meaningful because the integrals are well defined. This indicates an advantage of the variational formulation over the differential equation formulation: the variational formulation covers a larger set of data.

8.1.3. An equivalent minimization problem

In Chapter 7, we saw that certain classes of linear algebraic systems can be reformulated as minimization problems. Analogously, the variational formulation (8.6) of the boundary value problem (8.2) can be reformulated as the following equivalent *minimization problem*: find the function $u \in V$ such that

$$F(u) \leq F(w) \quad \text{for all } w \in V, \tag{8.8}$$

where

$$F(w) = \frac{1}{2} \int_0^1 a(w')^2 \, dx - \int_0^1 fw \, dx.$$

The quantity $F(w)$ is the *total energy* of the function $w(x)$ which is the sum of the *internal energy* $\frac{1}{2} \int_0^1 a(w')^2 \, dx$ and the *load potential* $-\int_0^1 fw \, dx$. The space V introduced above may be described as the space of functions w on $(0, 1)$ satisfying the boundary condition $w(0) =$

$w(1) = 0$ and having finite total energy $F(w)$. The detailed motivation behind this statement is given in Chapter 21.

To see that (8.6) and (8.8) are equivalent, we assume first that $u \in V$ satisfies (8.6). For $w \in V$ chosen arbitrarily, we set $v = w - u \in V$ and compute

$$F(w) = F(u + v) = F(u) + \int_0^1 au'v'\, dx - \int_0^1 fv\, dx + \frac{1}{2} \int_0^1 av'v'\, dx$$

$$= F(u) + \frac{1}{2} \int_0^1 av'v'\, dx,$$

proving that $F(w) \geq F(u)$. Conversely, if $u \in V$ satisfies (8.8) then the function $g(\epsilon) = F(u + \epsilon v)$, where $v \in V$ is fixed but arbitrary and $\epsilon \in \mathbb{R}$, has a minimum at $\epsilon = 0$. The derivative of $g(\epsilon)$ at $\epsilon = 0$ must be zero since that is a minimum point. Computing, we get $g'(0) = \int_0^1 au'v'\, dx - \int_0^1 fv\, dx$ from which (8.6) follows. For a differential equation which has an equivalent formulation as a minimization of an energy, questions such as existence, uniqueness and stability of solutions may be easier to tackle than for a general equation. For existence the idea is to construct a sequence of approximate solutions with energies decreasing to the minimal energy, and then prove that the sequence is a Cauchy sequence in some suitable space. We use this technique to prove existence of solutions of e.g. Poisson's equation in Chapter 21. Also, the corresponding discrete system obtained by Galerkin's method often may be solved efficiently by making use of a minimization formulation.

8.1.4. The finite element method

We consider the Galerkin finite element method for (8.2) based on continuous piecewise linear approximation, which we call the cG(1) finite element method (continuous Galerkin method of order one). The method is constructed following the same pattern we used to discretize (6.9). Let $\mathcal{T}_h : 0 = x_0 < x_1 < \ldots < x_M < x_{M+1} = 1$ be a partition of $(0, 1)$ and let V_h be the corresponding finite element space of continuous piecewise linear functions vanishing at $x = 0$ and $x = 1$. The finite element method is obtained by applying Galerkin's method to (8.6) with the finite dimensional space V_h replacing V and reads: find $U \in V_h$ such that

$$\int_0^1 aU'v'\, dx = \int_0^1 fv\, dx \quad \text{for all } v \in V_h. \tag{8.9}$$

Subtracting (8.9) from (8.6), and using $V_h \subset V$, we obtain the relation

$$\int_0^1 a(u - U)' v' \, dx = 0 \quad \text{for all } v \in V_h, \tag{8.10}$$

which expresses the Galerkin orthogonality property of the error $u - U$ in variational form.

As before, we write U in terms of the nodal basis functions φ_j, $j = 1, ..., M$,

$$U(x) = \sum_{j=1}^{M} \xi_j \varphi_j(x),$$

and substitute this expression into (8.9). Choosing v to be each one of the basis functions φ_i in turn, we obtain the systems of equations $A\xi = b$ for the coefficients $\xi = (\xi_j)$ of U, where $A = (a_{ij})$ is the $M \times M$ stiffness matrix and $b = (b_i)$ is the load vector with elements

$$a_{ij} = \int_0^1 a\varphi_j' \varphi_i' \, dx, \qquad b_i = \int_0^1 f \varphi_i \, dx.$$

The stiffness matrix A is symmetric because the trial and test spaces are the same, positive-definite since a is strictly positive, and tridiagonal because the supports of the nodal basis functions overlap only for nearby nodes.

Problem 8.5. Prove these properties of A.

Problem 8.6. Compute the stiffness matrix and load vector for the cG(1) method on a uniform triangulation for (8.2) with $a(x) = 1 + x$ and $f(x) = \sin(x)$.

Problem 8.7. Formulate the cG(1) method for (8.1). Compute the stiffness matrix when c is a constant. Is the stiffness matrix still symmetric, positive-definite, and tridiagonal?

Problem 8.8. Prove that the Galerkin finite element method can be reformulated in terms of the minimization problem: find $U \in V_h$ such that

$$F(U) \leq F(v) \quad \text{for all } v \in V_h.$$

Show that this problem takes the matrix form: find the vector $\xi = (\xi_j) \in \mathbb{R}^M$ that minimizes the quadratic function $\frac{1}{2}\eta^{\top} A\eta - b^{\top}\eta$ for $\eta \in \mathbb{R}^M$.

Problem 8.9. Show that the cG(1) solution U of (8.2) is exact at the nodes x_j if $a \equiv 1$. Hint: show that the error $e = u - U$ can be written

$$e(z) = \int_0^1 g_z'(x) e'(x) \, dx,$$

where

$$g_z(x) = \begin{cases} (1-z)x, & 0 \le x \le z, \\ z(1-x), & z \le x \le 1, \end{cases}$$

and then use Galerkin orthogonality and the fact that $g_{x_j} \in V_h$ for $j = 1, ..., M$. Does the result extend to variable a?

Problem 8.10. Prove that the function $g_z(x)$ defined in the previous problem satisfies the equation $-g_z'' = \delta_z$ in $(0,1)$, $g_z(0) = g_z(1) = 0$, where δ_z is the delta function at $x = z$. Show that a function u solving $-u'' = f$ in $(0,1)$, $u(0) = u(1) = 0$ can be represented in terms of the right-hand side f via g_z as follows:

$$u(z) = \int_0^1 g_z(x) f(x) \, dx.$$

The function $g_z(x)$ is the *Green's function* for the boundary value problem (8.2) with $a = 1$.

8.1.5.   Different boundary conditions

The boundary conditions of (8.2) are called *homogeneous Dirichlet* boundary conditions. A commonly encountered alternative is a *homogeneous Neumann* boundary condition

$$a(x)u'(x) = 0 \text{ for } x = 0 \text{ or } 1. \tag{8.11}$$

In the models we have discussed, this corresponds to prescribing zero flux $q(x) = -a(x)u'(x)$ or stress $\sigma(x) = a(x)u'(x)$ at $x = 0$ or 1. Effectively, this condition reduces to $u'(0) = 0$ or $u'(1) = 0$ since we assume that $a(x)$ is positive on $[0,1]$. More generally, we can impose a *Robin* boundary condition at $x = 1$ of the form

$$a(1)u'(1) + \gamma(u(1) - u_1) = g_1, \tag{8.12}$$

where $\gamma \ge 0$ is a given boundary heat conductivity, u_1 is a given outside temperature and g_1 a given heat flux. For example if $g_1 = 0$, then (8.12) says that the heat flux $-a(1)u'(1)$ at the boundary is proportional to the temperature difference $u(1) - u_1$ between the inside and outside temperatures $u(1)$ and u_1. We can experience this kind of boundary condition in a poorly insulated house on a cold winter day. The size of heat conductivity γ is a "hot issue" in the real estate business in the north of Sweden. When $\gamma = 0$, (8.12) reduces to a Neumann boundary condition.

Changing the boundary conditions requires a change in the variational formulation associated to the two-point boundary value problem. As an example, we consider the problem

$$\begin{cases} -(au')' = f & \text{in } (0,1), \\ u(0) = 0, \ a(1)u'(1) = g_1. \end{cases} \tag{8.13}$$

To derive the variational formulation of this problem, we multiply the differential equation $-(au')' = f$ by a test function v and integrate by parts to get

$$\int_0^1 fv\,dx = -\int_0^1 (au')'v\,dx = \int_0^1 au'v'\,dx - a(1)u'(1)v(1) + a(0)u'(0)v(0).$$

Now $a(1)u'(1)$ is specified but $a(0)u'(0)$ is unknown. If we assume that v satisfies the homogeneous Dirichlet condition $v(0) = 0$ and replace $a(1)u'(1)$ by g_1, then we are led to the following variational formulation of (8.13): find $u \in V$ such that

$$\int_0^1 au'v'\,dx = \int_0^1 fv\,dx + g_1v(1) \quad \text{for all } v \in V, \tag{8.14}$$

where

$$V = \left\{ v : \int_0^1 v^2\,dx < \infty, \ \int_0^1 (v')^2\,dx < \infty, \ v(0) = 0 \right\}. \tag{8.15}$$

Note that we only require that the functions in V satisfy the homogeneous Dirichlet boundary condition $v(0) = 0$.

To see that a solution of the variational problem (8.14) that is twice continuously differentiable solves the two-point boundary value problem (8.13), we have to show the solution satisfies the differential equation and the boundary conditions. The Dirichlet condition at $x = 0$ is satisfied of course because of the choice of V, but it is not immediately obvious that the Neumann condition is satisfied. To see this, we argue in the same way as above for the problem with Dirichlet boundary conditions. We integrate by parts in (8.14) to get

$$\int_0^1 fv\,dx + g_1 v(1) = -\int_0^1 (au')'v\,dx + a(1)u'(1)v(1). \tag{8.16}$$

Compared to (8.7), there are two additional boundary terms $g_1 v(1)$ and $a(1)u'(1)\,v(1)$. We first choose v as in Fig. 8.2 with $v(1) = 0$ so the boundary terms drop out, which shows that $-(au')' = f$ in $(0,1)$. Consequently, (8.16) reduces to $g_1 v(1) = a(1)u'(1)v(1)$ for all v in V. Now we choose a test function v with $v(1) = 1$, which shows that $a(1)u'(1) = g_1$. In short, the Neumann boundary condition is included implicitly in the variational formulation (8.14) through the free variation of the test functions v on the boundary with the Neumann condition.

To construct the cG(1) finite element approximation for (8.13), we define V_h to be the space of continuous functions v that are piecewise linear with respect to a partition $\mathcal{T}_h$ of $(0,1)$ such that $v(0) = 0$. Since the value of $v \in V_h$ at $x_{M+1} = 1$ is unspecified, we include the "half-hat" function φ_{M+1} in the set of nodal basis functions for V_h. The finite element method reads: compute $U \in V_h$ such that

$$\int_0^1 aU'v'\,dx = \int_0^1 fv\,dx + g_1 v(1) \quad \text{for all } v \in V_h. \tag{8.17}$$

We substitute $U(x) = \sum_{i=1}^{M+1} \xi_i \varphi_i(x)$, now with an undetermined value at the node x_{M+1}, into (8.17) and choose $v = \varphi_1, \cdots, \varphi_{M+1}$ in turn to get a $(M+1) \times (M+1)$ system of equations for ξ. We show the form of the resulting stiffness matrix with $a = 1$ and load vector in Fig. 8.3.

> **Problem 8.11.** Compute the coefficients of the stiffness matrix for the cG(1) finite element method for (8.17) using a uniform partition and assuming $a = f = 1$ and $g_1 = 1$. Check if the discrete equation corresponding to the basis function φ_{M+1} at $x = 1$ looks like a discrete analog of the Neumann condition.

To conclude, Neumann and Robin boundary conditions, unlike Dirichlet conditions, are not explicitly enforced in the trial and test spaces used

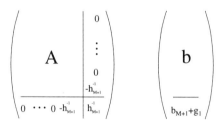

Figure 8.3: The stiffness matrix and load vector computed from (8.17) in the case that $a \equiv 1$. A and b are the stiffness matrix and load vector previously obtained in the problem with Dirichlet boundary conditions (8.6) and $b_{M+1} = \int_0^1 f\varphi_{M+1}\, dx$.

in the variational formulation. Instead, these conditions are automatically satisfied in the variational formulation. The Dirichlet boundary condition is called an *essential* boundary condition and Neumann and Robin conditions are called *natural* boundary conditions. An essential boundary condition is imposed explicitly in the definition of the trial and test spaces, i.e. it is a *strongly imposed* boundary condition, while natural boundary condition is automatically satisfied in the variational formulation, and is said to be a *weakly imposed* boundary condition.

We consider briefly also the case of non-homogeneous Dirichlet boundary conditions. As an example, we consider the equation $-(au')' = f$ in $(0,1)$ with the boundary conditions $u(0) = u_0$ and $u(1) = 0$, where $u_0 \neq 0$. The variational formulation of this problem is: find u in

$$V_{u_0} = \left\{ w : \int_0^1 w^2\, dx < \infty,\ \int_0^1 (w')^2\, dx < \infty,\ w(0) = u_0,\ w(1) = 0 \right\},$$

such that

$$\int_0^1 au'v'\, dx = \int_0^1 fv\, dx \quad \text{for all } v \in V_0,$$

where test functions in $v \in V_0$ satisfy the homogenous Dirichlet boundary condition $v(0) = 0$. In this case the space of trial functions V_{u_0} and the space of test functions V_0 are not the same. The reason a test function v satisfies the homogeneous Dirichlet boundary condition $v(0) = 0$ is that $v = u - w$, where both $w(0) = u(0) = u_0$, cf. Section 8.1.3.

Problem 8.12. Show that the Neumann problem (8.14) takes the following variational form if the Neumann condition is generalized to the Robin condition (8.12): find $u \in V$ such that

$$\int_0^1 aU'v' \, dx + \gamma u(1)v(1) = \int_0^1 fv \, dx + g_1 v(1) + \gamma u_1 v(1) \quad \text{for all } v \in V,$$

where the space V is still defined by (8.15). Compute the coefficients of the stiffness matrix of the cG(1) method applied to this problem with $a = 1$ on a uniform partition and check how the Robin boundary condition is approximated in the discrete system.

Problem 8.13. Determine the discrete system if cG(1) is applied to the boundary value problem (8.2) with $a = 1$ and the boundary conditions $u(0) = 1$, $u(1) = 0$.

8.1.6. Quadrature

In order to construct the discrete equations for U, we have to compute integrals involving a and f. For example, a typical component of b is

$$b_i = \int_{x_{i-1}}^{x_i} f(x) \frac{(x - x_{i-1})}{h_i} \, dx + \int_{x_i}^{x_{i+1}} f(x) \frac{(x - x_{i+1})}{-h_{i+1}} \, dx.$$

In practice, it is often inconvenient or even impossible to compute these integrals exactly. Hence, we often resort to evaluating the integrals giving the coefficients of A and b by quadrature formulas. We saw in Section 6.3 that the use of quadrature can have a strong effect on the accuracy of a Galerkin approximation. A basic principle is to choose a quadrature of sufficient accuracy that the order of convergence of the resulting method is the same as the order of convergence of the method computed without quadrature. Since we expect that the piecewise linear finite element approximation, like the piecewise linear interpolant, should be second order accurate, we use a quadrature on integrals involving f that is at least second order accurate. To verify that this does not affect the convergence rate, we later prove error estimates for methods with and without quadrature.

As an example, using the trapezoidal rule,

$$\int_{x_{i-1}}^{x_i} g(x) \, dx \approx \frac{1}{2} \big(g(x_i) + g(x_{i-1}) \big) h_i,$$

the components of the approximate load vector are

$$f(x_i)\frac{1}{2}\big(h_i + h_{i-1}\big).$$

Problem 8.14. Verify this formula.

In general, we prefer to use the midpoint Gauss rule rather than the trapezoidal rule because we get the same order of accuracy with fewer function evaluations. However in this case, the trapezoidal rule requires the same number of function evaluations as the midpoint rule because of the properties of the basis functions φ_i.

Problem 8.15. Derive a formula for the approximate load vector $\tilde{b}$ that is computed using the midpoint Gauss rule to evaluate the integrals.

Problem 8.16. Compute the coefficients of A and b in the equations for the Galerkin finite element approximation of the problem described in Problem 8.6 using the trapezoidal rule to evaluate integrals involving a and f.

8.2. Error estimates and adaptive error control

When conducting scientific experiments in a laboratory or building a suspension bridge, for example, there is always a lot of worry about the errors in the process. In fact, if we were to summarize the philosophy behind the scientific revolution, a main component would be the modern emphasis on the quantitative analysis of error in measurements during experiments and the reporting of the errors along with the results. The same issue comes up in computational mathematical modeling: whenever we make a computation on a practical problem, we must be concerned with the accuracy of the results and the related issue of how to compute efficiently. These issues naturally fit into a wider framework which also addresses how well the differential equation models the underlying physical situation and what effect errors in data and the model have on the conclusions we can draw from the results.

We address these issues by deriving two kinds of error estimates for the error $u - U$ of the finite element approximation. First we prove an *a priori* error estimate which shows that the Galerkin finite element method for (8.2) produces the best possible approximation of the solution u in V_h in a certain sense. If u has continuous second derivatives, then we know that V_h contains good approximations of u, for example the piecewise linear interpolant. So the a priori estimate implies that the error of the finite element approximation can be made arbitrarily small by refining the mesh provided that the solution u is sufficiently smooth to allow the interpolation error to go to zero as the mesh is refined. This kind of result is called an *a priori* error estimate because the error bound does not depend on the approximate solution to be computed. One the other, it does requires knowledge about the derivatives of the (unknown) exact solution.

After that, we prove an *a posteriori* error bound that bounds the error of the finite element approximation in terms of its residual error. This error bound can be evaluated once the finite element solution has been computed and used to estimate the error. Through the a posteriori error estimate, it is possible to estimate and adaptively control the finite element error to a desired tolerance level by suitably refining the mesh.

To measure the size of the error $e = u - U$, we use the energy norm $\|v\|_E$ corresponding to (8.2) defined for $v \in V$ by

$$\|v\|_E = \left(\int_0^1 a(x)(v'(x))^2 \, dx \right)^{1/2} .$$

This norm arises naturally in the analysis of the variational formulation. If we introduce the *weighted L_2 norm*

$$\|w\|_a = \left(\int_0^1 a\,w^2 \, dx \right)^{1/2} ,$$

with *weight a*, then the energy norm satisfies $\|v\|_E = \|v'\|_a$. In other words, the energy norm of a function v in V is the a-weighted L_2 norm of the first derivative v'. We will use the following variations of Cauchy's inequality with the weight a present:

$$\left| \int_0^1 a v' w' \, dx \right| \leq \|v'\|_a \|w'\|_a \quad \text{and} \quad \left| \int_0^1 v w \, dx \right| \leq \|v\|_a \|w\|_{a^{-1}} .$$
$$(8.18)$$

Problem 8.17. Prove these inequalities.

Assuming that a is bounded away from zero, a function v in V that is small in the energy norm, like the error of the finite element approximation, is also small pointwise. We illustrate this in

Problem 8.18. Prove the following inequality for functions $v(x)$ on $I = (0, 1)$ with $v(0) = 0$,

$$v(y) = \int_0^y v'(x)\, dx \le \left(\int_0^y a^{-1}\, dx \right)^{1/2} \left(\int_0^y a(v')^2\, dx \right)^{1/2}, \quad 0 < y < 1. \tag{8.19}$$

Use this inequality to show that if a^{-1} is integrable on $I = (0, 1)$ so that $\int_I a^{-1}\, dx < \infty$, then a function v is small in the maximum norm on $[0, 1]$ if $\|v\|_E$ is small and $v(0) = 0$.

8.2.1. An a priori error estimate

To estimate the error of the Galerkin finite element approximation, we compare it to the error of other approximations of the solution in V_h. The a priori error estimate for (8.9) states that the finite element approximation $U \in V_h$ is the best approximation of u in V_h with respect to the energy norm. This is a consequence of the Galerkin orthogonality built into the finite element method as expressed by (8.10).

We have for any $v \in V_h$,

$$\|(u - U)'\|_a^2 = \int_0^1 a(u - U)'(u - U)'\, dx$$

$$= \int_0^1 a(u - U)'(u - v)'\, dx + \int_0^1 a(u - U)'(v - U)'\, dx$$

$$= \int_0^1 a(u - U)'(u - v)'\, dx,$$

where the last line follows because $v - U \in V_h$. Estimating using Cauchy's inequality, we get

$$\|(u - U)'\|_a^2 \le \|(u - U)'\|_a \|(u - v)'\|_a,$$

so that

$$\|(u - U)'\|_a \le \|(u - v)'\|_a \quad \text{for all } v \in V_h.$$

This is the best approximation property of U. We now choose in particular $v = \pi_h u$, where $\pi_h u \in V_h$ is the nodal interpolant of u, and use the following weighted analog of (5.19)

$$\|(u - \pi_h u)'\|_a \le C_i \|h u''\|_a,$$

where C_i is an interpolation constant that depends only on (the variation of) a, to obtain the following error estimate.

Theorem 8.1. *The finite element approximation U satisfies $\|u-U\|_E \leq \|u - v\|_E$ for all $v \in V_h$. In particular, there is a constant C_i depending only on a such that*

$$\|u - U\|_E = \|u' - U'\|_a \leq C_i \|hu''\|_a.$$

Problem 8.19. Compute the bound in Theorem 8.1 for the solution of the example in Section 6.2.

This energy norm estimate says that the derivative of the error of the finite element approximation converges to zero at a first order rate in the mesh size h. It does not directly indicate that the error itself, say pointwise or in the L_2 norm, tends to zero. Arguing as in Problem 8.18, we can derive a pointwise bound on the error by direct integration using the derivative estimate, but this does not give the best possible results since this approach yields only first order convergence, when in fact the finite element method is second order convergent. We return to the problem of proving precise L_2 error estimates for the error itself in Section 15.5.

8.2.2. An a posteriori error estimate

Recall that we used an a posteriori error estimate to estimate the error of the numerical solution of a system of linear algebraic equations. That error estimate was based on the residual error which is the remainder left after substituting the numerical solution back into the original equation. Similarly, we measure the error of the finite element approximation in terms of the residual error of the approximation.

We start by computing the energy norm of the error using the variational equation (8.6)

$$\|e'\|_a^2 = \int_0^1 ae'e'\,dx = \int_0^1 au'e'\,dx - \int_0^1 aU'e'\,dx$$
$$= \int_0^1 fe\,dx - \int_0^1 aU'e'\,dx.$$

We use the Galerkin orthogonality (8.9) with $v = \pi_h e$ denoting the nodal interpolant of e in V_h to obtain

$$\|e'\|_a^2 = \int_0^1 f\,(e - \pi_h e)\,dx - \int_0^1 aU'(e - \pi_h e)'\,dx$$

$$= \int_0^1 f\,(e - \pi_h e)\,dx - \sum_{j=1}^{M+1} \int_{I_j} aU'(e - \pi_h e)'\,dx.$$

Now, we integrate by parts over each sub-interval I_j in the last term and use the fact that all the boundary terms disappear because $(e - \pi_h e)(x_j) = 0$, to get the *error representation formula*

$$\|e'\|_a^2 = \int_0^1 R(U)(e - \pi_h e)\,dx, \tag{8.20}$$

where $R(U)$ is the residual error which is a (discontinuous) function defined on $(0,1)$ by

$$R(U) = f + (aU')' \quad \text{on each sub-interval } I_j.$$

From the weighted Cauchy inequality (8.18), we get

$$\|e'\|_a^2 \le \|hR(U)\|_{a^{-1}}\|h^{-1}(e - \pi_h e)\|_a.$$

Using the results of Chapter 5, we can show that the interpolation error is bounded in the weighted L_2 norm as

$$\|h^{-1}(e - \pi_h e)\|_a \le C_i\|e'\|_a,$$

where C_i is an interpolation constant depending on a. Notice that here the mesh function $h(x)$ appears on the left-hand side. This proves the basic a posteriori error estimate:

Theorem 8.2. *There is an interpolation constant C_i depending only on a such that the finite element approximation U satisfies*

$$\|u' - U'\|_a \le C_i\|hR(U)\|_{a^{-1}}. \tag{8.21}$$

Problem 8.20. Prove that

$$\|hR(U)\|_{a^{-1}} \le CC_i\|hu''\|_a,$$

where C is a constant depending on a. This estimate indicates that the a posteriori error estimate is optimal in the same sense as the a priori estimate.

Problem 8.21. Write down the a posteriori estimate (8.21) explicitly for the example in Section 6.2.

Problem 8.22. Prove a priori and a posteriori error estimates for the cG(1) method applied to (8.13).

Problem 8.23. Prove a priori and a posteriori error estimates for the cG(1) method applied to the boundary value problem $-u'' + bu' + u = f$, $u(0) = u(1) = 0$. Consider first the case $b = 0$ and then try $b \neq 0$. Consider also the case of non-homogeneous Dirichlet boundary conditions.

Problem 8.24. If you have written some code that computes the cG(1) approximation of (8.2), then add a module that computes the a posteriori error bound and reports an estimate of the error after a computation. Note that in the case $a \equiv 1$, $R(U) = f$ on I_j. The constant C_i can be difficult to estimate precisely, but we can *calibrate* the a posteriori error bound in the following way: construct a problem with a known solution, then for several different meshes, compute the error and the a posteriori error bound. Average the ratio of these two to get an approximate value for C_i. Test the code out by constructing another problem with a known solution.

Problem 8.25. Consider the problem of estimating the error in corresponding stress or flux using the cG(1) method.

8.2.3. Adaptive error control

Since the a posteriori error estimate (8.21) indicates the size of the error of an approximation on a given mesh, it is natural to try to use this information to generate a better mesh that gives more accuracy. This is the basis of adaptive error control.

The computational problem that arises once a two-point boundary value problem is specified is to find a mesh such that the finite element approximation achieves a given level of accuracy, or in other words, such that the error of the approximation is bounded by an *error tolerance* TOL. In practice, we are also concerned with efficiency, which means in this case, that we want to determine a mesh with the fewest number of elements that yields an approximation with the desired accuracy. We try to reach this optimal mesh by starting with a coarse mesh and successively refining based on the size of the a posteriori error estimate. By starting with a coarse mesh, we try to keep the number of elements as small as possible.

More precisely, we choose an initial mesh $\mathcal{T}_h$, compute the corresponding cG(1) approximation U, and then check whether or not

$$C_i \|hR(U)\|_{a^{-1}} \leq \text{TOL}.$$

This is the *stopping criterion*, which guarantees that $\|u' - U'\|_a \leq \text{TOL}$ by (8.21), and therefore when it is satisfied, U is sufficiently accurate. If the stopping criterion is not satisfied, we try to construct a new mesh $\mathcal{T}_{\tilde{h}}$ of mesh size $\tilde{h}$ with as few elements as possible such that

$$C_i \|\tilde{h}R(U)\|_{a^{-1}} = \text{TOL}.$$

This is the *mesh modification criterion* from which the new mesh size $\tilde{h}$ is computed from the residual error $R(U)$ of the approximation on the old mesh. In order to minimize the number of mesh points, it turns out that the mesh size should be chosen to equidistribute the residual error in the sense that the contribution from each element to the integral giving the total residual error is roughly the same. In practice, this means that elements with large residual errors are refined, while elements in intervals where the residual error is small are combined together to form bigger elements.

We repeat the mesh modification followed by solution on the new mesh until the stopping criterion is satisfied. By the a priori error estimate, we know that if u'' is bounded, then the error tends to zero as the mesh is refined. Hence, the stopping criterion will be satisfied eventually. In practice, the adaptive error control rarely requires more than a few iterations.

This is a basic outline of the adaptive error control aimed at producing an approximation with a given accuracy using as few elements as possible. This adaptive method is implemented in Femlab and we give more details in Chapter 15 and in the advanced companion volume.

Problem 8.26. Following Problem 8.21, use the idea of equidistribution to compute the mesh used in the example in Section 6.2 by solving for h_1, h_2, ... This requires the use of Newton's method or something similar, since we are not computing the mesh by refining from a coarse mesh using the procedure just described. If you can, compare your results to the mesh generated by Femlab.

8.2.4. Supporting a sagging tent

We consider a circular horizontal elastic membrane, for example an elastic tent roof, loaded by its own weight and supported by a rim on its
outer boundary of radius 1.01 and by a pole of radius $\epsilon = 0.01$ in its
center. We choose the variables so that the height is zero at the level
of the supports. Since the tent is circular and symmetric through the
center pole, we can use polar coordinates to reduce the problem to one
dimension. With x representing the distance to the center, we use $u(x)$
to denote the vertical deflection of the membrane from the zero height,
so that $u(x) = 0$ at the rim at $x = 1.01$ and at the pole at $x = 0.01$.
Assuming the deflection $u(x)$ is small, it satisfies (8.2) with $a(x) = x + \epsilon$
and $f(x) = x + \epsilon$. This model is analogous to the elastic string model
with the specific coefficient $a(x)$ and the load $f(x)$ entering from the use
of polar coordinates with x representing the radial coordinate. We use
the adaptive code Femlab1d, which implements the adaptive error control described above, choosing TOL=0.05 and plot the results in Fig. 8.4.
We see that the "valley" is steeper close to the pole. We plot the residual
error $R(U)$, which is large near the pole, and the mesh size $h(x)$, which
is small close to the pole.

Problem 8.27. Solve this problem using Femlab1d.

Problem 8.28. *(For tent manufacturers)* Consider (8.2) with $a(x) = x^\alpha$,
where α is a positive real number. In this case, we have $a(0) = 0$ while
$a(x) > 0$ for $x > 0$. The variational formulation (8.6) formally extends to
this case, but we have to be careful about the boundary conditions and we
now make an investigation into this problem: determine the values of α
so that the boundary value problem (8.2) is meaningful in the sense that
we can expect a unique solution to exist. In particular investigate the case
$\alpha = 1$ corresponding to the above tent problem with $\epsilon = 0$ and seek to
answer the following question: can a very thin pole help support a membrane (or will the pole just penetrate the membrane and be of no use)?
Hint: compute the cG(1) finite element approximation with Femlab1d using different values of α to see what happens. (A non-adaptive code can
be used as long as you monitor the errors carefully using the a posteriori
error bound). From a theoretical point of view the basic question takes the
form: for what values of α is there a constant C such that

$$|v(0)| \leq C \left(\int_I x^\alpha (v')^2 \, dx \right)^{\frac{1}{2}} = C\|v\|_E$$

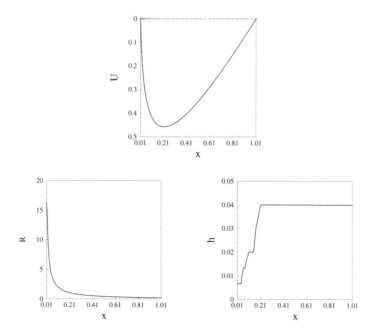

Figure 8.4: Finite element approximation U, residual error R, and mesh size h for the model of a tent computed with adaptive error control.

for all functions with $v(1) = 0$. This inequality is directly related to the pertinent question if it is possible to impose the boundary condition $u(0) = 0$ in the boundary value problem (8.2), that is if the elastic energy is strong enough to allow a "point load". To prove the inequality, start from the representation $v(0) = -\int_I v'\, dx$ and use Cauchy's inequality in a suitable way.

Problem 8.29. Justify the statement that the definition of V in (8.6) is meaningful in the sense that if v and v' belong to $L_2(0, 1)$ then the boundary values $v(0)$ and $v(1)$ are well defined. Hint: look at the previous problem.

8.3. Data and modeling errors

In practice, we often compute the finite element function using approximations $\hat{a}(x)$ and $\hat{f}(x)$ of $a(x)$ and $f(x)$. For example, a and f might be known only through physical measurements, or we might have to use quadrature to evaluate integrals involving a and f, or we might simply be concerned with the effects of round-off error. This means that we compute a modified finite element approximation $\hat{U} \in V_h$ using the orthogonality relation

$$\int_0^1 a\hat{U}'v'\,dx = \int_0^1 \hat{f}\,dx \quad \text{for all } v \in V_h. \tag{8.22}$$

We now seek to estimate the effect on the energy norm of the error coming from the perturbations in a and f. We refer to the solution error resulting from the perturbation in the coefficient a as the the *modeling error* and that from the perturbation of the data f as the *data error*.

We seek an a posteriori error estimate of the total error $\|u - \hat{U}\|_E$ that includes both the Galerkin discretization error, and the modeling and data errors. We start by obtaining a modified form of the error representation (8.20), for $e = u - \hat{U}$,

$$\|(u - \hat{U})'\|_a^2 = \int_0^1 (au'e' - a\hat{U}'e')\,dx = \int_0^1 (fe - a\hat{U}'e')dx$$

$$= \int_0^1 \hat{f}(e - \pi_h e)\,dx - \sum_{j=1}^{M+1} \int_{I_j} a\hat{U}'(e - \pi_h e)'\,dx$$

$$+ \int_0^1 (f - \hat{f})e\,dx - \sum_{j=1}^{M+1} \int_{I_j} (a - \hat{a})\hat{U}'e'\,dx = I + II - III,$$

where I represents the first two terms, II the third, and III the fourth term. The first term I is estimated as above. For the new term III, we have

$$III \leq \|(a - \hat{a})\hat{U}'\|_{a^{-1}}\|e'\|_a.$$

Similarly, integration by parts gives

$$II \leq \|\hat{F} - F\|_{a^{-1}}\|e'\|_a,$$

where $F' = f$, $\hat{F}' = \hat{f}$ and $F(0) = \hat{F}(0) = 0$. Altogether, we obtain:

Theorem 8.3. *There is a constant C_i only depending on a, such that the finite element approximation $\hat{U}$ satisfies*

$$\|u - \hat{U}\|_E \leq C_i\|h\hat{R}(\hat{U})\|_{a^{-1}} + \|\hat{F} - F\|_{a^{-1}} + \|(\hat{a} - a)\hat{U}'\|_{a^{-1}}, \tag{8.23}$$

where $\hat{R}(\hat{U}) = (\hat{a}\hat{U}')' + \hat{f}$ on each sub-interval.

This estimate can be used to adaptively control the total error with contributions from Galerkin discretization and perturbations in data and modeling. It can also be used to analyze the error resulting from the use of a quadrature formula to evaluate integrals that define U.

Problem 8.30. Use the a posteriori error estimate (8.23) to estimate the effect of using the composite trapezoidal quadrature rule to approximate the coefficients of the load vector $b_i = \int_0^1 f\varphi_i \, dx$.

Problem 8.31. Show that if the relative error in $a(x)$ is less than or equal to δ for $x \in I$, then $\|(\hat{a} - a)\hat{U}'\|_{a^{-1}} \leq \delta\|\hat{U}'\|_a$.

8.4.   Higher order finite element methods

The finite element method can be formulated with piecewise polynomials of any degree. We illustrate this by describing the cG(2) method which uses a discrete space V_h consisting of continuous piecewise *quadratic* approximating function, on a partition $\mathcal{T}_h$ of $[0, 1]$ into sub-intervals as for the cG(1) method. Using the idea of the Lagrange basis described in Chapter 5, we can describe a quadratic polynomial on a given interval uniquely by its values at three distinct points. On the sub-interval I_i of the partition, we use the two endpoints, x_{i-1} and x_i, and in addition, we use the midpoint $x_{i-1/2} = (x_i + x_{i-1})/2 = x_{i-1} + h_i/2$. We illustrate the Lagrange basis functions for I_i based on these points in Fig. 8.5. A

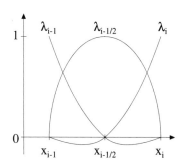

Figure 8.5: The Lagrange basis functions for $\mathcal{P}^2(I_i)$.

quadratic polynomial $p(x)$ on I_i can be written

$$p(x) = p(x_{i-1}) \frac{(x - x_{i-1/2})(x - x_i)}{-h_i/2 \cdot -h_i} + p(x_{i-1/2}) \frac{(x - x_{i-1})(x - x_i)}{h_i/2 \cdot -h_i/2}$$
$$+ p(x_i) \frac{(x - x_{i-1})(x - x_{i-1/2})}{h_i \cdot h_i/2}.$$

Just as in the case of piecewise linear approximations, we use the continuity assumption to piece together the Lagrange basis functions on each interval to form the basis functions for the finite element space V_h. We find that

$$\varphi_{i-1/2}(x) = \begin{cases} \dfrac{4(x_i - x)(x - x_{i-1})}{h_i^2}, & x \in I_i, \\ 0, & \text{otherwise}, \end{cases} \qquad i = 1, 2, ..., M+1,$$

and

$$\varphi_i(x) = \begin{cases} \dfrac{2(x - x_{i+1/2})(x - x_{i+1})}{h_{i+1}^2}, & x \in I_{i+1}, \\ \dfrac{2(x - x_{i-1/2})(x - x_{i-1})}{h_i^2}, & x \in I_i, \\ 0, & \text{otherwise}, \end{cases} \qquad i = 1, 2, ..., M.$$

is the nodal basis for V_h. We illustrate the different possible shapes of the basis functions in Fig. 8.6.

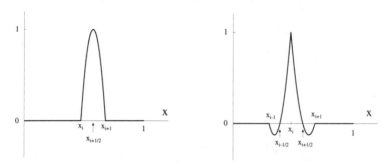

Figure 8.6: Basis functions for the cG(2) method.

Problem 8.32. Construct the nodal basis functions for V_h and reproduce the plot in Fig. 8.6.

This is a *nodal basis* since $v \in V_h$ can be written uniquely as

$$v(x) = \sum_{i=1}^{M+1} v(x_{i-1/2})\varphi_{i-1/2}(x) + \sum_{i=1}^{M} v(x_i)\varphi_i(x).$$

Problem 8.33. Prove this.

We obtain a system of equations $A\xi = b$ for the $2M + 1$ coefficients of U in the expansion

$$U(x) = \xi_{\frac{1}{2}}\varphi_{\frac{1}{2}}(x) + \xi_1\varphi_1(x) + \ldots + \xi_{M+1\frac{1}{2}}\varphi_{M+\frac{1}{2}}(x),$$

where the sum runs over the half-integers $1/2, 1, \ldots, M + 1/2$. By substituting the expansion into (8.6) and taking the test function v to be each one of the basis functions φ_i in turn. The stiffness matrix has coefficients

$$a_{ij} = \int_0^1 \varphi_i'\varphi_j'\, dx, \quad i, j = 1/2, 1, \ldots, M + 1/2.$$

When $i = k + 1/2$ for some integer $0 \leq k \leq M$, then $\varphi_i'\varphi_j'$ is zero except possibly for $j = k$, $k + 1/2$, and $k + 1$. Likewise, when $i = k$ for some integer $1 \leq k \leq M$, $\varphi_i'\varphi_j'$ is zero except possibly for $j = k - 1$, $k - 1/2$, k, $k + 1/2$, and $k + 1$. Therefore, A is banded with 5 non-zero diagonals. Further, A is positive-definite and symmetric. Computing the coefficients of the stiffness matrix is straightforward, though a little tedious, so we give it as an exercise. The data vector b has coefficients $b_i = \int_0^1 f\varphi_i\, dx$ for $i = 1/2, 1, \ldots, M + 1/2$.

An advantage of the cG(2) method compared to cG(1) is that it is higher order accurate, which means that the cG(2) approximation is generally more accurate than the cG(1) approximation computed on the same mesh. We expect this to be true because a piecewise quadratic interpolant is more accurate than a piecewise linear interpolant.

Problem 8.34. Prove a version of Theorem 8.1 for the cG(2) method.

Problem 8.35. Write a program that computes the cG(2) finite element approximation of the two-point boundary value problem (6.9) assuming that the user supplies the data vector b. Make sure that the code is as efficient as possible using the material from Chapter 7. Test the code on the equation $-u'' = 6x$ which has solution $u = x - x^3$. Compare with Femlab1d.

Problem 8.36. We can numerically measure the order of convergence of
an approximation on a uniform mesh in the following way. If we assume
that the norm of the error e_h corresponding to the mesh with meshsize h
is a function of h of the form $\|e_h\| = Ch^q$ for some constants C and q, then
by taking logarithms, we get $\log(\|e_h\|) = \log(C) + q\log(h)$. We create a
test problem with a known solution and compute the error for a sequence
of 5 or 6 meshsizes. We then compute the best fit line (or least squares
line fit) through the pairs of data $(\log h, \log(\|e_h\|))$. The slope of this line
is approximately the order of convergence q. Test this procedure on the
problem in Problem 8.35 using both the cG(1) and cG(2) approximations.
Compare the respective orders.

8.5. The elastic beam

The techniques used to construct and analyze a finite element method
for the second order two-point boundary value problem (8.2) can also be
used to study higher order problems. For example, we consider a model
for a horizontal clamped beam made of an elastic material subject to a
vertical load of intensity $f(x)$, see Fig. 8.7. The vertical deflection $u(x)$

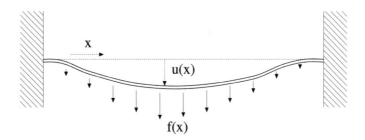

Figure 8.7: A clamped beam under a vertical load f.

of the beam satisfies

$$\begin{cases} D^4 u(x) = f(x) & \text{for } 0 < x < 1, \\ u(0) = u'(0) = u(1) = u'(1) = 0, \end{cases} \tag{8.24}$$

where $D = \frac{d}{dx}$ and $f(x)$ is the intensity of the transversal load. The
boundary conditions correspond to the beam being *clamped* at its end
points, with both the deflection u and the rotation u' equal to zero.

The equation $D^4 u = f$ results from combining an equilibrium equation of the form $m'' = f$, where $m(x)$ represents the *bending moment*, and a constitutive equation of the form $m = \kappa u''$, where $\kappa > 0$ is *bending stiffness* parameter, which we set equal to one for the sake of simplicity.

As the first step in the Galerkin discretization of (8.24), we give the following variational formulation of the problem: find $u \in W$ such that

$$\int_0^1 u'' v'' \, dx = \int_0^1 fv \, dx \quad \text{for all } v \in W, \tag{8.25}$$

where

$$W = \left\{ v : \int_0^1 \left(v^2 + (v')^2 + (v'')^2 \right) dx < \infty, \right.$$

$$\left. v(0) = v'(0) = v(1) = v'(1) = 0 \right\}.$$

We note that W is different from the space V used for the second order differential equation above.

Problem 8.37. Derive (8.25) from (8.24).

Problem 8.38. (a) Define a suitable finite element space W_h for (8.25) consisting of piecewise cubics. Hint: prove first that a cubic polynomial p on the interval $[x_{i-1}, x_i]$ is uniquely determined by the values $p(x_{i-1})$, $p'(x_{i-1})$, $p(x_i)$, and $p'(x_i)$. (b) Construct a *nodal* basis for $\mathcal{P}^3(x_{i-1}, x_i)$ for which the coefficients of a polynomial p are $p(x_{i-1})$, $p'(x_{i-1})$, $p(x_i)$, and $p'(x_i)$. (c) Explain why it is impossible in general to use piecewise linears or quadratics to construct a suitable discrete space of continuously diffentiable function for the discretization of (8.25).

Problem 8.39. Construct the global basis functions for W_h. What is the dimension of W_h? What are the degrees of freedom of a function in W_h with respect to this basis? Formulate a finite element method for (8.24) based on W_h. Compute the corresponding stiffness matrix and load vector assuming a uniform partition.

Problem 8.40. Prove a priori and a posteriori error estimates for the finite element method for the beam problem.

Problem 8.41. Extend (8.24) to suitable combinations of boundary conditions of the form: (a) $u(0) = u''(0) = 0$ corresponding to a *freely supported* beam at $x = 0$; (b) $u''(0) = D^3 u(0) = 0$ corresponding to a free beam end at $x = 0$; and (c) $u'(0) = D^3 u(0) = 0$ corresponding to the rotation and transversal shear force being zero.

8.6.   A comment on a priori and a posteriori analysis

In philosophy, a distinction is made between *analytic* and *synthetic* statements. A synthetic statement is one that can be verified a posteriori by experiment or observation, while the truth of an analytic statement is to be verified a priori by a purely logical analysis proceeding from definitions and axioms. For example, "A bachelor is an unmarried man" is a true analytic statement (which simply gives the definition of a bachelor). Statements of mathematics tend to be analytic in nature; in principle, to be verified a priori by logical analysis, although the work by Gödel set limits to this approach. However, also mathematical statements may sometimes be verified a posteriori by explicitly providing concrete examples. For example, the statement "There is a prime number larger than 17" can be verified by first giving the number 19 and then proving mathematically that this number is prime by trying the factors 2, 3, 5, and 7 without success. This statement is different from the statement "There are infinitely large prime numbers", which cannot be checked by this concrete case by case a posteriori method, but turns out to be provable by an a priori argument.

Traditionally, the error analysis of numerical methods for differential equations has been of a priori type with a dependence on the unknown exact solution and appears to require a "complete understanding" of the problem in order to be useful from a quantitative point of view. In the posteriori error analysis the "understanding" is replaced by computing, which opens the possibility of quantitative error control also for complex problems. In the spirit of synthesis, we conclude that a priori and a posteriori analysis serve different purposes and are both fundamental.

> Kurtele, if I compare your lecture with the others, there is no comparison. (Adele Gödel)

9

Scalar Initial Value Problems

So long as man marked his life only by the cycles of nature – the changing seasons, the waxing or waning moon – he remained a prisoner of nature. If he was to go his own way and fill his world with human novelties, he would have to make his own measures of time. (D. Boorstin)

This is a very fine theorem and one that is not at all obvious. (Leibniz)

In this chapter, we consider the numerical solution of the initial value problem for a linear scalar first order differential equation:

$$\begin{cases} \dot{u}(t) + a(t)u(t) = f(t) & \text{for } 0 < t \leq T, \\ u(0) = u_0, \end{cases} \tag{9.1}$$

where $a(t)$ is a bounded *coefficient*, the right-hand side $f(t)$ is a *production* or *forcing* term, and u_0 is an *initial value*. In an initial value problem, the independent variable t represents time and evolves in a forward direction so that the solution $u(t)$ at a time $t > 0$ is determined by the initial value $u(0)$ and the forcing $f(s)$ for $s \leq t$. In particular, the value $u(s)$ influences the value $u(t)$ if $s < t$, but not vice versa. For the numerical solution of initial value problems, it is natural to use "time-stepping" methods in which the approximation is computed successively on one time interval after another.

Initial value problems are often classified according to how a perturbation of the solution at one time affects the solution later on. A property that is concerned with behavior, such as growth or decay, of perturbations of a solution as time passes is generally called a *stability* property. If we think of numerical discretization as introducing a kind of perturbation of the true solution, then the stability properties of an initial value problem influence the error in a numerical solution of an initial value problem. In general, we expect the error of a numerical solution to grow with time, since the error at one time is affected by the *accumulation* of errors made at earlier times.

There is a class of problems with particular stability properties to which we shall pay special attention. This is the class of *parabolic problems* characterized by the condition

$$a(t) \geq 0 \quad \text{for } 0 \leq t \leq T. \tag{9.2}$$

If $a(t) \geq \alpha$ where $\alpha > 0$ is constant, then perturbations of solutions of (9.1) decay as time passes, something akin to a "fading memory", and perturbations from different time levels do not accumulate. Such problems are also called *dissipative*. We shall see that even the weaker condition $a(t) \geq 0$ implies a kind of "non-accumulation" of perturbations, which allows the problem (9.1) to be solved numerically over long time. Parabolic problems arise in applications where diffusion plays a dominant role such as heat conduction and viscous flow as we will see in Chapters 16 and 21.

We first derive (9.1) as a model in population dynamics and give a formula for the solution that we use to investigate its properties. In particular, we prove some stability estimates in the parabolic case which are used later. We then introduce two Galerkin methods for (9.1) based on piecewise polynomial approximation: a discontinuous Galerkin method based on piecewise constant approximation, and a continuous Galerkin method based on continuous piecewise linear approximation. We give a detailed error analysis of these methods and present some computational examples.

This chapter presents the foundation for the discretization in time of any initial value problem. Studying the material in detail is a good investment and preparation for the extension to systems of differential equations in the next chapter and the extensions to linear partial differential equations later in the book, and to nonlinear problems in the advanced companion volume.

9.1. A model in population dynamics

We now extend the simple model of the population of rabbits in West Virginia discussed briefly in Chapter 6 to cover more complicated situations. That model was derived by assuming that the birth rate is proportional to the population. We use $a_1(t)$ to denote the factor of proportionality, which is now allowed to vary with time. Similarly, the death rate is also roughly proportional to the population and we use $a_2(t)$ to denote the factor of proportionality. To model emigration and immigration, we use a function $f(t)$ that gives the net rate at which rabbits enter West Virginia at time t, i.e. the difference between the immigration and emigration rates. A principle of "conservation of rabbits" implies that the density $u(t)$ must satisfy

$$\begin{cases} \dot{u}(t) + a(t)u(t) = f(t) & \text{for } 0 < t \leq T, \\ u(0) = u_0, \end{cases} \tag{9.3}$$

where $a(t) = a_2(t) - a_1(t)$. The solution of (9.3) is given by the *variation of constants* formula

$$u(t) = e^{-A(t)}u_0 + \int_0^t e^{-(A(t)-A(s))}f(s)\,ds, \tag{9.4}$$

where $A(t)$ is the primitive function of $a(t)$ satisfying $A'(t) = a(t)$, and $A(0) = 0$. We verify this formula by computing the time derivative of $u(t)$:

$$\dot{u}(t) = -a(t)e^{-A(t)}u_0 + f(t) + \int_0^t (-a(t))e^{-(A(t)-A(s))}f(s)\,ds$$

$$= -a(t)\left(e^{-A(t)}u_0 + \int_0^t e^{-(A(t)-A(s))}f(s)\,ds\right) + f(t).$$

Problem 9.1. Show using the solution formula (9.4) that the solution of (9.3) with $f(t) = 2\sin(t)$, $a = 1$, and $u_0 = -1$ is $u(t) = \sin(t) - \cos(t) = \sqrt{2}\sin\left(t - \frac{\pi}{4}\right)$.

Problem 9.2. Compute the solution when $a = 0$ and $f(t) = 200\cos(t)$.

Problem 9.3. Compute the solutions corresponding to (a) $a(t) = 4$ and (b) $a(t) = -t$ with $f(t) = t^2$ and $u_0 = 1$.

Problem 9.4. Derive the solution formula (9.4). Hint: multiply the differential equation (9.3) by the *integrating factor* $e^{A(t)}$ and rewrite the equation in terms of $w = e^{A(t)}u(t)$ using the product rule.

9.1.1. Stability features of the parabolic case

In general, the exponential factors in (9.4) can grow and shrink with time but in the parabolic case with $a(t) \geq 0$ for all t, that is if the death rate is greater than or equal to the birth rate, then $A(t) \geq 0$ and $A(t) - A(s) \geq 0$ for all $t \geq s$. In this case, both u_0 and f are multiplied by quantities that are less than one if $a(t) > 0$ and less than or equal to one if $a(t) \geq 0$. We now state these qualitative observations in more precise quantitative form. First, if $a(t) \geq \alpha$, where $\alpha > 0$ is constant, then the solution u of (9.4) satisfies for $t > 0$,

$$|u(t)| \leq e^{-\alpha t}|u_0| + \frac{1}{\alpha}\left(1 - e^{-\alpha t}\right) \max_{0 \leq s \leq t} |f(s)|. \tag{9.5}$$

Problem 9.5. Prove this estimate.

We see that the effect of the initial data u_0 decays exponentially with time and that the effect of the right-hand side f does not depend on the length of the time interval t, only on the maximum value of f and on α. Second, under the more general condition $a(t) \geq 0$, the solution u of (9.4) satisfies for $t > 0$ (prove this)

$$|u(t)| \leq |u_0| + \int_0^t |f(s)|\, ds. \tag{9.6}$$

In this case the influence of u_0 stays bounded in time, and the integral of f expresses an accumulation in time. We conclude in particular that the contributions from the right-hand side f can accumulate in time in the second case, but not in the first.

Problem 9.6. Suppose u solves (9.3), where $a(t) \geq \alpha > 0$ for $t > 0$ and f is continuous, non-decreasing and positive. Determine $\lim_{t \to \infty} u(t)$.

Problem 9.7. Suppose u solves (9.3) with initial data u_0 and v solves (9.3) starting with initial data v_0. If $a \geq 0$, show that $|u(t) - v(t)| \leq |u_0 - v_0|$ for all t. If $a \geq \alpha > 0$, show that $|u(t) - v(t)| \to 0$ as $t \to \infty$.

Problem 9.8. Let $f(t) = 3t$ and plot the solutions corresponding to $a = .01$ and $a = -.01$ over a long time interval. Interpret your results in terms of the population of rabbits.

Note that an inequality like (9.6) does not mean that a solution necessarily must grow as t increases. For example, if $\dot{u} = -\sin(t)$, where $a \equiv 0$, then $u(t) = \cos(t)$ and u is bounded for all t. In this case, the cancellation that occurs in the integral of f does not occur in the integral of $|f|$.

Problem 9.9. Assume that

$$\int_{I_j} f(s)\,ds = 0 \quad \text{for } j = 1, 2, ..., \tag{9.7}$$

where $I_j = (t_{j-1}, t_j)$, $t_j = jk$ with k a positive constant. Prove that if $a(t) \geq 0$, then the solution of (9.3) satisfies

$$|u(t)| \leq e^{-A(t)}|u_0| + \max_{0 \leq s \leq t} |kf(s)|. \tag{9.8}$$

We note that in this estimate the effect of f is independent of t, and that there is also a factor k multiplying f. We will meet this situation in the error analysis of Galerkin methods with k representing a time step. Hint: use (9.7) to see that

$$u(t) = e^{-A(t)}u_0 + \int_0^t (\varphi - \pi_k \varphi)f(s)\,ds,$$

where $\varphi(s) = e^{-(A(t)-A(s))}$ and π_k is the L_2 projection into the piecewise constants on the partition $\{t_j\}$, together with the estimate

$$\int_0^t |\varphi - \pi_k \varphi|\,ds \leq \int_0^t ka(s)e^{-(A(t)-A(s))}\,ds = k(1 - e^{-A(t)}) \leq k.$$

9.2. Galerkin finite element methods

We consider two Galerkin finite element methods for (9.3) based on piecewise polynomial approximation. The first is defined using continuous trial functions of degree 1 and discontinuous test functions of degree 0. We call this method the *continuous Galerkin method of degree* 1, or the cG(1) method. The other method is defined using discontinuous trial and test functions of degree 0, and we call this method the *discontinuous Galerkin method of degree* 0, or the dG(0) method. These methods generalize to piecewise polynomial approximation of general order q and are then referred to as the cG(q) and dG(q) methods, respectively.

The methods we consider are based on two variations of the Galerkin method used in Chapter 6, which applied to (9.3) takes the form: find $U \in \mathcal{P}^q(0, T)$ satisfying $U(0) = u_0$ such that

$$\int_0^T (\dot{U} + aU)v \, dt = \int_0^T fv \, dt \qquad (9.9)$$

for all v in $\mathcal{P}^q(0, T)$ with $v(0) = 0$. In other words, the residual $R = \dot{U} + aU - f$ is orthogonal to all polynomials in $\mathcal{P}^q(0, T)$ with $v(0) = 0$. We called this a *global* Galerkin method because the interval $[0, T]$ is taken as a whole. The piecewise polynomial methods cG(q) and dG(q) are constructed by applying variations of this global Galerkin method successively on each sub-interval $I_n = (t_{n-1}, t_n)$ of a partition $\mathcal{T}_k : 0 = t_0 < t_1 < .. < t_N = T$ of $(0, T)$.

Problem 9.10. Without using (9.4), prove that if $a(t) \geq 0$ then a continuously differentiable solution of (9.3) is unique. Hint: let u and w be two functions that satisfy (9.3) such that $w(0) = u(0) = u_0$ and find an equation for the difference $e = u - w$. Multiply by $v(t)$ and integrate to obtain a variational equation for e corresponding to (9.9) and show that the choice $v = e$ leads to an equation that implies that $e(t) = 0$ for all t.

9.2.1. The continuous Galerkin method

The continuous Galerkin cG(q) method is based on the following variation of the global Galerkin method: find U in $\mathcal{P}^q(0, T)$ with $U(0) = u_0$ such that

$$\int_0^T (\dot{U} + aU)v \, dt = \int_0^T fv \, dt \quad \text{for all } v \in \mathcal{P}^{q-1}(0, T).$$
$$(9.10)$$

We introduce this variation because it yields a more accurate approximation of degree q than the original global Galerkin method described above. The difference between the two methods is the choice of test space. Recall that forcing the Galerkin approximation to satisfy the initial value exactly reduces the number of degrees of freedom by one. Consequently, in the original global Galerkin method, we test using the q polynomials t^j, $j = 1, ..., q$. In the variation, we test against the q

polynomials t^j with $j = 0, 1, ..., q-1$. With $q = 1$, we thus choose $v = 1$ and we get:

$$U(T) - U(0) + \int_0^T a\left(U(T)\frac{t}{T} + U(0)\frac{(T-t)}{T}\right) dt = \int_0^T f \, dt. \tag{9.11}$$

Since $U(0) = u_0$, this gives an equation for $U(T)$ that in turn determines the linear function U on $[0, T]$.

We now formulate the cG(q) method for (9.3) on the partition $\mathcal{T}_k$ by applying (9.10) successively on each time interval $(t_{n-1}, t_n]$ to obtain an approximation in the space $V_k^{(q)}$ of continuous piecewise polynomials of degree q on $\mathcal{T}_k$. We first obtain the formula on $(0, t_1]$ by using (9.11) with $T = t_1$ and $U_0 = U(t_0) = u_0$. Continuing interval by interval, assuming that U has already been computed on $(t_{n-2}, t_{n-1}]$, and letting U_{n-1} denote the value of $U(t_{n-1})$, we compute the cG(q) approximation U on the next interval $(t_{n-1}, t_n]$ as the function $U \in \mathcal{P}^q(t_{n-1}, t_n)$ that satisfies $U(t_{n-1}) = U_{n-1}$ and

$$\int_{t_{n-1}}^{t_n} (\dot{U} + aU)v \, dt = \int_{t_{n-1}}^{t_n} fv \, dt \quad \text{for all } v \in \mathcal{P}^{q-1}(t_{n-1}, t_n).$$

In the case $q = 1$, the piecewise linear function U satisfies

$$U_n - U_{n-1} + \int_{t_{n-1}}^{t_n} a(t)U(t) \, dt = \int_{t_{n-1}}^{t_n} f(t) \, dt \quad \text{for } n = 1, ..., N, \tag{9.12}$$

where $U(t)$ is determined on (t_{n-1}, t_n) by its nodal values U_{n-1} and U_n. For example, the cG(1) approximation for $\dot{u} + \cos(t)u = t^3$ satisfies

$$\left(1 + \sin(t_n) + \frac{\cos(t_n) - \cos(t_{n-1})}{k_n}\right)U_n$$

$$= \left(1 + \sin(t_{n-1}) + \frac{\cos(t_n) - \cos(t_{n-1})}{k_n}\right)U_{n-1} + (t_n^4 - t_{n-1}^4)/4.$$

Given U_{n-1} from the previous interval, (9.12) determines U_n uniquely for any k_n provided $a(t) \geq 0$, and for any sufficiently small time step k_n provided that $\int_{I_n} |a| \, dt < 1$ for general a. Under these conditions on the step sizes, we can thus compute the nodal values U_n successively for $n = 1, 2, ...$

Equivalently, we can write the equation for the cG(q) approximation U in global form as: find $U \in V_k^{(q)}$ such that

$$\int_0^{t_N} (\dot{U} + aU)v \, dt = \int_0^{t_N} fv \, dt \quad \text{for all } v \in W_k^{(q-1)},$$

$$(9.13)$$

where $W_k^{(q-1)}$ is the space of discontinuous piecewise polynomials of degree $q - 1$ on $\mathcal{T}_k$ and $U_0 = u_0$.

Problem 9.11. Show that if a is constant, then (9.12) reduces to

$$U_n - U_{n-1} + \frac{U_{n-1} + U_n}{2} a k_n = \int_{I_n} f \, dt.$$

Problem 9.12. Compute the cG(1) approximations for the differential equations specified in Problem 9.3. In each case, determine the condition on the step size that guarantees that U exists.

Problem 9.13. Formulate a continuous Galerkin method using piecewise polynomials based on the original global Galerkin method.

Problem 9.14. Prove that (9.13) and (9.12) determine the same function. Hint: choose successively for $n = 1, ..., N$, the test function v to be the basis function for $W_k^{(0)}$ that is equal to one on I_n and zero elsewhere.

9.2.2. The discontinuous Galerkin method

The discontinuous Galerkin dG(q) method is based on the following variation of the global Galerkin method: find $U \in \mathcal{P}^q(0, T)$ such that

$$\int_0^T (\dot{U} + aU)v \, dt + (U(0) - u(0))v(0) = \int_0^T fv \, dt \quad \text{for all } v \in \mathcal{P}^q(0, T).$$

$$(9.14)$$

In this method the trial and test spaces are the same, which turns out to be an advantage in the error analysis and which also gives improved stability properties for parabolic problems in comparison to the continuous Galerkin method. To obtain these advantages, we give up the requirement that U satisfy the initial condition exactly, since otherwise the coefficients of U would be over-determined. Instead, the initial condition is imposed in a variational sense through the presence of the term

$(U(0) - u(0))v(0)$. The equation (9.14) expresses the Galerkin orthogonality of the approximation U in the sense that the sum of the residual $\dot{U} + aU - f$ inside the interval $(0, T)$ and the "residual" $U(0) - u(0)$ at the initial time is "orthogonal" to all discrete test functions.

Actually, we can generalize (9.14) to: find $U \in \mathcal{P}^q(0,1)$ such that

$$\int_0^T (\dot{U} + aU)v\, dt + \alpha(U(0) - u(0))v(0) = \int_0^T fv\, dt \quad \text{for all } v \in \mathcal{P}^q(0, T),$$

where α is a coefficient that weights the relative importance of the residual error $U(0) - u(0)$ at the initial time against the residual error $\dot{U} + aU - f$ in solving the differential equation. The method (9.14) corresponds to $\alpha = 1$ which turns out to give the best accuracy and stability.

The dG(q) method is applied using the method (9.14) successively in the intervals $I_n = (t_{n-1}, t_n)$ of the partition $\mathcal{T}_k$. We therefore compute the dG(q) solution U in the space $W_k^{(q)}$ of discontinuous piecewise polynomials of degree q on $\mathcal{T}_k$. To account for discontinuities of the functions in $W_k^{(q)}$ at the time nodes t_n, we introduce the notation

$$v_n^+ = \lim_{s \to 0^+} v(t_n + s), \quad v_n^- = \lim_{s \to 0^+} v(t_n - s), \quad \text{and} \quad [v_n] = v_n^+ - v_n^-.$$

That is, v_n^+ is the limit "from above", v_n^- is the limit "from below", and $[v_n] = v_n^+ - v_n^-$ is the "jump" in $v(t)$ at time t_n. We illustrate this in Fig. 9.1.

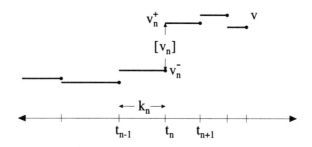

Figure 9.1: The notation for dG(0).

We formulate the dG(q) method as follows: for $n = 1, ..., N$, find

$U \in \mathcal{P}^q(t_{n-1}, t_n)$ such that

$$\int_{t_{n-1}}^{t_n} (\dot{U} + aU)v \, dt + U_{n-1}^+ v_{n-1}^+ = \int_{t_{n-1}}^{t_n} fv \, dt + U_{n-1}^- v_{n-1}^+$$

$$\text{for all } v \in \mathcal{P}^q(t_{n-1}, t_n). \quad (9.15)$$

In the piecewise constant case $q = 0$, we choose $v = 1$ to get

$$\int_{t_{n-1}}^{t_n} aU \, dt + U_{n-1}^+ = \int_{t_{n-1}}^{t_n} f \, dt + U_{n-1}^-. \quad (9.16)$$

Since $v \in W_k^{(0)}$ is constant on each sub-interval, it is natural to denote v's value on the interval I_n by v_n so that $v_n = v_n^- = v_{n-1}^+$. This notation will be used repeatedly below and should be noticed. Using this notation, the dG(0) method (9.16) takes the form

$$\int_{t_{n-1}}^{t_n} aU_n \, dt + U_n = \int_{t_{n-1}}^{t_n} f \, dt + U_{n-1}, \quad (9.17)$$

where we set $U_0 = u_0$. For example, the dG(0) approximation for $\dot{u} + \cos(t)u = t^3$, satisfies

$$U_n + (\sin(t_n) - \sin(t_{n-1}))U_n = U_{n-1} + (t_n^4 - t_{n-1}^4)/4.$$

As above, (9.17) determines U_n uniquely for any k_n provided $a(t) \geq 0$, and for any time step k_n small enough that $\int_{I_n} |a| \, dt < 1$ for general a.

Equivalently, we can write (9.15) in global form as: find $U \in W_k^{(q)}$ such that

$$\sum_{n=1}^N \int_{t_{n-1}}^{t_n} (\dot{U} + aU)v \, dt + \sum_{n=1}^N [U_{n-1}]v_{n-1}^+ = \int_0^{t_N} fv \, dt \quad \text{for all } v \in W_k^{(q)}, \quad (9.18)$$

where, in $[U_0] = U_0^+ - U_0^-$, $U_0^- = u_0$.

Problem 9.15. Formulate the dG(1) method for the differential equations specified in Problem 9.3.

Problem 9.16. Verify that (9.15) and (9.18) are equivalent.

Problem 9.17. Construct a Galerkin finite element method for U in $W_k^{(0)}$ using testfunctions in $V_k^{(1)}$, and compute explicit equations for the nodal values of U in the case a is constant.

9.2.3.   Comparing the accuracy and stability of the cG(1) and dG(0) methods

The error of an approximate solution at the final time, or *global error*, is determined by the *accuracy* and *stability* properties of the method. Accuracy is related to the size of the perturbation introduced by the method of discretization at each time step and stability is related to the growth, decay, and accumulation of perturbations over many time steps. Accuracy and stability properties are determined both by the corresponding properties of the continuous differential equation and the choice of discretization method. For example, parabolic problems have special stability properties that allow long time approximation of solutions by some choices of discretization methods (for example, dG(0)). In this section, we make a preliminary comparison of the stability and accuracy properties of the dG(0) and cG(1) finite element methods for (9.3).

   In the first comparison, we demonstrate that the error of the cG(1) approximation is smaller than that of the dG(0) approximation for a given a and f over a fixed interval granted that the time steps are sufficiently small. In particular, the cG(1) method converges more rapidly than the dG(0) method as the mesh is refined to zero. If the time steps $k_n = k$ are constant, the comparison can be made assuming that the error at the final time T is proportional to k^p, that is $|\text{error}| \approx Ck^p$ with the constant of proportionality C depending on the exact solution of the differential equation, the discretization method, and the final time T. p is called the *order of convergence*. We prove error estimates of this form below. To determine p experimentally, we remark that by taking logarithms:

$$\log(|\text{error}|) \approx \log C + p \log(k),$$

we can determine p as the slope of a line that passes through data points $(\log(k), \log(|\text{error}|))$. We construct a problem with a known solution and then compute the numerical solution to time T using a set of different time steps k. We compute a best fit line through the points $(\log(k), \log(|\text{error}|))$ using least squares and finally compute the slope of the line, which should approximate p.

   We illustrate this technique on the problem $\dot{u} - .1u = -\sin(t)$, $u(0) = 1$ with solution $u(t) = e^{.1t}/101 + 100\cos(t)/101 + 10\sin(t)/101$. We compute to time $T = 1$ using the dG(0) and cG(1) methods with time steps .25, .125, .0625, ..., .00390625. We plot the logarithms of the errors versus the logarithms of the corresponding steps in Fig. 9.2. The slopes of the least squares lines are 1.02 for the dG(0) method and 2.00 for cG(1).

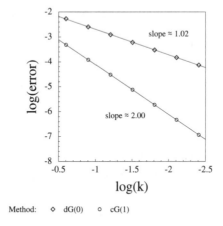

Figure 9.2: Errors at time $T = 1$ for the dG(0) and cG(1) method and least square fitted lines.

Problem 9.18. (a) Repeat this experiment using the equation $\dot{u} + u = e^{-t}$, $u(0) = 0$ with solution $u = te^{-t}$. (b) Do the same for $\dot{u} - u = -4(1-t)^{1/3}e^t/3$, $u(0) = 1$ with solution $u(t) = (1-t)^{4/3}e^t$.

In the second comparison, we show that the stability properties of the dG(0) method may mean that it gives better results for parabolic problems than the cG(1) method. To explain this, we write out the formulas for the nodal values of the dG(0) and the cG(1) approximations, denoted by U and $\tilde{U}$ respectively. In the case that $a > 0$ is constant and $f \equiv 0$ and the time step k is constant, we get

$$U_n = \left(\frac{1}{1+ak}\right)^n u_0 \quad \text{and} \quad \tilde{U}_n = \left(\frac{1-ak/2}{1+ak/2}\right)^n u_0.$$

Problem 9.19. Verify these formulas.

We fix $t_n = nk$ and examine the behavior of the error as the size of a varies. In the figure on the left in Fig. 9.3, we plot the errors of U and

$\tilde{U}$ over the time interval $[0,5]$ with the moderate value $a = 1$ and we see that the cG(1) solution $\tilde{U}$ is more accurate in this case. In the figure on the right, we plot the errors at time $t = 2$ computed using $k = .02$ as a increases in value. We see that cG(1) is more accurate for moderate values of a but gets worse as a increases. In fact, $\tilde{U} \to \pm u_0$ as $a \to \infty$, while both u and U tend to zero. We conclude that the dG(0) method is more robust than the cG(1) method for large a due to its stability properties.

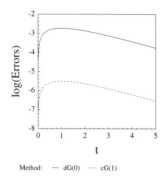 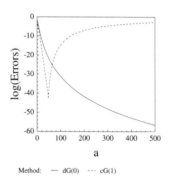

Figure 9.3: The plot on the left shows the log of the errors of the dG(0) and cG(1) methods versus time for $a = 1$ using a fixed step $k = .01$. The plot on the right shows the log of the errors at $t = 2$ versus a using the fixed step $k = .02$.

Problem 9.20. Compare dG(1) with dG(0) and cG(1) on the same two test problems using these techniques.

Problem 9.21. Use these techniques to compare the accuracies of the methods in Problems 9.13 and 9.17 to the dG(0) and cG(1) methods on the same test problems.

9.3. A posteriori error analysis and adaptive error control

"Don't gamble! Take all your savings and buy some good stocks, and hold it till it goes up. Then sell it."
"If it don't go up, don't buy it." (W. Rogers)

In this section, we prove a posteriori error estimates for dG(0) and cG(1) and develop corresponding adaptive algorithms. The a posteriori error

estimate bounds the error in terms of the residual of the computed solution at earlier times and a quantity called the stability factor that
measures the accumulation of errors. Recall that we use such estimates
for error estimation and adaptive error control. In the next section, we
prove an a priori estimate that shows that the methods converge.

9.3.1. An a posteriori error estimate for the cG(1) method

We begin by deriving an a posteriori error estimate for the cG(1) method.
The analysis is based on representing the error in terms of the solution
of a continuous *dual problem* related to (9.3), which is used to determine
the effects of the accumulation of errors, and in terms of the residual
of the computed solution, which measures the step by step production
of error. Specifically, using the dual problem, we can get information
about the final value of a solution using the fact that the solution solves
the differential equation over the time interval. To see this, recall that
we can rewrite the differential equation $\dot{u} + au = f$ for all $0 < t < T$ in
variational form as

$$\int_0^T (\dot{u} + au)\, v\, dt = \int_0^T f v\, dt,$$

for all test functions v. Integration by parts gives the equivalent equation,

$$u(T)v(T) - u(0)v(0) + \int_0^T u(t)\left(-\dot{v} + av\right) dt = \int_0^T f v\, dt.$$

If we now choose v to solve $-\dot{v} + av = 0$ in $(0, T)$, which is the dual
problem, then this relation simplifies to

$$u(T)v(T) = u(0)v(0) + \int_0^T f v\, dt.$$

In other words, by using the fact that u solves the differential equation
$\dot{u} + au = f$ on $(0, T)$ and using the solution v of a dual problem, we
can get information about the final value $u(T)$ in terms of the initial
value $u(0)$ and the right hand side f. This type of representation underlies the error analysis for (9.3) which we shall now present, and also
the corresponding analysis for more general problems developed later in

the book. We will provide further motivation for the dual problem in Chapter 15.

The dual problem for (9.1) reads: find $\varphi(t)$ such that

$$\begin{cases} -\dot{\varphi} + a\varphi = 0 & \text{for } t_N > t \geq 0, \\ \varphi(t_N) = e_N, \end{cases} \tag{9.19}$$

where $e_N = u_N - U_N$ is the error at time t_N that we are trying to estimate. Note that (9.19) runs "backwards" in time starting at time t_N and that the time derivative term has a minus sign. Starting from the identity

$$e_N^2 = e_N^2 + \int_0^{t_N} e\left(-\dot{\varphi} + a\varphi\right) dt,$$

where $e = u - U$, we integrate by parts and use the fact that $e(0) = 0$ to get

$$e_N^2 = \int_0^{t_N} (\dot{e} + ae)\varphi \, dt.$$

Since u solves the differential equation (9.3), this simplifies to give an equation for the error

$$e_N^2 = \int_0^{t_N} (f - aU - \dot{U})\varphi \, dt = -\int_0^{t_N} r(U)\varphi \, dt, \tag{9.20}$$

in terms of the residual error $r(U) = \dot{U} + aU - f$ and the solution of the dual problem φ.

Problem 9.22. Supply the details for this discussion. Give an alternative motivation for the equation for the error e_N, by first showing that the error $e(t)$ satisfies

$$\dot{e} + ae = -r(U) \quad \text{in } (0, t_N), \quad e(0) = 0,$$

and then use (9.4) to see that

$$e_N = -\int_0^{t_N} e^{-(A(t_N) - A(s))} r(U) \, ds = -e_N^{-1} \int_0^{t_N} \varphi r(U) \, ds.$$

Next, we use the Galerkin orthogonality (9.13) expressing that

$$\int_0^{t_N} r(U)v \, dt \quad \text{for } v \in W_k^{(0)},$$

choosing $v = \pi_k \varphi$ to be the L_2 projection of φ into the space of piecewise constants $W_k^{(0)}$, that is $\pi_k \varphi = k_n^{-1} \int_{I_n} \varphi \, ds$ on I_n, to obtain from (9.20) the *error representation formula*:

$$e_N^2 = - \int_0^{t_N} r(U)(\varphi - \pi_k \varphi) \, dt. \qquad (9.21)$$

We emphasize that this is an exact representation of the error from which we can obtain a good estimate of the error. Recall that the error of the L_2 projection π_k is bounded by

$$\int_{I_n} |\varphi - \pi_k \varphi| \, dt \le k_n \int_{I_n} |\dot{\varphi}| \, dt.$$

Therefore, using the notation $|v|_J = \max_{t \in J} |v(t)|$, we have

$$e_N^2 \le \sum_{n=1}^N |r(U)|_{I_n} \int_{I_n} |\varphi - \pi_k \varphi| dt \le \sum_{n=1}^N |r(U)|_{I_n} k_n \int_{I_n} |\dot{\varphi}| dt. \qquad (9.22)$$

There are several ways to proceed from this point. Here, we bring out the max of $k_n |r(U)|_{I_n}$ over n to obtain

$$e_N^2 \le \max_{1 \le n \le N} (k_n |r(U)|_{I_n}) \int_0^{t_N} |\dot{\varphi}| \, dt \le S(t_N) |e_N| \| kr(U) \|_{[0, t_N]}, \qquad (9.23)$$

where the *stability factor* $S(t_N)$ is defined by

$$S(t_N) = \frac{\int_0^{t_N} |\dot{\varphi}| \, dt}{|e_N|},$$

and the *step function* $k(t)$ is the piecewise constant function with the value k_n on I_n. The stability factor measures the effects of the accumulation of error in the approximation and to give the analysis a quantitative meaning, we have to give a quantitative bound of this factor. The following lemma gives an estimate for $S(t_N)$ both in the general case and the parabolic case when $a(t) \ge 0$ for all t.

Lemma 9.1. *If $|a(t)| \le A$ for $0 \le t \le t_N$ then the solution φ of (9.19) satisfies*

$$|\varphi(t)| \le \exp(A t_N) |e_N| \quad \text{for all } 0 \le t \le t_N, \qquad (9.24)$$

and

$$S(t_N) \leq \exp(\mathcal{A} t_N). \tag{9.25}$$

If $a(t) \geq 0$ for all t, then φ satisfies:

$$|\varphi(t)| \leq |e_N| \quad \text{for all } 0 \leq t \leq t_N. \tag{9.26}$$

and

$$S(t_N) \leq 1. \tag{9.27}$$

Proof. With the change of variable $s = t_N - t$ the dual problem takes the form $d\psi/ds + a(t_N - s)\psi(s) = 0$ with initial value $\psi(0) = e_N$, where $\psi(s) = \varphi(t_N - s)$. This problem has the same form as the original problem and therefore we can write an exact formula for the solution of the dual problem,

$$\varphi(t) = e^{A(t) - A(t_N)} e_N. \tag{9.28}$$

Problem 9.23. Carry out the indicated change of variables and derive (9.28).

Considering the parabolic case $a(t) \geq 0$, (9.26) follows from (9.28) using the fact that $A(t) - A(t_N) \leq 0$ for $t \leq t_N$. Further, since $a \geq 0$,

$$\int_0^{t_N} |\dot{\varphi}|\, dt = |e_N| \int_0^{t_N} a(t) \exp\big(A(t) - A(t_N)\big)\, dt$$
$$= |e_N|\big(1 - \exp(A(0) - A(t_N))\big) \leq |e_N|.$$

Problem 9.24. Prove (9.24) and (9.25).

∎

By inserting the stability estimates (9.25) or (9.27), as the case may be, into (9.22), we obtain an a posteriori error estimate for the cG(1) method:

Theorem 9.2. *For $N = 1, 2, ...$, the cG(1) finite element solution U satisfies*

$$|u(t_N) - U_N| \leq S(t_N)|kr(U)|_{[0,t_N]}, \tag{9.29}$$

where $k = k_n$ for $t_{n-1} < t \leq t_n$ and $r(U) = \dot{U} + aU - f$. If $|a(t)| \leq \mathcal{A}$ for $0 \leq t \leq t_N$ then $S(t_N) \leq \exp(\mathcal{A} t_N)$. If $a(t) \geq 0$ for all t then $S(t_N) \leq 1$.

Based on the numerical experiments performed above, we expect to get a second order estimate for the cG(1) approximation. In fact, (9.29) is second order because $|r(U)| \leq Ck$ is f is smooth. To see the second order accuracy, we derive a slightly different bound starting from (9.21). Writing out $r(U)$, we note that since $g - \pi_k g$ is orthogonal to constant functions for any function g and $\dot{U}$ is constant on I_n,

$$
\begin{aligned}
e_N^2 &= -\int_0^{t_N} (aU - f)\,(\varphi - \pi_k \varphi)\, dt \\
&= -\int_0^{t_N} ((aU - f) - \pi_k(aU - f))\,(\varphi - \pi_k \varphi)\, dt.
\end{aligned}
\tag{9.30}
$$

Estimating as above, we obtain

$$
|e_N| \leq S(t_N) \left| k^2 \frac{d}{dt}(aU - f) \right|_{[0,t_N]},
\tag{9.31}
$$

and the second order accuracy follows if $\frac{d}{dt}(aU - f)$ is bounded. When it actually comes to estimating the error, we normally use (9.29) instead of (9.31), because (9.31) involves computing the time derivative of the residual.

Problem 9.25. Derive (9.31).

Problem 9.26. Because of round-off error, if nothing else, we usually cannot specify that $U_0 - u_0 = 0$ exactly. State and prove a modified a posteriori error bound for the cG(1) method assuming that $e_0 = \delta$.

9.3.2.   An a posteriori error estimate for the dG(0) method

We now derive an a posteriori error bound for the dG(0) approximation. The proof has the same structure as the proof for the cG(1) method, but the residual error has a different form and in particular includes "jump terms" arising from the discontinuities at the time nodes in the approximation's values.

Once more starting with the dual problem and the identity

$$
e_N^2 = e_N^2 + \sum_{n=1}^{N} \int_{t_{n-1}}^{t_n} e\,(-\dot{\varphi} + a\varphi)\, dt,
$$

we integrate by parts over each sub-interval I_n to obtain

$$e_N^2 = \sum_{n=1}^{N} \int_{t_{n-1}}^{t_n} (\dot{e} + ae)\varphi \, dt + \sum_{n=1}^{N-1} [e_n]\varphi_n^+ + (u_0 - U_0^+)\varphi_0^+. \tag{9.32}$$

Problem 9.27. If v and w are piecewise differentiable functions on the partition $\mathcal{T}_k$, prove that

$$\sum_{n=1}^{N} \int_{t_{n-1}}^{t_n} \dot{v}w \, dt + \sum_{n=1}^{N-1} [v_n]w_n^+ + v_0^+ w_0^+$$

$$= -\sum_{n=1}^{N} \int_{t_{n-1}}^{t_n} v\dot{w} \, dt - \sum_{n=1}^{N-1} v_n^- [w_n] + v_N^- w_N^-.$$

Write down the special version of this formula when w is a smooth function and apply this to prove (9.32).

Using the facts that u solves the differential equation (9.3), $\dot{U} \equiv 0$ on I_n, and $U_0^- = u_0$, (9.32) simplifies to

$$e_N^2 = \sum_{n=1}^{N} \left(\int_{t_{n-1}}^{t_n} (f - aU)\varphi \, dt - [U_{n-1}]\varphi_{n-1}^+ \right).$$

Problem 9.28. Verify this equation.

Once again we use the Galerkin orthogonality (9.18) by choosing $v = \pi_k\varphi$ to be the L_2 projection into the space of piecewise constants $W_k^{(0)}$ and obtain the *error representation formula*:

$$e_N^2 = \sum_{n=1}^{N} \left(\int_{t_{n-1}}^{t_n} (f - aU)(\varphi - \pi_k\varphi) \, dt - [U_{n-1}](\varphi - \pi_k\varphi)_{n-1}^+ \right). \tag{9.33}$$

We may now finish as above using the stability estimates (9.25) or (9.27), and obtain

Theorem 9.3. *For $N = 1, 2, \ldots$, the dG(0) finite element solution U satisfies*

$$|u(t_N) - U_N| \leq S(t_N)|kR(U)|_{[0,t_N]}, \tag{9.34}$$

where

$$R(U) = \frac{|U_n - U_{n-1}|}{k_n} + |f - aU| \quad \text{for } t_{n-1} < t \leq t_n.$$

If $|a(t)| \leq A$ for $0 \leq t \leq t_N$ then $S(t_N) \leq \exp(A t_N)$ and if $a(t) \geq 0$ for all t then $S(t_N) \leq 1$.

Problem 9.29. Supply the details to finish the proof of this theorem.

Based on the experiments above, we expect to get a first order estimate on the error of the dG(0) approximation, which means that $R(U)$ stays bounded as the mesh size tends to zero if f is bounded. In fact, this follows from the a priori error estimate that we prove in the next section. It is interesting that the residual errors $R(U)$ for the dG(0) method and $r(U)$ for the cG(1) method have a very similar appearance, but $r(U)$ is much smaller in general. Note that the trick (9.30) we used for the cG(1) method, indicating that $|r(U)| = O(k)$, will not work for the dG(0) method unless the discrete solution is continuous (i.e. constant!), because the part of the residual error arising from the jump terms does not cancel in the same way as the part inside the integral, and we cannot conclude (9.31) for the dG(0) method.

9.3.3. Order and estimating the error

We have seen that the a posteriori analysis gives a second order estimate (9.29) for the error of the cG(1) method and a first order estimate (9.34) for the dG(0) method. Moreover, the triangle inequality means that for the cG(1) approximation U, $|r(U)| \leq |R(U)|$, and therefore (9.29) can also be interpreted as a first order estimate like (9.34). When using the cG(1) method, we can maximize the size of the time steps, and reduce the computational work, by using whichever order bound gives the smaller bound on the error. A situation in which the first order bound can be smaller than the second order occurs for example when the derivative of the residual is singular, but the residual itself is continuous. In some situations when the cG(1) method is converging only at a first order rate, it is more efficient to switch to the dG(0) method since that also gives first order convergence, but at a lower cost per step. An efficient general code for solving time dependent differential equations uses a synthesis of methods and orders and as a part of the error control makes decisions not only on the step sizes, but also on the order and choice of method.

9.3.4. Adaptive error control

To guarantee that the dG(0) approximation U satisfies

$$|u(t_N) - U_N| \le TOL,$$

where t_N is a given time, we seek to determine the time steps k_n so that

$$S(t_N)|kR(U)|_{I_n} = TOL \quad \text{for } n = 1, 2, ..., N. \tag{9.35}$$

The approach is the same for the cG(1) method after replacing $R(U)$ by $r(U)$. Using the language introduced for adaptive integration, we recognize (9.35) as the stopping criterion and we try to satisfy it by using a "predictor-corrector" strategy. We first compute U_n from U_{n-1} using a predicted step k_n, then we compute $|kR(U)|_{I_n}$ and check to see if (9.35) is satisfied for this choice of k_n. If so then we accept the solution U_n and continue to the next time step, and if not then we recompute with a smaller step k_n. We give more details of this procedure in the advanced companion volume.

We also have to estimate the stability factor $S(t_N)$. Of course, we could use the exponential bound $\exp(At_N)$ on $S(t_N)$ in Lemma 9.1, but this is usually too crude for most problems. This is particularly true of course in the parabolic case. In the case of a linear scalar problem, we have a precise formula for the solution of the dual problem that we can evaluate. In more general problems, i.e. nonlinear or set in higher dimensions, we have to solve the dual problem numerically and compute an approximation to $S(t_N)$ in an auxiliary computation. This raises some interesting questions that we address in the advanced companion book. In the present case however, these don't come up because we have the exact solution formula.

We illustrate the adaptive error control on three examples. All of these computations were performed using the Cards code in which the the stability factor is computed in an auxiliary computation.

For the first example, we return to our rabbits considering a situation in which the death rate dominates the birth rate and there is a periodic pattern in emigration and immigration. We model this with the parabolic problem $\dot{u} + u = \sin(t)$ with $u(0) = 1$, which has solution $u(t) = 1.5e^{-t} + .5(\sin(t) - \cos(t))$. We compute with dG(0), controlling the error by means of the a posteriori error bound using an error tolerance of .001. The solution and the approximation along with the stability factor are shown in Fig. 9.4. Note that $S(t)$ tends to 1 as t increases, indicating that the numerical error does not grow significantly with time, and accurate computations can be made over arbitrarily long time intervals.

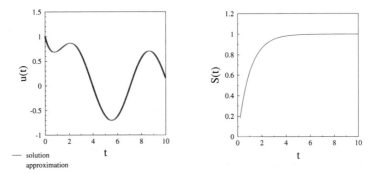

Figure 9.4: Results for the dG(0) approximation of $\dot{u} + u = \sin(t)$ with $u(0) = 1$ computed with an error smaller than .001. The plot on the left shows the approximation and the solution. The plot on the right shows the stability factor $S(t)$.

Problem 9.30. Compute $S(t)$ and the a posteriori error bound explicitly for this example and compute the steps that will be generated by the error control. Discuss what happens to the step sizes as time passes.

Next we consider a situation with no immigration or emigration in which the birth rate dominates the death rate. This situation occurred in the rabbit population of Australia, causing terrible troubles until a plague of myxomatosis killed off many rabbits. We model this using the equation $\dot{u} - u = 0$ with $u(0) = 1$, which has solution $u(t) = e^t$. We compute with dG(0) keeping the error below .025. Since the problem is not dissipative, we expect to see the error grow. We plot the error $U(t) - u(t)$ in Fig. 9.5 and the exponential growth rate is clearly visible. The stability factor is also plotted, and we note that it reflects the error growth precisely.

Problem 9.31. Repeat Problem 9.30 for this example. Then compare the sequence of step sizes produced by the adaptive error control for the two examples.

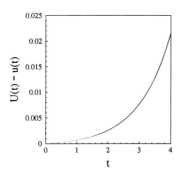 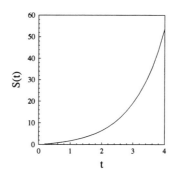

Figure 9.5: Results for the dG(0) approximation of $\dot{u} - u = 0$ with $u(0) = 1$ computed with an error smaller than .025. The plot on the left shows the error. The plot on the right shows the stability factor $S(t)$.

Finally, we consider a situation with no emigration or immigration in which the death rate dominates the birth rate in an oscillatory fashion. We model this with the problem

$$\begin{cases} \dot{u} + (.25 + 2\pi \sin(2\pi t))u = 0, & t > 0, \\ u(0) = 1, \end{cases} \tag{9.36}$$

with solution

$$u(t) = \exp(-.25t + \cos(2\pi t) - 1).$$

The population oscillates, increasing and decreasing as time passes. However, the size of the oscillations dampens as time passes. In Fig. 9.6, we plot the solution together with the dG(0) approximation computed with error below .12. We also plot the time steps used for the computation. We see that the steps are adjusted for each oscillation and in addition that there is an overall trend to increasing the steps as the size of the solution decreases.

In addition, the solution has changing stability characteristics. In Fig. 9.7, we plot the stability factor versus time, and this clearly reflects the fact that the numerical error decreases and increases in an oscillatory fashion. If a crude "exponential" bound on the stability factor like the one in the a priori estimate is used instead of a computational estimate, then the error is greatly overestimated. To demonstrate the efficiency

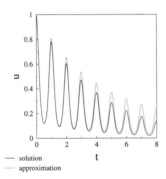 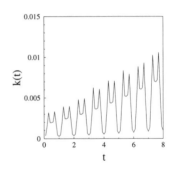

Figure 9.6: Results for the dG(0) approximation of (9.36) computed
with an error smaller than .12. The plot on the left shows
the solution and approximation. The plot on the right
shows the time steps.

of the a posteriori estimate for use in error control, we also plot the
ratio of the true error to the computed bound versus time in Fig. 9.7.
The ratio quickly settles down to a constant, implying that the bound
is predicting the behavior of the error in spite of the fact that the error
oscillates a good deal.

Problem 9.32. Compute $S(t)$ explicitly for this example.

9.4. A priori error analysis

In this section we give an a priori error analysis of dG(0). We start with
the general case and then consider the parabolic case with $a(t) \geq 0$. The
a priori analysis of cG(1) in the general case is similar with $k^2 \ddot{u}$ replacing
$k \dot{u}$.

9.4.1. The general case

We begin by deriving an a priori error estimate in the simplified case
when $a(t) = a$ is constant so that the dG(0) approximation is given by

$$U_n - U_{n-1} + k_n a U_n = \int_{I_n} f \, dt \quad \text{for } n = 1, 2, \ldots,$$

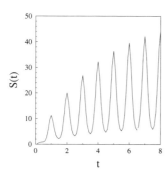
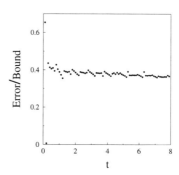

Figure 9.7: Results for the dG(0) approximation of (9.36) computed with an error smaller than .12. The plot on the left shows the stability factor $S(t)$. The plot on the right shows the ratio of the error to the a posteriori error bound.

where $U_0 = u_0$. By integrating the equation for the exact solution u, we see that the exact values $u_n = u(t_n)$ satisfy the following variant of this relation

$$u_n - u_{n-1} + k_n a u_n = \int_{I_n} f \, dt + k_n a u_n - \int_{I_n} a u \, dt \quad \text{for } n = 1, 2, ..,$$

so that subtracting, we get the equation $e_n + k_n a e_n = e_{n-1} + \rho_n$ for the error $e_n = u_n - U_n$, that is

$$e_n = (1 + k_n a)^{-1}(e_{n-1} + \rho_n), \tag{9.37}$$

where

$$|\rho_n| = \left| k_n a u_n - \int_{I_n} a u \, dt \right| \leq \frac{1}{2} |a| k_n^2 |\dot{u}|_{I_n}. \tag{9.38}$$

Problem 9.33. Prove (9.38) using the results in Chapter 5.

We see that the error e_n at time t_n is determined by the error e_{n-1} at the previous time t_{n-1} and the contribution ρ_n made over the interval I_n. The task is to "unwind" the recursive relationship (9.37). Assuming that $k_n |a| \leq 1/2$, $n \geq 1$, and using the inequality (prove it!) $1/(1 - x) \leq \exp(2x)$ for $0 \leq x \leq 1/2$, it follows that

$$|e_N| \leq (1 - k_N|a|)^{-1}|e_{N-1}| + (1 - k_N|a|)^{-1}|\rho_N|$$
$$\leq e^{2|a|k_N}|e_{N-1}| + e^{2|a|k_N}|\rho_N|$$

Iterating this estimate and noting that $e_0 = 0$, we find that

$$|e_N| \leq e^{2|a|k_{N-1}}e^{2|a|k_N}|e_{N-2}| + e^{2|a|k_{N-1}}e^{2|a|k_N}|\rho_{N-1}| + e^{2|a|k_N}|\rho_N|$$

$$\vdots$$

$$\leq \sum_{n=1}^{N} e^{2|a|\sum_{m=n}^{N} k_m}|\rho_n|.$$

Problem 9.34. Verify this estimate.

Noting that $\sum_{m=n}^{N} k_m = t_N - t_{n-1}$ and using (9.38), we conclude that

$$|e_N| \leq \frac{1}{2}\sum_{n=1}^{N} e^{2|a|(t_N - t_{n-1})}|a|k_n \max_{1 \leq n \leq N} k_n|\dot{u}|_{I_n}. \qquad (9.39)$$

To simplify the sum on the right, we set $\tau_n = t_N - t_{n-1}$, $1 \leq n \leq N+1$, and note that under the assumption on k_n,

$$2|a|\tau_n = 2|a|\tau_{n+1} + 2|a|k_n \leq 2|a|\tau_{n+1} + 1 \quad \text{for } 1 \leq n \leq N,$$

while

$$2|a|\tau_{n+1} \leq 2|a|\tau \quad \text{for } \tau_{n+1} \leq \tau \leq \tau_n.$$

Hence,

$$e^{2|a|\tau_n}k_n = \int_{\tau_{n+1}}^{\tau_n} e^{2|a|\tau_n} d\tau \leq e^1 \int_{\tau_{n+1}}^{\tau_n} e^{2|a|\tau} d\tau,$$

and thus,

$$\sum_{n=1}^{N} e^{2|a|\tau_n}|a|k_n \leq e\int_{\tau_{N+1}}^{\tau_1} |a|e^{2|a|\tau} d\tau = e\int_0^{t_N} |a|e^{2|a|\tau} dt.$$

Evaluating the integral and placing the result in (9.39), we obtain

Theorem 9.4. *Assuming that* $k_n|a| \leq 1/2$, $n \geq 1$, *the error of the dG(0) approximation U satisfies*

$$|u(t_N) - U_N| \leq \frac{e}{4}\left(e^{2|a|t_N} - 1\right)\max_{1 \leq n \leq N} k_n|\dot{u}|_{I_n}. \qquad (9.40)$$

Problem 9.35. Provide all the details for this proof.

Problem 9.36. Arguing from (9.39) directly, prove the error bound

$$|e_N| \leq |a| t_N e^{2|a| t_N} \max_{1 \leq n \leq N} k_n |\dot{u}|_{I_n}.$$

Compare this to the result in the theorem and explain if or when it is "worse".

Problem 9.37. Write out the a priori error estimate for the equations specified in Problem 9.3.

This result proves that the error of U_N approaches zero at a first order rate as the time steps tend to zero as long as $\dot{u}$ is bounded on $[0, T]$. However, to force the error to be small using this bound, we have to choose the steps small enough to overcome the size of $e^{2|a| t_N}$ as well as $|\dot{u}|$, which could be a severe restriction if a, t_N, or $|\dot{u}|$ is large. The exponential factor, which we call the *stability factor*, is precisely a bound on the accumulation of errors. Thus, (9.40) is a priori in two ways: it measures the errors made in each interval using the size of $\dot{u}$ and it measures the accumulation of error by taking the worst case when all the errors are of the same sign.

This proof does not make any use of the special properties of parabolic problems. Since the goal is to obtain accurate error estimates, this is not satisfactory. We present a more sophisticated analysis below.

Problem 9.38. *(Hard!)* Combine the techniques used in the proof of the Fundamental Theorem of Calculus given in Chapter 3 and the proof above to show that the dG(0) approximation forms a Cauchy sequence. Conclude that the dG(0) approximation converges to the solution of (9.3).

Problem 9.39. Use the a priori error bound to show that $R(U) = O(1)$ as the mesh size is refined to zero.

9.4.2. The parabolic case

We conclude this section with a more sophisticated a priori error estimate for dG(0) for (9.3) that applies to the general case and also gives a sharp estimate in the parabolic case $a(t) \geq 0$. The weaker stability properties of cG(1) do not allow a corresponding estimate for cG(1).

Theorem 9.5. *If* $|a(t)| \leq A$ *for all* t, *then the dG(0) approximation* U *satisfies for* $N = 1, 2, ..$,

$$|u(t_N) - U_N| \leq 3e^{2At_N}|k\dot{u}|_{[0,t_N]},$$

and if $a(t) \geq 0$ *for all* t,

$$|u(t_N) - U_N| \leq 3|k\dot{u}|_{[0,t_N]}. \tag{9.41}$$

Proof. It suffices to estimate the "discrete" error $\bar{e} \equiv \pi_k u - U$ in $W_k^{(0)}$, where $\pi_k u$ is the L_2 projection into $W_k^{(0)}$, since $u - \pi_k u$ can be estimated using the results in Chapter 5.

The proof is analogous to the proof of the posteriori estimate and uses the following *discrete dual* problem: find Φ in $W_k^{(0)}$ such that for $n = N, N - 1, .., 1$,

$$\int_{t_{n-1}}^{t_n} (-\dot{\Phi} + a(t)\Phi)v \, dt - [\Phi_n]v_n = 0, \quad \text{for all } v \in W_k^{(0)}, \tag{9.42}$$

where $\Phi_N^+ = \Phi_{N+1} = (\pi_k u - U)_N$.

Problem 9.40. Determine the relation between the discrete and continuous dual problems (9.42) and (9.19).

Choosing $v = \bar{e}$, we get

$$|\bar{e}_N|^2 = \sum_{n=1}^{N} \int_{t_{n-1}}^{t_n} (-\dot{\Phi} + a(t)\Phi)\bar{e} \, dt - \sum_{n=1}^{N-1} [\Phi_n]\bar{e}_n + \Phi_N \bar{e}_N.$$

We use the Galerkin orthogonality to replace U by u and we obtain an error representation formula:

$$|\bar{e}_N|^2 = \sum_{n=1}^{N} \int_{t_{n-1}}^{t_n} (-\dot{\Phi} + a(t)\Phi)(\pi_k u - u) \, dt - \sum_{n=1}^{N-1} [\Phi_n](\pi_k u - u)_n$$

$$+ \Phi_N(\pi_k u - u)_N$$

$$= -\int_0^{t_N} (a\Phi(u - \pi_k u) \, dt + \sum_{n=1}^{N-1} [\Phi_n](u - \pi_k u)_n - \Phi_N(u - \pi_k u)_N,$$

where we use the fact that $\dot{\Phi} \equiv 0$ on each time interval. Combining estimates on the interpolation error with the following lemma expressing the weak and strong stability of the discrete dual problem (9.42), we reach the desired conclusion. ∎

Problem 9.41. Write out the details of these calculations.

Lemma 9.6. *If* $|a(t)| \leq A$ *for all* $t \in (0, t_N)$ *and* $k_j |a|_{I_j}$ *for* $j = 1, ..., N$, *then the solution of (9.42) satisfies*

$$|\Phi_n| \leq \exp(2A(t_N - t_{n-1}))|\bar{e}_N|, \tag{9.43}$$

$$\sum_{n=1}^{N-1} |[\Phi_n]| \leq \exp(2At_N)|\bar{e}_N|, \tag{9.44}$$

$$\sum_{n=1}^{N} \int_{t_{n-1}}^{t_n} |a\Phi_n| \, dt \leq \exp(2At_N)|\bar{e}_N|. \tag{9.45}$$

If $a(t) \geq 0$ *for all* t, *then*

$$|\Phi_n| \leq |\bar{e}_N|, \tag{9.46}$$

$$\sum_{n=1}^{N-1} |[\Phi_n]| \leq |\bar{e}_N|, \tag{9.47}$$

$$\sum_{n=1}^{N} \int_{t_{n-1}}^{t_n} a|\Phi_n| \, dt \leq |\bar{e}_N|. \tag{9.48}$$

Proof. The discrete dual problem (9.42) can be written

$$\begin{cases} -\Phi_{n+1} + \Phi_n + \Phi_n \int_{t_{n-1}}^{t_n} a(t) \, dt = 0, \quad n = N, N-1, ...1, \\ \Phi_{N+1} = \bar{e}_N^-, \end{cases}$$

where Φ_n denotes the value of Φ on I_n, so that

$$\Phi_n = \prod_{j=n}^{N} \left(1 + \int_{I_j} a \, dt\right)^{-1} \Phi_{N+1}.$$

In the case that a is bounded, (9.43)-(9.45) follow from standard estimates.

Problem 9.42. Prove this claim.

When $a \geq 0$, (9.46) follows immediately. To prove (9.47), we assume without loss of generality that Φ_{N+1} is positive, so the sequence Φ_n decreases when n decreases, and

$$\sum_{n=1}^{N} |[\Phi_n]| = \sum_{n=1}^{N} [\Phi_n] = \Phi_{N+1} - \Phi_1 \leq |\Phi_{N+1}|.$$

Finally (9.47) follows from the discrete equation.   ∎

Problem 9.43.   Prove the following stability estimate for the dG(0) method (9.18) in the case $f = 0$, $a \geq 0$:

$$|U_N|^2 + \sum_{n=0}^{N-1} |[U_n]|^2 \leq |u_0|^2.$$

Hint: multiply the equation for U_n by U_n and sum.

9.5.   Quadrature errors

A Galerkin method for (9.3) generally contains integrals involving the functions a and f. If these integrals are computed approximately then the resulting quadrature error also contributes to the total error. We met quadrature errors in the proof of the Fundamental Theorem of Calculus and we discussed quadrature further in Chapter 5. When solving (9.3) numerically, it is natural to control the quadrature error on the same tolerance level as the Galerkin discretization error controlled by (9.2).

To analyze the effect of quadrature errors, we consider the midpoint and endpoint rules, which on the interval $I_n = (t_{n-1}, t_n)$ take the form

$$\int_{I_n} g \, dt \approx g(t_{n-\frac{1}{2}})k_n, \quad \int_{I_n} g \, dt = g(t_n)k_n, \tag{9.49}$$

where $t_{n-\frac{1}{2}} = (t_{n-1} + t_n)/2$. We recall the analysis of quadrature formulas in Chapter 5, and in particular the fact that the midpoint rule is more accurate than the rectangle rule. To simplify the discussion, we assume that a is constant and analyze the error when a quadrature rule is used on the integral involving f. We compare dG(0) approximations computed with the two quadrature rules and conclude that the endpoint rule is less accurate on many problems, while both methods have the same cost of one function evaluation per step. The analysis shows the advantage of separating the Galerkin and quadrature errors since they accumulate at different rates.

The dG(0) method combined with the endpoint rule is equivalent to the classic *backward Euler difference scheme*.

For the midpoint rule, the quadrature error on a single interval is bounded by

$$\left| \int_{I_n} g\, dt - g(t_{n-\frac{1}{2}})k_n \right| \leq \min\left\{ \int_{I_n} |k\dot{g}|\, dt, \frac{1}{2}\int_{I_n} |k^2 \ddot{g}|\, dt \right\}. \tag{9.50}$$

The corresponding error estimate for the endpoint rule reads

$$\left| \int_{I_n} g\, dt - g(t_n)k_n \right| \leq \int_{I_n} |k\dot{g}|\, dt. \tag{9.51}$$

As already noticed, the midpoint rule is more accurate unless $|\ddot{g}| \gg |\dot{g}|$, while the cost of the two rules is the same.

We present an a posteriori error analysis that includes the effects of using quadrature. We start with the modified form of the error representation formula

$$e_N^2 = \sum_{n=1}^{N} \left(\int_{t_{n-1}}^{t_n} (f - aU)(\varphi - \pi_k\varphi)\, dt - [U_{n-1}](\varphi - \pi_k\varphi)_{n-1}^+ \right.$$
$$\left. + \int_{t_{n-1}}^{t_n} f\pi_k\varphi\, dt - (\overline{f\pi_k\varphi})_n k_n \right), \tag{9.52}$$

where

$$\overline{g}_n = \begin{cases} g(t_n) & \text{for the endpoint rule,} \\ g(t_{n-1/2}) & \text{for the midpoint rule.} \end{cases}$$

Problem 9.44. Compare (9.21) and (9.52).

The accumulation of quadrature error is measured by a different stability factor. We introduce the *weak stability factor*

$$\tilde{S}(t_N) = \frac{\int_0^{t_N} |\dot{\varphi}|\, dt}{|e_N|},$$

where we recall that φ satisfies the dual problem (9.19).

Problem 9.45. Prove that $\tilde{S}(t_N) \leq t_N(1 + S(t_N))$. Hint: $\varphi(t) = \int_{t_N}^t \dot{\varphi}\, ds + e_N$.

Problem 9.46. Assume that $a > 0$ is constant. Prove that $\tilde{S}(t_N) \gg S(t_N)$ if a is small.

Using the facts that $\pi_k\varphi$ is piecewise constant and $\int_{I_n} |\pi_k\varphi(t)|\, dt \leq \int_{I_n} |\varphi(t)|\, dt$, we obtain a modified a posteriori error estimate that includes the quadrature errors.

Theorem 9.7. *The dG(0) approximation U computed using quadrature on terms involving f satisfies for $N = 1, 2, ..$,*

$$|u(t_N) - U_N| \leq S(t_N)|kR(U)|_{(0,t_N)} + \tilde{S}(t_N)C_j|k^j f^{(j)}|_{(0,t_N)},$$

where

$$R(U) = \frac{|U_n - U_{n-1}|}{k_n} + |f - aU|, \quad \text{on } I_n,$$

and $j = 1$ for the rectangle rule, $j = 2$ for the midpoint rule, $C_1 = 1$, $C_2 = 1/2$, $f^{(1)} = \dot{f}$, and $f^{(2)} = \ddot{f}$.

In general, the two stability factors $S(t_N)$ and $\tilde{S}(t_N)$ grow at different rates. In cases when $\tilde{S}(t_N)$ grows more quickly, it is natural to use the midpoint rule since the size of $\tilde{S}(t_N)$ then is compensated by the second order accuracy. In general, the computational cost of the quadrature is usually small compared to the Galerkin computational work (which requires the solution of a system of equations) and the precision of the quadrature may be increased without significantly increasing the overall work. Thus, it is possible to choose a higher order quadrature formula to compensate for rapid accumulation of quadrature errors. This illustrates the importance of separating Galerkin discretization and quadrature errors since they accumulate differently.

As an illustration, we consider a situation in which the birth rate dominates the death rate and there is an increasing rate of immigration. We model this with the equation $\dot{u} - .1u = t^3$ and $u(0) = 1$. This problem is not dissipative, so we expect error accumulation. We compute using the dG(0), the dG(0) with the rectangle rule quadrature, and the dG(0) with the midpoint point rule quadrature.

Problem 9.47. Write down explicit equations for the three approximations.

We plot the dG(0) approximation in Fig. 9.8 together with the errors of the three approximations. The dG(0) with exact and midpoint quadrature are very close in accuracy, while the error in the rectangle rule backward Euler computation accumulates at a much faster rate.

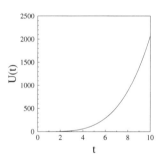
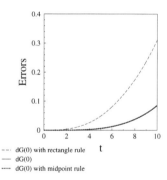

--- dG(0) with rectangle rule
— dG(0)
···· dG(0) with midpoint rule

Figure 9.8: Results for three approximations of $\dot{u} - .1u = t^3$ and $u(0) = 1$ based on the dG(0) method. The plot on the left shows the dG(0) approximation while the plot on the right shows the errors of the three approximations. Note that the results for the dG(0) and dG(0) with midpoint quadrature are nearly identical.

Problem 9.48. Discuss an alternative way of estimating the quadrature error based on an integral of $|k^j f^{(j)}|$ in time instead of the above maximum, and define the corresponding stability factor.

9.6.   The existence of solutions

We gave an expression for the solution of (9.3) in terms of the data u_0 and f using the variation of constants formula (9.4):

$$u(t) = e^{-A(t)}u_0 + \int_0^t e^{-(A(t)-A(s))} f(s)\, ds,$$

where $A(t)$ is the primitive function of a given by $\int_0^t a(s)\, ds$ and the exponential function $e^{\pm t} = \exp(\pm t)$ is defined to be the solution of the initial value problem

$$\begin{cases} \dot{u}(t) = \pm u & \text{for } t > 0, \\ u(0) = 1. \end{cases} \tag{9.53}$$

It is easy to verify that the function defined by (9.4) satisfies (9.3), given certain properties of the integral and the exponential function. Therefore, in order to prove the existence of the solution of (9.3), it suffices to prove the existence of the integral of a and some of its properties, and to prove the existence and some of the properties of the solution of (9.53).

We already proved the existence of the integral in Chapter 3, and at the end of that chapter, we promised to outline a proof of the existence of the solution of (9.53). We do this now. The proof amounts to an extension of the proof of the Fundamental Theorem of Calculus.

We show the existence of the solution of (9.53) by showing that the dG(0) approximation converges to a unique function and then showing this function satisfies the differential equation. We restrict attention to the parabolic case $\dot{u} = -u$ since the case $\dot{u} = u$ is covered noting that $v = \frac{1}{u}$ satisfies $\dot{v} = v$ if $\dot{u} = -u$. We show that the dG(0) approximation converges by showing that it generates a Cauchy sequence. This follows from an estimate of the difference $U^{(N)} - U^{(M)}$, where $U^{(N)}$ is the dG(0) approximate solution of (9.53) on a partition $\mathcal{T}_k^{(N)}$ of $(0, T)$ with uniform time steps 2^{-N}, assuming $T = 1$ for simplicity, and $U^{(M)}$ a corresponding dG(0) approximation on a finer partition $\mathcal{T}_k^{(M)}$ with time steps 2^{-M}, with $M > N$. This is the same setup as in the proof of the Fundamental Theorem; see Fig. 3.10. The first step is to prove the following close analog of the a priori error estimate (9.41) with the exact solution u replaced by the fine mesh approximation $U^{(M)}$

$$|U^{(M)} - U^{(N)}|_{[0,T]} \le 3 \cdot 2^{-N} |\dot{U}^{(M)}|_{[0,T]}, \qquad (9.54)$$

where $\dot{U}^{(M)}$ is the piecewise constant function defined on the partition $\mathcal{T}_k^{(M)} = \{t_n\}$ by

$$\dot{U}^{(M)} = \frac{(U_n^{(M)} - U_{n-1}^{(M)})}{2^{-M}} \quad \text{on } (t_{n-1}, t_n].$$

Problem 9.49. Prove this result modifying the proof of (9.41).

The next step is to prove that $|\dot{U}^{(M)}|_{[0,T]}$ is bounded independently of M. By the definition of $\dot{U}^{(M)}$, we have

$$\dot{U}^{(M)} + U^{(M)} = 0 \quad \text{on } (t_{n-1}, t_n],$$

and thus, $\dot{U}^{(M)}$ is bounded if $U^{(M)}$ is bounded. But we know the following formula for $U_n^{(M)}$

$$U_n^{(M)} = (1 + 2^{-M})^{-n} \le 1,$$

and the desired bound follows. This proves that $\{U^{(N)}\}$ is a Cauchy sequence on $[0, T]$ and therefore converges to a unique function u. By the construction it follows that u satisfies $\dot{u} + u = 0$. We have now proved the existence of a solution of the initial value problem $\dot{u} + u = 0$ in $(0, T]$, $u(0) = 1$, and we denote it by $u(t) = \exp(-t)$.

Problem 9.50. Prove that $\lim_{n \to \infty} (1 + \frac{1}{n})^{-n} = \exp(-1) = e^{-1}$. More generally, prove that $\lim_{n \to \infty} (1 + \frac{x}{n})^{-n} = \exp(-x)$.

Problem 9.51. Give a direct proof of existence of a solution to the problem (9.1) following the same strategy as in the special case above.

Problem 9.52. Prove that $\exp(t + s) = \exp(t) \exp(s)$ for $t, s \geq 0$.

This proof extends rather directly to systems of differential equations, as we show in the next chapter. The proof illustrates the close connection between mathematical questions like existence and uniqueness and convergence of numerical solutions.

> Nature that fram'd us of four elements
> Warring within our breasts for regiment,
> Doth teach us all to have aspiring minds:
> Our souls, whose faculties can comprehend
> The wondrous architecture of the world,
> And measure every wandring planet's course,
> Still climbing after knowledge infinite,
> And always moving as the restless spheres,
> Will us to wear ourselves, and never rest,
> Until we reach the ripest fruit of all,
> That perfect bliss and sole felicity,
> The sweet fruition of an earthly crown. (C. Marlowe, 1564-1593)

Figure 9.9: The house in Hannover where Leibniz lived from 1698 to his death.

10

Initial Value Problems for Systems

The search for general methods for integrating ordinary differential equations ended about 1755. (Kline)

In this chapter, we study the generalization of the scalar initial value problem (9.3) to the case of a system of linear ordinary differential equations of dimension $d \geq 1$: find $u(t) \in \mathbb{R}^d$ such that

$$\begin{cases} \dot{u}(t) + A(t)u(t) = f(t) & \text{for } 0 < t \leq T, \\ u(0) = u_0, \end{cases} \tag{10.1}$$

where $A(t)$ is a $d \times d$ matrix, $f(t) \in \mathbb{R}^d$ is a vector forcing function, $u_0 \in R^d$ is an initial vector, and $(0, T]$ a given time interval. Systems of equations are more complicated than scalar equations because of the possibility of interaction between the different components of the system, and in general numerical computation must be used to obtain detailed information about the solutions of particular problems.

We begin by presenting some models from mechanics and a simple model of the climate in the ocean of the form (10.1) and we discuss the general structure of the solution of such systems. We distinguish between *autonomous* systems where $A(t) = A$ is a constant matrix and *non-autonomous* systems where $A(t)$ depends on time t. Non-autonomous systems generally allow richer behavior in their solutions than autonomous systems while also being more difficult to analyze. We then extend the Galerkin methods introduced in the previous chapter to systems, derive error bounds, and present some examples. We pay particular attention to two classes of systems of the form (10.1) referred to as parabolic and hyperbolic systems. Finally, we conclude with a short presentation of an initial value problem for a nonlinear system of ordinary differential equations that describes the motion of a satellite. Nonlinear systems are treated in detail in the advanced companion volume.

10.0.1.   Linear problems and the principle of superposition

In this book we restrict the discussion to linear differential equations. These are of fundamental importance and in fact much of the study of nonlinear problems is based on analysis of associated linear problems. A basic consequence of linearity is the *principle of superposition* stating that a linear combination of solutions of a linear differential equation is a solution of the equation for the corresponding linear combination of data. In other words, if u satisfies $\dot{u} + Au = f$ with initial data $u(0)$ and v satisfies $\dot{v} + Av = g$ with initial data $v(0)$, then $w = \alpha u + \beta v$ satisfies $\dot{w} + Aw = \alpha f + \beta g$ with initial data $\alpha u(0) + \beta v(0)$ for any constants α and β. We have already used this property many times in the book, for example it underlies Fourier's method for solving two point boundary value problems. This property does not hold for nonlinear problems.

The following consequence of the principle of superposition is often used: a solution u of $\dot{u} + Au = f$ with initial data $u(0)$, can be written as $u = u_p + u_h$ where u_p is any particular solution of $\dot{u}_p + Au_p = f$ and u_h solves the *homogeneous* problem $\dot{u}_h + Au_h = 0$ with the initial data $u(0) - u_p(0)$. This allows the reduction of the general problem to the homogeneous problem $\dot{u} + Au = 0$ if a particular solution u_p has been found.

10.1.   Model problems

To provide some examples of autonomous problems, we generalize the models of stationary systems of masses, springs, and dashpots introduced in Chapter 7 to non-stationary *dynamical systems* including inertial forces from the motion of the masses. After that, we introduce a simple non-autonomous model of the climate in the ocean.

10.1.1. Dynamical systems of masses, springs and dashpots

We start by considering a body of mass m resting on a frictionless horizontal surface and connected to a rigid wall by a spring. If the body is moved a distance u from its rest position, where the spring force is zero, then the spring reacts with a restoring force σ on the body. By Hooke's law, σ is proportional to the displacement: $\sigma = ku$, where k is the spring constant ; see Fig. 10.1. We assume that the mass is also acted upon

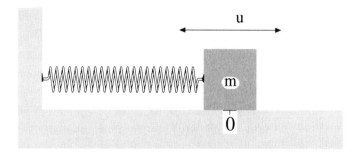

Figure 10.1: A mass-spring system.

by an external force $f(t)$ that varies with time t. Newton's second law states that the rate of change with respect to time of the momentum $m\dot{u}$ of the body is equal to the total force, which in this case is the difference of the applied force and the spring force. We thus obtain the following model for the dynamics of the mass-spring system:

$$\begin{cases} \sigma(t) = ku(t) & \text{for } 0 < t \le T, \\ m\ddot{u}(t) = f(t) - \sigma(t) & \text{for } 0 < t \le T, \\ u(0) = u_0, \quad \dot{u}(0) = \dot{u}_0, \end{cases} \tag{10.2}$$

where we denote the initial position and velocity by u_0 and $\dot{u}_0$, respectively. Eliminating σ, we obtain an initial value problem for a linear second order differential equation:

$$\begin{cases} m\ddot{u} + ku = f & \text{in } (0, T], \\ u(0) = u_0, \quad \dot{u}(0) = \dot{u}_0. \end{cases} \tag{10.3}$$

We recall that in the case $f = 0$ and $m = 1$, the solution has the form

$$u(t) = a\cos(\sqrt{k}t) + b\sin(\sqrt{k}t) = \alpha\cos(\sqrt{k}(t - \beta)), \tag{10.4}$$

where the constants a and b, or α and β, are determined by the initial conditions. We conclude that if $f = 0$ and $m = 1$, then the motion of the mass is periodic with frequency $\sqrt{k}$ and phase shift β and amplitude α depending on the initial data.

Problem 10.1. Verify (10.4) and determine the values of a and b, or α and β, in terms of u_0 and $\dot{u}_0$.

Problem 10.2. (a) Using the notation of Chapter 7, show that the dynamics of a system of masses and springs modeled by the relation $\sigma = DBu$ between the vector of spring forces σ and the vector of displacements u and the equilibrium equation $B^{\mathsf{T}}\sigma = f$ at rest, where f is a vector of exterior forces acting on the masses, is modeled by

$$\begin{cases} \sigma = DBu & \text{in } (0, T] \\ M\ddot{u} + B^{\mathsf{T}}\sigma = f & \text{in } (0, T] \\ u(0) = u_0, \quad \dot{u}(0) = \dot{u}_0, \end{cases}$$

where M is the diagonal matrix with diagonal entry m_j equal to mass j. (b) Show this system can be written as

$$\begin{cases} M\ddot{u} + B^{\mathsf{T}}DBu = f & \text{in } (0, T], \\ u(0) = u_0, \quad \dot{u}(0) = \dot{u}_0. \end{cases} \tag{10.5}$$

Problem 10.3. Consider particular configurations of the form (10.5) with several masses and springs.

Introducing the new variables $u_1 = \dot{u}$ and $u_2 = u$, and for simplicity assuming that $m = 1$, the second order equation (10.3) can be written as the first order system

$$\begin{cases} \dot{u}_1 + ku_2 = f & \text{in } (0, T], \\ \dot{u}_2 - u_1 = 0 & \text{in } (0, T], \\ u_1(0) = \dot{u}_0, \qquad u_2(0) = u_0, \end{cases} \tag{10.6}$$

which has the form (10.1) with $u = (u_1, u_2)$ and

$$A = \begin{pmatrix} 0 & k \\ -1 & 0 \end{pmatrix}.$$

We next consider a model with the spring replaced by a *dashpot*, which is a kind of shock absorber consisting of a piston that moves

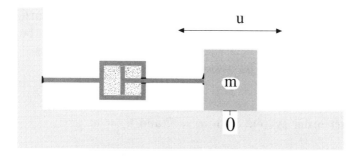

Figure 10.2: Cross section of a dashpot connected to a mass.

inside a cylinder filled with oil or some other viscous fluid see Fig. 10.2. The resistance of the fluid to flowing past the piston as it moves gives a force $\sigma = \mu \dot{u}$ that is proportional to the rate of change of displacement $\dot{u}$ with the constant of proportionality $\mu \geq 0$ representing the *viscosity coefficient* of the dashpot. Assuming that the mass is acted upon also by an exterior force $f(t)$, we obtain the following model by using Newton's law,

$$\begin{cases} m\ddot{u} + \mu\dot{u} = f & \text{in } (0, T], \\ u(0) = u_0, \quad \dot{u}(0) = \dot{u}_0. \end{cases} \tag{10.7}$$

If $f = 0$ and $m = 1$, then the solution has the form $u(t) = a\exp(-\mu t) + b$, in which case the mass approaches a fixed position $u = b$ determined by the initial data as time increases. Written as a first order system, (10.7) with $m = 1$ takes the form

$$\begin{cases} \dot{u}_1 + \mu u_1 = f & \text{in } (0, T], \\ \dot{u}_2 - u_1 = 0 & \text{in } (0, T], \\ u_1(0) = \dot{u}_0, & u_2(0) = u_0. \end{cases} \tag{10.8}$$

Using these ingredients, we can model systems of masses, springs and dashpots in different configurations. For example, a mass connected to both a spring and a dashpot coupled in parallel is modeled by

$$\begin{cases} \sigma = ku + \mu\dot{u}, & \text{in } (0, T], \\ m\ddot{u} + \sigma = f, & \text{in } (0, T], \\ u(0) = u_0, \quad \dot{u}(0) = \dot{u}_0. \end{cases}$$

where the total force σ is the sum of the forces from the spring and the dashpot. Eliminating σ and assuming that $m = 1$, this can be rewritten

$$\begin{cases} \ddot{u} + \mu\dot{u} + ku = f, & \text{in } (0, T], \\ u(0) = u_0, \quad \dot{u}(0) = \dot{u}_0, \end{cases} \tag{10.9}$$

or as a first order system with $u_1 = \dot{u}$ and $u_2 = u$:

$$\begin{cases} \dot{u}_1 + \mu u_1 + k u_2 = f & \text{in } (0, T], \\ \dot{u}_2 - u_1 = 0, & \text{in } (0, T], \\ u_1(0) = \dot{u}_0, & u_2(0) = u_0. \end{cases} \tag{10.10}$$

In the case $f = 0$, the solution of (10.9) is a linear combination of terms of the form $\exp(\lambda t)$, where λ solves the associated polynomial equation $\lambda^2 + \mu\lambda + k = 0$, that is $\lambda = (\mu \pm \sqrt{\mu^2 - 4k})/2$. If $\mu^2 - 4k > 0$, then the roots are real and the solution is

$$u(t) = a \exp\left(-\frac{1}{2}\left(\mu + \sqrt{\mu^2 - 4k}\right)t\right) + b \exp\left(-\frac{1}{2}\left(\mu - \sqrt{\mu^2 - 4k}\right)t\right),$$

with the constants a and b determined by the initial condition. In this case, when the viscous damping of the dashpot dominates, the solution converges exponentially to a rest position, which is equal $u = 0$ if $k > 0$. If $\mu^2 - 4k < 0$, then the roots are complex and the solution is

$$u(t) = ae^{-\frac{1}{2}\mu t} \cos\left(\frac{1}{2}\sqrt{4k - \mu^2}\, t\right) + be^{-\frac{1}{2}\mu t} \sin\left(\frac{1}{2}\sqrt{4k - \mu^2}\, t\right).$$

The solution again converges to the zero rest position as time passes if $\mu > 0$, but now it does so in an oscillatory fashion. Finally, in the limit case $\mu^2 - 4k = 0$ the solution is

$$u(t) = (a + bt)e^{-\frac{1}{2}\mu t},$$

and exhibits some non-exponential growth for some time before eventually converging to the zero rest position as time increases. We illustrate the three possible behaviors in Fig. 10.3.

Problem 10.4. Verify the solution formulas for the three solutions shown in Fig. 10.3.

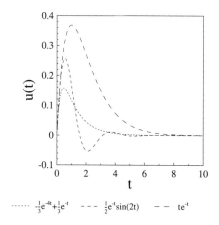

$$-\tfrac{1}{3}e^{-4t}+\tfrac{1}{3}e^{-t} \qquad \tfrac{1}{2}e^{-t}\sin(2t) \qquad te^{-t}$$

Figure 10.3: Three solutions of the mass-spring-dashpot model (10.9) satisfying the initial conditions $u(0) = 0$ and $\dot{u}(0) = 1$. The first solution corresponds to $\mu = 5$ and $k = 4$, the second to $\mu = 2$ and $k = 5$, and the third to $\mu = 2$ and $k = 1$.

Problem 10.5. (a) Show that a mass connected to a spring and dashpot coupled in series is modeled by

$$\begin{cases} \dot{u} = \frac{1}{\mu}\sigma + \frac{1}{k}\dot{\sigma}, & \text{in } (0, T], \\ m\ddot{u} + \sigma = f, & \text{in } (0, T], \\ u(0) = u_0, \quad \dot{u}(0) = \dot{u}_0., \end{cases}$$

where the total displacement rate $\dot{u}$ is the sum of the displacement rates related to the spring and the dashpot. (b) Show that the system for $m = 1$ can be written on the form

$$\begin{cases} \dot{u}_1 - u_2 = f, & \text{in } (0, T], \\ \dot{u}_2 + \frac{k}{\mu}u_2 + ku_1 = 0 & \text{in } (0, T], \end{cases} \tag{10.11}$$

with suitable initial conditions. (c) Explain the range of behavior that is possible in the solutions if $f = 0$.

10.1.2. Models of electrical circuits

There is an analogy between models of masses, dashpots, and springs in mechanics and models of electrical circuits involving inductors, resistors, and capacitors, respectively. The basic model for an electric circuit with these components has the form

$$L\ddot{u}(t) + R\dot{u}(t) + \frac{1}{C}u(t) = f(t) \quad \text{for} \quad t > 0, \qquad (10.12)$$

together with initial conditions for u and $\dot{u}$, where $f(t)$ represents an applied voltage and u represents a primitive function of the current I. This model says that the applied voltage is equal to the sum of the voltage changes $V = L\dot{I}$, $V = RI$ and $V = u/C$ from the inductor, resistor, and capacitor, where L, R and C are the coefficients of inductance, resistance and capacitance, respectively. The system (10.12) takes the same form as (10.9) and the discussion above (including systems with components coupled in parallel or series) applies.

10.1.3. A simple model of the climate in the ocean

> I was born a simple man, I got no command of the written word.
> I can only try and tell you the things I've seen and heard.
> Listen to the picture forever etched on my mind.
> The day that the hell broke loose just north of Marietta,
> all along the Kennesaw line.
> The day that the hell broke loose just north of Marietta.
> Oh, the sun rose high above us that morning on a clear and cloud-less day.
> Well, the heat blistered down through the leaves on the trees.
> The air seemed hot enough to catch fire.
> The heavens seemed to be made of brass as the sun rose higher and higher. (D. Oja-Dunnaway)

We consider a model of the temperature variations of the atmosphere and the ocean taking into account different periodic effects due to the sun. The model is based on *Newton's law of cooling* which states that the rate of change of heat between two regions is proportional to the difference between their temperatures. We choose units so that the temperature of space is 0 while the temperature of the earth is 1, and we let u_1 denote the temperature of the ocean and u_2 the temperature of the atmosphere. See Fig. 10.4. We model the influence of the sun on the atmosphere and the ocean through two periodic forcing terms with different time scales (corresponding to daily and yearly variations). We also allow the factors of proportionality in Newton's law to vary with time. In total, the rate of change of heat of the ocean is given by the

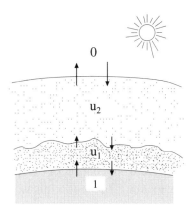

Figure 10.4: An illustration of the exchange of heat between the earth, ocean, and atmosphere.

exchange with the earth and with the atmosphere plus the effect of a forcing term and that of the atmosphere is determined similarly. This gives a system of the form

$$\begin{cases} \dot{u}_1 = \alpha(t)(u_2 - u_1) + \beta_1(t)(1 - u_1) + S_1 \sin(P_1 t), & 0 < t \le T, \\ \dot{u}_2 = \alpha(t)(u_1 - u_2) - \beta_2(t)u_2 + S_2 \sin(P_2 t) & 0 < t \le T, \\ u(0) = u_0, & (10.13) \end{cases}$$

where α, β_1, and β_2 are positive, continuous functions. We can rewrite the system in the form $\dot{u} + Au = f$ with a positive-definite, symmetric matrix

$$A(t) = \begin{pmatrix} \alpha(t) + \beta_1(t) & -\alpha(t) \\ -\alpha(t) & \alpha(t) + \beta_2(t) \end{pmatrix}.$$

10.2. The existence of solutions and Duhamel's formula

To prove existence of a solution to (10.1) in the autonomous case and with $f = 0$, we follow the approach used for the scalar problem $\dot{u} + u = 0$ in Chapter 9. We define an approximate solution $U = U^{(N)}$ by the dG(0) method on a uniform partition of $(0, T]$ with time steps $k = T2^{-N}$ and value U_n on $((n-1)k, nk)$, as follows:

$$U_n + kAU_n = U_{n-1}, \quad \text{for } n = 1, 2, ..., \quad (10.14)$$

where $U_0 = u_0$. We first deal with the simpler case when A is positive semi-definite (but not necessarily symmetric) so that $(Av, v) \geq 0$ for all $v \in \mathbb{R}^d$. Taking the dot product with U_n in (10.14) gives (denoting the Euclidean scalar product and norm by $(\cdot, \cdot)$ and $|\cdot|$),

$$|U_n|^2 + k(AU_n, U_n) = (U_{n-1}, U_n) \leq |U_n| |U_{n-1}|.$$

By the positive-definiteness of A this proves that $|U_n| \leq |U_{n-1}| \leq \dots \leq |u_0|$. It follows that a solution of (10.14) is necessarily unique (since $u_0 = 0$ implies $U_n = 0$), and since (10.14) is a linear system for U_n, it implies that U_n also exists. In the same way as in Chapter 9 we now prove that $\{U^{(N)}\}$ forms a Cauchy sequence on $[0, T]$ and therefore converges to a solution u of (10.1) with $f = 0$. We denote the solution by $u(t) = \exp(-tA)u_0$, where we think of $\exp(-tA)$ as a matrix which when applied to the initial data u_0 gives the solution $u(t)$. The columns of $\exp(-tA)$ are the solutions of (10.1) with $f = 0$ corresponding to choosing u_0 equal to the standard basis vectors e_i in $\mathbb{R}^d$.

Problem 10.6. Compute the solution matrix $\exp(-tA)$ for (10.6), (10.8), and (10.10) with $f \equiv 0$ by choosing the indicated data.

Problem 10.7. (*Harder*) Prove that $\{U^{(N)}\}$ is a Cauchy sequence. Hint: follow the corresponding proof in the scalar case closely.

The case of a general matrix A can be reduced to the positive semi-definite case by introducing the change of variables $v(t) = \exp(\kappa t)u(t)$ for a constant $\kappa > 0$, which changes $\dot{u} + Au = 0$ into $\dot{v} + (A + \kappa I)v = 0$, where I is the $d \times d$ identity matrix. Now, $A + \kappa I$ is positive semi-definite if κ is sufficiently large and so we know that v, and therefore u, exists.

Once we have the general solution of (10.1) with $f = 0$, we can write down the following formula for the solution of (10.1) with $f \neq 0$:

$$u(t) = \exp(-tA)u_0 + \int_0^t \exp(-(t-s)A)f(s)\, ds.$$
$$(10.15)$$

This is referred to as *Duhamel's formula*, and is analogous to the variation of constants formula for a scalar problem.

Problem 10.8. Using Problem 10.6, write down formulas for the solutions of (10.6), (10.8), and (10.10) with the forcing function $f(t) = t$.

Problem 10.9. Prove Duhamel's formula using the properties of $\exp(-tA)$ just established.

10.3.   Solutions of autonomous problems

We now describe the structure of the solutions of (10.1) in the autonomous case with $f = 0$ in terms of properties of the eigenvalues
and eigenvectors of A. It is convenient to allow scalars to be complex temporarily. If A is a real matrix, we obtain real-valued solutions
by taking real and imaginary parts of complex-valued solutions. We
start observing that if v is an eigenvector of A corresponding to the
eigenvalue λ so that $Av = \lambda v$, then $u(t) = \exp(-\lambda t)v$ is a solution
of $\dot{u} + Au = 0$, which we refer to as an *eigenmode*. If $\lambda = \alpha + i\theta$,
where $\alpha \in \mathbb{R}$ and $\theta \in \mathbb{R}$ are the real and imaginary parts of λ, so
that $\exp(-\lambda t) = \exp(-\alpha t)(\cos(\theta t) + i\sin(\theta t))$, then the modulus of the
eigenmode is exponentially decaying (increasing) if $\alpha > 0$ ($\alpha < 0$), and
if $\theta \neq 0$ then the eigenmode is oscillating with frequency θ. We conclude
that if there is a basis of eigenvectors of A, then it should be possible to
describe the solutions of (10.1) with $f = 0$ in terms of eigenmodes. We
make this precise below.

   We recall that if the eigenvectors of A form a basis then A is diagonalizable, that is there is an invertible matrix P such that $P^{-1}AP = \Lambda$,
where Λ is a diagonal matrix with non-zero entries only on the diagonal from the upper left corner to the lower right corner. The diagonal
elements of Λ are the eigenvalues of A and the columns of P are the
corresponding eigenvectors. If P is orthogonal so that $P^{-1} = P^*$, then
A is said to be normal. Recall that A^* denotes the adjoint of $A = (a_{ij})$
with elements $\bar{a}_{ji}$, where the bar denotes the complex conjugate so that
$\bar{\lambda} = \alpha - i\theta$ if $\lambda = \alpha + i\theta$ with $\alpha, \theta \in \mathbb{R}$. If A is real, then $A^* = A^{\mathsf{T}}$. A
basic fact of linear algebra states that A is normal if $A^*A = AA^*$. An
important special case is given by a selfadjoint (or symmetric) matrix
satisfying $A^* = A$.

   We shall see that in the autonomous case the structure of the solutions of (10.1) with $f = 0$ is entirely determined by the eigenvalues if A
is normal, while if A is not normal then the structure of the eigenvectors
of A is also important.

   Note that the eigenvalues and the structure of the eigenvectors of
$A(t)$ do not determine the structure of solutions of non-autonomous
problems in general.

10.3.1. Diagonalizable and normal systems

We assume that A is diagonalizable with $P^{-1}AP = \Lambda$ diagonal. Changing to the new variable v defined by $u = Pv$, we rewrite (10.1) with $f = 0$ as

$$P\dot{v} + APv = 0,$$

which upon pre-multiplication by P^{-1} gives a diagonal system for v:

$$\begin{cases} \dot{v} + \Lambda v = 0 & \text{in } (0, T], \\ v(0) = v_0 = P^{-1}u_0, \end{cases}$$

with solution

$$v(t) = \big(v_{0,1}\exp(-\lambda_1 t), \cdots, v_{0,d}\exp(-\lambda_d t)\big)^{\top} = \exp(-\Lambda t)v_0,$$

where the $v_{0,i}$ are the components of the initial value $v(0)$ and $\exp(-\Lambda t)$ denotes the diagonal matrix with diagonal elements $\exp(-\lambda_i t)$. It follows that the solution u of (10.1) with $f = 0$ can be written

$$u(t) = P\exp(-\Lambda t)P^{-1}u_0, \tag{10.16}$$

which gives a complete description of the structure of solutions. If A is normal so that P is orthogonal, then the relation between the norms of $u = Pv$ and v is particularly simple, since $|u| = |v|$.

   If A is real, then the real and imaginary parts of $u = Pv$ are real-valued solutions of $\dot{u} + Au = 0$. This means that each component of a real-valued solution u of $\dot{u} + Au = 0$ is a linear combination of terms of the form

$$\exp(-\alpha_j t)\cos(-\theta_j t) \quad \text{and} \quad \exp(-\alpha_j t)\sin(-\theta_j t),$$

where $\alpha_j \in \mathbb{R}$ and $\theta_j \in \mathbb{R}$ are the real and imaginary parts of the eigenvalue λ_j. The factor $\exp(-\alpha_j t)$ is exponentially decaying if $\alpha_j > 0$, is constant if $\alpha_j = 0$, and is exponentially growing if $\alpha_j < 0$. Further, there are oscillatory factors $\cos(-\theta_j t)$ and $\sin(-\theta_j t)$ of frequency θ_j.

   We apply this framework to the models above. The matrix for the mass-spring system (10.5) is diagonalizable with two purely imaginary eigenvalues $\pm i\sqrt{k}$ and the solution accordingly has the form $a\cos(\sqrt{k}t) + b\sin(\sqrt{k}t)$. The corresponding matrix A is normal if $k = 1$. Next, the matrix of the mass-dashpot system (10.8) is diagonalizable but not normal with two real eigenvalues 0 and μ, and the solution components have the form $a+b\exp(-\mu t)$. Finally, the coefficient matrix for the mass-spring-dashpot system (10.10) has eigenvalues $\lambda_j = \mu/2 \pm \sqrt{\mu^2/4 - k}$ and is diagonalizable if $\mu^2/4 - k \neq 0$. If $\mu^2/4 - k > 0$, then there are two positive eigenvalues and the solution components are a linear combination of two exponentially decaying functions. If $\mu^2/4 - k < 0$, then there are two complex conjugate eigenvalues with positive real part equal to $\mu/2$ and the corresponding solution components both decay exponentially and oscillate. The matrix is normal only if $k = 1$ and $\mu = 0$. The degenerate case $\mu^2/4 - k = 0$ is treated in the next section. We illustrated the range of possible behavior in Fig. 10.3.

Problem 10.10. Verify the statements of the preceeding paragraph.

Problem 10.11. (a) Show that the eigenvalues of the general mass-spring model (10.5) are purely imaginary assuming that $M = I$. (b) Compute the eigenvalues of the mass-spring-dashpot system (10.11) and discuss the possible behavior of the solutions.

10.3.2. Non-diagonalizable systems

If there is a multiple root λ of the characteristic equation $\det(A-\lambda I) = 0$ for the eigenvalues, i.e. if λ is an eigenvalue of multiplicity $m > 1$, and there are less than m linearly independent eigenvectors associated to λ, then A is not diagonalizable. In this case, the components of the solution of (10.1) with $f = 0$ have the form $p_j(t)\exp(-\lambda_j t)$, where the p_j are polynomials in t of degree at most $m - 1$. An example is given by $A = \begin{pmatrix} 0 & 1 \\ 0 & 0 \end{pmatrix}$ which has the eigenvalue zero of multiplicity 2 with a one-dimensional eigenspace spanned by $(1,0)^\top$. The corresponding solution of (10.1) with $f = 0$ is given by $u(t) = (u_{0,1} - tu_{0,2}, u_{0,2})^\top$. Another example is given by the mass-spring-dashpot system (10.10) in the degenerate case $\mu^2/4 - k = 0$, where the matrix has an eigenvalue $\mu/2$ of multiplicity 2 and a corresponding one-dimensional eigenspace spanned by $(-\mu/2, 1)^\top$. The corresponding solution of (10.1) with $f = 0$ is

$$u(t) = \exp\left(-\frac{\mu}{2}t\right)\begin{pmatrix} u_{0,1} - \frac{\mu}{2}(u_{0,1} + \frac{\mu}{2}u_{0,2})t \\ u_{0,2} + (u_{0,1} + \frac{\mu}{2}u_{0,2})t \end{pmatrix}.$$

$$(10.17)$$

Problem 10.12. Verify (10.17).

The degeneracy of the eigenspace may be viewed as a limit case of almost colinear eigenvectors. This cannot occur if A is normal because the eigenvectors are orthogonal. We give an example:

Problem 10.13.   Consider the initial value problem associated to the coefficient matrix $A = \begin{pmatrix} \mu & -1 \\ -\epsilon & \mu \end{pmatrix}$ with $\epsilon > 0$. Determine the eigenvalues and eigenvectors and show how degeneracy of the eigenspace develops as $\epsilon \to 0$.

10.4.   Solutions of non-autonomous problems

The above proof of existence of solution of (10.1) directly extends to the non-autonomous case assuming e.g, that $A(t)$ depends continuously on t. There is also a corresponding generalization of Duhamel's formula, namely

$$u(t) = X(t)u_0 + X(t) \int_0^t X(s)^{-1} f(s)\, ds, \qquad (10.18)$$

where $X(t)$ is the *fundamental solution* of the matrix differential equation

$$\begin{cases} \dot{X}(t) + A(t)X(t) = 0 & \text{in } (0, T], \\ X(0) = I, \end{cases} \qquad (10.19)$$

where I is the identity matrix. This simply means that the columns of $X(t)$ are the solutions of (10.1) with $f = 0$ and initial data u_0 equal to the standard basis vectors e_i in $\mathbb{R}^d$.

Problem 10.14. Verify the solution formula (10.18).

Problem 10.15.   Use (10.18) to compute the solution of (10.13) in the case that α, β_1, β_2 are constant.

We remark that the general solution of a non-autonomous differential equation cannot be described in terms of the eigenvalues of the coefficient matrix $A(t)$ even if $A(t)$ is diagonalizable for each t. This can be understood by attempting the change of variables used in the constant coefficient case. Assuming that $P(t)$ diagonalizes $A(t)$ so that $P(t)^{-1}A(t)P(t) = \Lambda(t)$ for every t with $\Lambda(t)$ diagonal, and using the change of variables $u(t) = P(t)v(t)$, we obtain the following system for v:

$$\dot{v} = \Lambda v + P^{-1}\dot{P}v.$$

This system is not diagonal in general and the term $P^{-1}\dot{P}v$ may drastically change the nature of solutions as compared to solutions of $\dot{v} = \Lambda v$ that are determined by the eigenvalues of A.

Problem 10.16. Verify the differential equation in the new variable.

To illustrate, we present an example constructed by the mathematician Vinograd. Let u solve $\dot{u} + A(t)u = 0$ where

$$A(t) = \begin{pmatrix} 1 + 9\cos^2(6t) - 6\sin(12t) & -12\cos^2(6t) - \frac{9}{2}\sin(12t) \\ 12\sin^2(6t) - \frac{9}{2}\sin(12t) & 1 + 9\sin^2(6t) + 6\sin(12t) \end{pmatrix}. \tag{10.20}$$

The eigenvalues of A are 1 and 10, independent of time. In the constant coefficient case, we would conclude that the solution decays to zero as time passes. However, the solution is

$$u(t) = C_1 e^{2t}\begin{pmatrix} \cos(6t) + 2\sin(6t) \\ 2\cos(6t) - \sin(6t) \end{pmatrix} + C_2 e^{-13t}\begin{pmatrix} \sin(6t) - 2\cos(6t) \\ 2\sin(6t) + \cos(6t) \end{pmatrix},$$

where C_1 and C_2 are constants, which in fact increases exponentially in size as t becomes large for a non-zero C_1.

Problem 10.17. (a) Compute the eigenvalues of the coefficient matrix for Vinograd's example, and verify the formula given for the solution. Plot the solution for $C_1 = C_2 = 1$. (b) *(Messy.)* Compute the change of variables map P and the matrix $P^{-1}\dot{P}$.

10.5. Stability

The *stability* properties of a solution of (10.1) refers to the sensitivity of the solution to perturbations in the data f and u_0. By the principle of superposition, we know that if u solves $\dot{u} + Au = f$ with initial value $u(0) = u_0$ and v solves $\dot{v} + Av = f + g$ with initial value $v(0) = v_0 = u_0 + w_0$, where g and w_0 are perturbations of data, then the corresponding solution perturbation $w = u - v$ solves $\dot{w} + Aw = g$ with initial data $w(0) = w_0$, i.e. a problem of the same form as (10.1). A stability estimate bounds the perturbation w in terms of the perturbations g and w_0 of the data, or equivalently bounds the solution u of (10.1) in terms of f and u_0.

The stability properties of the problem (10.1) may be expressed through different stability estimates on the solutions. Consideration of the stability properties is crucial when analyzing the error of a numerical solution. We shall give a rough classification of ordinary differential equations into (a) general problems (b) parabolic problems and (c) hyperbolic problems. Each class is characterized by a particular stability property. General problems allow errors to accumulate at an exponential rate; hyperbolic problems allow the errors to accumulate at a linear rate; and parabolic problems are characterized by no, or very slow, accumulation of errors. In the autonomous case, we may identify the following basic models of parabolic and hyperbolic type:

- parabolic: A is symmetric positive semidefinite, that is, A is normal with non-negative real eigenvalues

- hyperbolic: A is normal with purely imaginary eigenvalues.

The parabolic model case with A symmetric positive semidefinite corresponds to assuming the coefficient a in the scalar problem (9.3) to be a nonnegative constant, and the hyperbolic model case to assuming a to be a purely imaginary constant.

10.5.1. General problems

We assume that $|A(t)| \leq \mathcal{A}$ for $0 \leq t \leq T$, where $\mathcal{A}$ is a constant and $|\cdot|$ denotes the Euclidean norm and the corresponding matrix norm so that $|A(t)| = \|A(t)\|_2$ with the notation from Chapter 4. The basic stability estimates for the solution u of (10.1) with $f = 0$ read:

$$|u(t)| \leq \exp(\mathcal{A}t)|u_0| \quad \text{for } 0 \leq t \leq T, \tag{10.21}$$

and

$$\int_0^t |\dot{u}(s)|ds \leq e^{\mathcal{A}t}|u_0|. \tag{10.22}$$

We note the presence of the exponential stability factor $\exp(\mathcal{A}t)$ which grows very quickly with t unless $\mathcal{A}$ is small.

Problem 10.18. (a) Prove (10.21). Hint: multiply both sides of (10.1) with $f = 0$ by u, estimate, and then integrate. Use the fact that $(u, \dot{u}) = \frac{1}{2}\frac{d}{dt}|u|^2 = |u|\frac{d}{dt}|u|$. (b) Prove (10.22) using (10.21).

Problem 10.19. Compute bounds for the coefficient matrices in the three models (10.6), (10.8), and (10.10) and write out the stability bounds (10.21) for each problem. Compare the bounds to the actual sizes of the solutions of the three models (with $f \equiv 0$).

Problem 10.20. Compute a bound for the coefficient matrix of Vinograd's example (10.20) and compare the bound (10.21) to the size of the solution.

10.5.2. Parabolic model problem: A is symmetric positive semi-definite

We consider (10.1) with $f = 0$ and assume that A is a constant symmetric and positive semi-definite matrix, that is $A^\mathsf{T} = A$ and $(Av, v) \geq 0$ for all v.

We first prove that the norm of a solution cannot grow as time passes, that is

$$|u(t)| \leq |u_0| \quad \text{for } 0 \leq t \leq T. \tag{10.23}$$

To prove this, we multiply the differential equation $\dot{u} + Au = 0$ by u to get

$$0 = (\dot{u}, u) + (Au, u) = \frac{d}{dt}\frac{1}{2}|u|^2 + (Au, u).$$

It follows after integration that

$$|u(t)|^2 + 2 \int_0^t (Au, u)\, ds = |u_0|^2, \tag{10.24}$$

which by the positive semi-definiteness of A implies (10.23).

Problem 10.21. If we know that A is strictly positive-definite, then we can show that the norm of a solution actually decreases as time passes. For example, assuming that A is symmetric and there is a function $a(t) \geq 0$ for all t such that $(A(t)v, v) \geq a(t)|v|^2$ for all $v \in \mathbb{R}^d$ and $t \geq 0$, prove that a solution u of (10.1) with $f = 0$ satisfies

$$|u(t)| \leq |u(0)| \exp\left(-\int_0^t a(s)\, ds\right) \quad \text{for } t \geq 0.$$

Hint: first verify that $\frac{d}{dt}|u| + a(t)|u| \leq 0$ (or $\frac{1}{2}\frac{d}{dt}|u|^2 + a(t)|u|^2 \leq 0$) and multiply by an *integrating factor*. What can you conclude if there is a constant α such that $a(t) \geq \alpha > 0$ for all t?

Using the symmetry of A in addition, we can prove a *strong* stability estimate that bounds the derivative of u,

$$\int_0^t s|\dot{u}(s)|^2 \, ds \leq \frac{1}{4}|u_0|^2 \quad \text{for all } t \geq 0. \tag{10.25}$$

To prove (10.25), we multiply the differential equation $\dot{u} + Au = 0$ by $tAu(t)$ and use the symmetry of A to get

$$\frac{1}{2}\frac{d}{dt}(u(t), tAu(t)) + (Au(t), tAu(t)) = \frac{1}{2}(u(t), Au(t)).$$

Integrating and recalling (10.24), we obtain

$$\frac{1}{2}t(u(t), Au(t)) + \int_0^t s|Au(s)|^2 \, ds = \frac{1}{2}\int_0^t (u, Au) \, ds \leq \frac{1}{4}|u_0|^2, \tag{10.26}$$

from which (10.25) follows.

Note that from (10.25), it follows by Cauchy's inequality that for $0 < \epsilon < t$,

$$\int_\epsilon^t |\dot{u}(s)| \, ds \leq \left(\int_\epsilon^t \frac{ds}{s}\right)^{1/2} \left(\int_\epsilon^t s|\dot{u}(s)|^2 \, ds\right)^{1/2} = \frac{1}{2}(\log(t/\epsilon))^{1/2}|u_0|.$$

Furthermore, (10.23) implies

$$\int_0^\epsilon |\dot{u}(s)| \, ds \leq \epsilon|A||u_0|.$$

Adding these estimates together we find that

$$\int_0^t |\dot{u}(s)| \, ds \leq \left(|A|\tau + \frac{1}{2}(\log(\frac{t}{\tau}))^{1/2}\right)|u_0|, \tag{10.27}$$

where we took $\epsilon = \tau = \min\{t, |A|^{-1}\}$ to essentially minimize the resulting bound. Compared to (10.22), we get a slow logarithmic growth instead of the exponential growth of the general case.

Problem 10.22. Prove that the indicated choice of ϵ gives the best possible result essentially.

Note that the proof of the estimate (10.23) does not use the symmetry of A and directly generalizes to the non-autonomous case if $A(t)$ is positive semi-definite for $0 \le t \le T$. The proof of (10.25) on the other hand needs both the symmetry and time-independence of A.

Problem 10.23. Fill in the missing details of the proofs of (10.26) and (10.25).

Problem 10.24. (a) Show that (10.25) implies that $|\dot{u}| \to 0$ as $t \to \infty$. Hint: assume that there is some $\epsilon > 0$ such that $|\dot{u}| \ge \epsilon$ and reach a contradiction. (b) Show that for small t, (10.25) is satisfied even if $|\dot{u}(t)| = t^{-1/2}|u_0|$.

Problem 10.25. Even though the norm of u cannot increase when the system is symmetric and positive semi-definite, individual components u_i may increase initially. As an example, consider the linear homogeneous system $\dot{u} + Au = 0$ with $A = \frac{1}{2}\begin{pmatrix} a+b & a-b \\ a-b & a+b \end{pmatrix}$, where a and b are constants with $b > a > 0$. (a) Show that the eigenvalues of A are a and b and compute the general solution by computing the matrix P that diagonalizes A. (b) Write out the solution u corresponding to the initial data $(1, 3)^{\mathsf{T}}$. (c) By explicit computation, show that $\frac{d}{dt}|u(t)|^2 < 0$ always. What does this mean? (d) Show that one component of u decreases monotonically, but the other component increases for some time before beginning to decrease. Show that the time where the maximum value of this component is reached is inversely proportional to $b - a$, which is the gap between the eigenvalues of A.

Problem 10.26. Give a physical interpretation of a system of the form (10.1) with $A = \begin{pmatrix} 1 & -1 \\ -1 & 1 \end{pmatrix}$. Analyze the stability properties of the system.

The technique we used to prove the stability estimates are generally called the *energy method*. Characteristically, the equation $\dot{u} + Au = 0$ is multiplied by u or Au. We will use analogous arguments below for partial differential equations.

10.5.3.   Hyperbolic model problem: A is normal with purely imaginary eigenvalues

We now assume that A is normal with purely imaginary eigenvalues. In this case, the norm of a solution u of $\dot{u} + Au = 0$ is conserved in time:

$$|u(t)| = |u(0)| \quad \text{for all } t. \tag{10.28}$$

If $|u(t)|$ is the energy of $u(t)$, this corresponds to *conservation of energy*.

A different route to energy conservation comes from assuming that A is *skew-symmetric*, that is $(Av, w) = -(v, Aw)$, so that in particular $(Av, v) = 0$. An example of a skew-symmetric matrix is given by the matrix $A = \begin{pmatrix} 0 & 1 \\ -1 & 0 \end{pmatrix}$.

Below, we shall use the following strong stability estimate for $\dot{u}$ resulting from (10.28):

$$\int_0^t |\dot{u}| ds \leq t|A||u_0|. \tag{10.29}$$

Problem 10.27.   (a) Prove (10.28). Hint: diagonalize A. (b) Prove (10.28) assuming A is skew-symmetric.

Problem 10.28.   Prove (10.29).

10.5.4.   An example with A non-normal

Above, we considered two cases with A normal: the parabolic model case with the eigenvalues of A real non-negative, and the hyperbolic model case with the eigenvalues purely imaginary. We now consider the non-normal matrix $A = \begin{pmatrix} \mu & -1 \\ 0 & \mu \end{pmatrix}$, where $\mu > 0$. The matrix A has a single positive eigenvalue μ of multiplicity two with a corresponding one-dimensional eigenspace spanned by $(1, 0)^T$. The solution of (10.1) with $f = 0$ is given by $u(t) = e^{-\mu t}(u_{0,1} + tu_{0,2}, u_{0,2})^T$. Note the factor t in the first component. Choosing $u_0 = (0, 1)^T$ gives the solution $u(t) = e^{-\mu t}(t, 1)$ for which the first component reaches a maximum $(e\mu)^{-1}$, which is large if μ is small. We conclude that there is an initial transient period where the norm of the solution may grow significantly, before eventually tending to zero. In this case, energy conservation or decay is far from being satisfied (in the sense that $|u(t)| \leq |u_0|$), and the solution actually gains energy temporarily, although the sign of the eigenvalue would seem to indicate decay.

The above type of non-normal system arises in the study of stability of fluid flow and explains how perturbations of fluid flow may grow in transition to turbulent flow by "stealing energy from the mean flow". We will return to this topic in the advanced companion book.

> **Problem 10.29.** (a) Verify the computations in this example and plot the solution. (b) Determine approximately the "width" of the initial transient. Hint: using a plot, explain why the point of inflection of $|u(t)|^2$ considered as a function of t is related to the length of the transient period, then compute the point(s) of inflection. Extend the considerations to $A = \begin{pmatrix} \mu & -1 \\ \epsilon & \mu \end{pmatrix}$ where ϵ is small positive, and study the relation to the limit case with $\epsilon = 0$.

10.6.   Galerkin finite element methods

The finite element methods for the scalar problem (9.3) naturally extend to the system (10.1). We let $W_k^{(q)}$ denote the space of piecewise polynomial vector functions of degree q with respect to a partition $\mathcal{T}_k = \{t_n\}$ of $[0, T]$ with $t_N = T$, that is $v \in W_k^{(q)}$ if $v|_{I_n} \in \mathcal{P}^q(I_n)^d$ for each interval I_n. Similarly, $V_k^{(q)}$ denotes the space of continuous piecewise polynomials vector functions of degree q on $\mathcal{T}_k$. As in the scalar case, we use $v_n^\pm = \lim_{s \to \pm 0} v(t_n + s)$ to denote the two values of v in $W_k^{(q)}$ at the time node t_n and $[v_n] = v_n^+ - v_n^-$ to denote the "jump" in value at the node.

The discontinuous Galerkin method of degree q (dG(q)) method reads: compute $U \in W_k^{(q)}$ such that

$$\sum_{n=1}^{N} \int_{t_{n-1}}^{t_n} (\dot{U} + AU, v)\, dt + \sum_{n=1}^{N} ([U_{n-1}], v_{n-1}^+) = \sum_{n=1}^{N} \int_{t_{n-1}}^{t_n} (f, v)\, dt$$

$$\text{for all } v \in W_k^{(q)}, \quad (10.30)$$

or equivalently for $n = 1, ..., N$,

$$\int_{t_{n-1}}^{t_n} (\dot{U} + AU, v)\, dt + ([U_{n-1}], v_{n-1}^+) = \int_{t_{n-1}}^{t_n} (f, v)\, dt$$

$$\text{for all } v \in \mathcal{P}^q(I_n)^d, \quad (10.31)$$

where $U_0^- = u_0$. In particular, the dG(0) method takes the form: for $n = 1, ..., N$, find $U_n \in \mathcal{P}^0(I_n)^d$ such that

$$U_n + \int_{t_{n-1}}^{t_n} AU_n \, dt = U_{n-1} + \int_{t_{n-1}}^{t_n} f \, dt,$$

where $U_n = U_n^-$ and $U_0 = u_0$.

The continuous Galerkin method of degree q (cG(q)) reads: compute $U \in V_k^{(q)}$ such that

$$\sum_{n=1}^{N} \int_{t_{n-1}}^{t_n} (\dot{U} + AU, v) \, dt = \sum_{n=1}^{N} \int_{t_{n-1}}^{t_n} (f, v) \, dt \quad \text{for all } v \in W_k^{(q-1)}, \tag{10.32}$$

where $U(0) = u_0$. In particular, the cG(1) method takes the form: find $U \in V_k^{(1)}$ such that for $n = 1, ..., N$

$$U(t_n) + \int_{t_{n-1}}^{t_n} A(t)U(t) \, dt = U(t_{n-1}) + \int_{t_{n-1}}^{t_n} f \, dt.$$

Problem 10.30. (a) Show that (10.31) is equivalent to (10.30).

Problem 10.31. Show that (10.31) and (10.32) specify unique approximations provided that $\int_{I_n} |A| \, dt$ is sufficiently small for $n \geq 1$.

Problem 10.32. (a) Find a change of variables that brings (10.6) into the form

$$\begin{cases} \dot{u}_1 + au_2 = f_1 & \text{for } t > 0, \\ \dot{u}_2 - au_1 = f_2 & \text{for } t > 0, \\ u_1(0) = u_{0,1}, \; u_2(0) = u_{0,2} \end{cases} \tag{10.33}$$

for a constant a. (b) Using $U_n = (U_{n,1}, U_{n,2})^\top$ to denote the value of U on I_n, show that dG(0) for this problem takes the form

$$\begin{cases} U_{n,1} + ak_n U_{n,2} = U_{n-1,1} + \int_{I_n} f_1 \, dt, \\ -ak_n U_{n,1} + U_{n,2} = U_{n-1,2} + \int_{I_n} f_2 \, dt. \end{cases}$$

(c) Show that the cG(1) with the nodal values U_n takes the form

$$\begin{cases} U_{n,1} + ak_n U_{n,2}/2 = U_{n-1,1} - ak_n U_{n-1,2}/2 + \int_{I_n} f_1 \, dt, \\ -ak_n U_{n,1}/2 + U_{n,2} = U_{n-1,2} + ak_n U_{n-1,1}/2 + \int_{I_n} f_2 \, dt. \end{cases}$$

Problem 10.33. *(Messy!)* Show that coefficients of the matrix in the equation $BU_n = U_{n-1}$ for the dG(0) approximation for Vinograd's example (10.20) are

$$B_{1,1} = 1 + \frac{1}{2}\big(\cos(12t_n) - \cos(12t_{n-1})\big)$$
$$+ \frac{3}{4}\big(\sin(t_n)\cos(t_n) - \sin(t_{n-1})\cos(t_{n-1})\big) + \frac{11}{2}k_n$$

$$B_{1,2} = \frac{3}{8}\big(\cos(12t_n) - \cos(12t_{n-1})\big)$$
$$- \big(\sin(t_n)\cos(t_n) - \sin(t_{n-1})\cos(t_{n-1})\big) - 6k_n$$

$$B_{2,1} = \frac{3}{8}\big(\cos(12t_n) - \cos(12t_{n-1})\big)$$
$$- \big(\sin(t_n)\cos(t_n) - \sin(t_{n-1})\cos(t_{n-1})\big) + 6k_n$$

$$B_{2,2} = 1 + \frac{1}{2}\big(\cos(12t_n) - \cos(12t_{n-1})\big)$$
$$- \frac{3}{4}\big(\sin(t_n)\cos(t_n) - \sin(t_{n-1})\cos(t_{n-1})\big) + \frac{11}{2}k_n.$$

Problem 10.34. (a) Compute equations for the dG(0) and cG(1) approximations for (10.8). (b) Do the same for (10.13) in the case of constant coefficients.

Problem 10.35. *(Messy!)* Compute equations for the dG(1) approximation for (10.6), (10.8), and (10.13) (with constant coefficients).

Problem 10.36. *(Messy!)* Repeat Problems 10.32- 10.35 using quadrature formula of the appropriate orders. When discretizing (10.13), treat the non-constant coefficient case.

10.7. Error analysis and adaptive error control

The a priori and a posteriori error estimates for the finite element methods for scalar problems extend directly to systems. We let $\mathcal{A} = |A|_{[0,t_N]}$, where as before $|v|_{[a,b]} = \max_{a \le t \le b} |v(t)|$. We first treat the general case, then we examine a parabolic problem in which there is essentially no accumulation of errors and a hyperbolic problem with linear growth in time.

10.7.1.   The general problem

The a priori error estimate for the dG(0) method in the general case is:

Theorem 10.1. *Assuming that $k_n A \leq 1/2$ for $n \geq 1$, the dG(0) approximation U satisfies*

$$|u(t_N) - U_N| \leq \frac{e}{2} \left(e^{2At_N} - 1\right) \max_{1 \leq n \leq N} k_n |\dot{u}|_{I_n}.$$
$$(10.34)$$

Problem 10.37.   Prove this theorem. Hint: the proof follows the scalar case very closely.

For the a posteriori result, we let φ solve the dual problem

$$\begin{cases} -\dot{\varphi} + A(t)^\top \varphi = 0 & \text{for } t_N > t \geq 0, \\ \varphi(t_N) = e_N, \end{cases} \qquad (10.35)$$

and define the stability factor by

$$S(t_N) = \frac{\int_0^{t_N} |\dot{\varphi}| dt}{|e_N|}.$$

Note the transpose on the coefficient matrix. The a posteriori result for dG(0) in the general case is:

Theorem 10.2. *The dG(0) approximation U satisfies for $N = 1, 2, ..$,*

$$|u(t_N) - U_N| \leq S(t_N)|kR(U)|_{[0,t_N]},$$

where $k = k_n$ for $t_{n-1} < t \leq t_n$ and

$$R(U) = \frac{|U_n - U_{n-1}|}{k_n} + |f - aU_n| \quad \text{for } t_{n-1} < t \leq t_n.$$

If $|A(t)| \leq A$ for $0 \leq t \leq t_N$, then $S(t_N) \leq \exp(At_N)$.

Proof. While the proof follows the scalar case closely, we write out the details to show how the solution of the dual problem enters into the analysis in the present system case. By (10.35) we have trivially

$$|e_N|^2 = (e_N, e_N) = (e_N, e_N) + \sum_{n=1}^{N} \int_{t_{n-1}}^{t_n} (e, -\dot{\varphi} + A^\top \varphi) \, dt,$$

which, after integration by parts over each sub-interval I_n and using the definition of the transpose, gives

$$|e_N|^2 = \sum_{n=1}^{N} \int_{t_{n-1}}^{t_n} (\dot{e} + Ae, \varphi) \, dt + \sum_{n=1}^{N-1} ([e_n], \varphi_n^+) + (u_0 - U_0^+, \varphi_0^+).$$

Problem 10.38. Verify this formula.

As in the scalar case, using the Galerkin orthogonality (10.31) with v the L_2 projection $\pi_k \varphi$ into the piecewise constants $W_k^{(0)}$, this equation simplifies to the *error representation formula*:

$$|e_N|^2 = \sum_{n=1}^{N} \left(\int_{t_{n-1}}^{t_n} (f - AU, \varphi - \pi_k \varphi) \, dt - ([U_{n-1}], (\varphi - \pi_k \varphi)_{n-1}^+) \right).$$

Problem 10.39. Derive this equality.

We now use the following extension of the error estimate for the L_2 projection to vector functions:

$$|\varphi - \pi_k \varphi|_{I_n} \leq \int_{I_n} |\dot{\varphi}| \, dt,$$

and obtain

$$|e_N| \leq S(t_N) \max_{n=1,\ldots,N} \left(|[U_{n-1}]| + |k(f - AU)|_{I_n} \right),$$

and the proof is completed by estimating the stability factor. The following lemma states a bound in the general case.

∎

Lemma 10.3. *If $|A(t)| \leq \mathcal{A}$ for $0 \leq t \leq t_N$, then the solution φ of (9.19) satisfies*

$$|\varphi(t)| \leq e^{\mathcal{A}t} |e_N| \quad \text{for all } 0 < t < t_N, \tag{10.36}$$

and

$$S(t_N) \leq e^{\mathcal{A}t_N} - 1. \tag{10.37}$$

Proof. Taking the inner product of the differential equation in (10.35) with φ, we get

$$(\varphi, \dot\varphi) = (A\varphi, \varphi),$$

from which we obtain

$$\frac{d}{dt}|\varphi|^2 \le 2A|\varphi|^2.$$

Integrating this equation and taking a square root gives (10.36). From the differential equation in (10.35), we get

$$\int_0^{t_N} |\dot\varphi|\, dt \le A|e_N| \int_0^{t_N} |\varphi|\, dt,$$

which gives (10.37) after substituting using (10.36) and computing the resulting integral. ∎

Problem 10.40. State and prove a version of Theorem 9.7 that considers the additional effects of quadrature errors.

We base adaptive error control on Theorem 10.2 following the algorithm used for scalar problems. We saw that the exponential bound on the stability factor can be much too large for use in adaptive error control for a scalar problem and instead we compute an approximation of the stability factor during the course of a computation. In a system of equations, there is even more possibility for cancellation of errors because the error can change direction as different components interact.

We present a computation made on Vinograd's example (10.20) using the dG(1) method with an error tolerance of .05. In Fig. 10.5, we show the components of the approximate solution together with the stability factor $S(t)$. The slope of a line fitted using least squares to the logarithms of the time and stability factor is about .94, and there is a good fit, indicating that the errors indeed are accumulating at an exponential rate in this case.

10.7.2. A parabolic model problem

We now consider the parabolic model case with A symmetric and positive semi-definite. We may then apply the general result of Theorem 10.2 together with the following stability estimate:

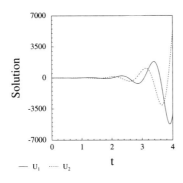 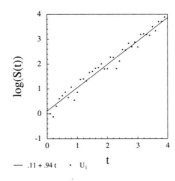

Figure 10.5: Results for a dG(1) approximation of the Vinograd example (10.20) with initial value $(-1\ 3)^\top$ computed with an error tolerance of .05. The plot on the left shows the two components of the numerical solution. The plot on the right shows a log-log plot of $S(t)$ versus time. A line fitted to the data using least squares is also drawn. The slope of this line is about .94

Lemma 10.4. *If A is symmetric positive semi-definite, then the solution φ of dual problem (9.19) satisfies*

$$|\varphi(t)| \le |e_N| \quad \text{for all } 0 \le t \le t_N, \tag{10.38}$$

and

$$S(t_N) \le |A|\tau + \frac{1}{2}\left(\log\left(\frac{t_N}{\tau}\right)\right)^{1/2}, \tag{10.39}$$

where $\tau = \min\{t_N, |A|^{-1}\}$. In particular $S(t_N) \le 1 + \frac{1}{2}(\log(t_N|A|))^{1/2}$ if $t_N \ge |A|^{-1}$.

This result says that $S(t_N)$ grows at most linearly with time for small times, i.e if $t_N < |A|^{-1}$, then $S(t_N) \le |A|t_N$. But for sufficiently large times, i.e. $t_N \ge |A|^{-1}$, then $S(t_N)$ grows at most at a logarithmic rate, which is very slow. This is the system generalization of the scalar parabolic estimate $S(t) \le 1$.

Proof. Changing the time variable and using the symmetry of A, the dual problem (10.35) can be written on the form $\dot\varphi + A\varphi = 0$ for $0 < t \le$

t_N with $\varphi(0) = e_N$. Applying the estimate (10.27) proves the desired result. ∎

Problem 10.41. Perform the indicated change of time variable.

We describe a computation on the atmosphere-ocean model (10.13) using constant parameters $\alpha = .1$ and $\beta_1 = \beta_2 = .01$, which amounts to assuming that the heat exchange between the water and the air is faster than that between water and land and between the air and space which are equal. We choose the forcing functions with parameters $S_1 = .005$, $P_1 = \pi/180$, $S_2 = .3$, and $P_2 = \pi$. With the time unit in days, this assumes that the temperature of the water varies bi-annually with the effects of the sun, while the temperature of the atmosphere is assumed to vary daily. We made a computation using the dG(1) method with an error tolerance of .018. We show the components of the approximation in the Fig. 10.6, and we can see that after a relatively short initial transient of a few days, the temperatures of the water and air settle to a regular pattern. The rapid daily effect of the sun on the air makes that plot appear almost solid as the temperature of the air oscillates on a frequency of a day. The slower effect of the sun on the water is visible in both components as the oscillation occurring at a frequency of approximately 360 days. In the plot on the left in Fig. 10.7, we show

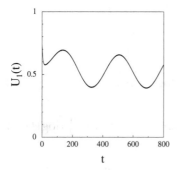 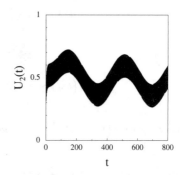

Figure 10.6: Results for a dG(1) approximation of the model of temperatures in the atmosphere and ocean (10.13) with initial value $(.75 \ .25)^\top$ computed with an error tolerance of .018. The two plots on top show the two components of the numerical solution.

the solution's components for the first fifty days. The initial transient and the daily oscillations are both visible. On the right, we plot the stability factor $S(t)$. Overall $S(t)$ is growly only slowly as time passes. In this problem, we thus can compute for a very long time without the accumulation of error ruining the accuracy.

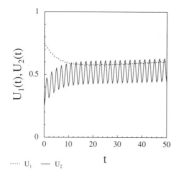 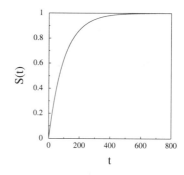

Figure 10.7: Results for a dG(1) approximation of the model of temperatures in the atmosphere and ocean (10.13) with initial value $(.75 \ .25)^\top$ computed with an error tolerance of .018. The plot on the left shows the two components of the numerical solution. The plot on the right shows a plot of $S(t)$ versus time.

It is possible to prove a more accurate a priori error bound in the parabolic case as well.

Problem 10.42. *(Hard!)* Prove the analog of the more precise priori error bound of Theorem 9.5 for (10.1) in the case A is symmetric positive semi-definite.

10.7.3. A hyperbolic model problem

We now consider the hyperbolic model problem model (10.33) with the normal matrix $A = [0 \ a, 0 \ -a]$ with purely imaginary eigenvalues $\pm a$. An important generalization of this problem is the wave equation, which we discuss in Chapter 17. We prove that the errors of the cG(1) method for (10.33) accumulate at most at a linear rate in time.

Lemma 10.5. *If $|a| \leq A$, then the cG(1) approximation of the solution of (10.33) satisfies*

$$|u(t_N) - U_N| \leq A \, t_N \, |kR|_{[0,t_N]}, \tag{10.40}$$

where the residual vector R is defined

$$R(t)|_{I_n} = \begin{pmatrix} \dot{U}_1(t) + aU_2(t) - f_1(t) \\ \dot{U}_2(t) - aU_1(t) - f_2(t) \end{pmatrix}.$$

Proof. The proof starts off in the same way as the proof of Theorem 10.2. Using Galerkin orthogonality, we obtain an error representation formula for the cG(1) approximation

$$|e_N|^2 = -\int_0^{t_N} (R, \varphi - \pi_k \varphi) \, dt, \tag{10.41}$$

where $\pi_k \varphi$ is the L_2 projection of φ into $W_k^{(0)}$, that is $\pi_k \varphi|_{I_n}$ is the average value of φ on I_n.

   Problem 10.43. Derive (10.41).

   The dual problem for (10.33) reads

$$\begin{cases} -\dot{\varphi}_1 - a\varphi_2 = 0 & \text{for } t_N > t \geq 0, \\ -\dot{\varphi}_2 + a\varphi_1 = 0 & \text{for } t_N > t \geq 0, \\ \varphi(t_N) = e_N. \end{cases} \tag{10.42}$$

To prove a stability estimate, we multiply the first two equations in (10.42) by φ_1 and φ_2, respectively. Adding them together, we obtain using that the terms $\pm a\varphi_1\varphi_2$ cancel,

$$\frac{d}{dt}(\varphi_1^2 + \varphi_2^2) \equiv 0.$$

Integrating, we obtain

$$|\varphi(t)| = |e_N| \quad \text{for } 0 \leq t \leq T,$$

and under the assumption on a, we conclude that

$$|\dot{\varphi}|_{[0,t_N]} \leq A|e_N|.$$

We now obtain (10.40) using the usual error bound for the error $\varphi - \pi_k \varphi$. ∎

We illustrate with computations on the problem:

$$\begin{cases} \dot{u}_1 + 2u_2 = \cos(\pi t/3), & t > 0, \\ \dot{u}_2 - 2u_1 = 0, & t > 0, \\ u_1(0) = 0, \quad u_2(0) = 1. \end{cases}$$

In Fig. 10.8, we first plot the two components of the cG(1) approximation computed to keep the error below .2. Note that the error control chooses the time steps to maintain accuracy in all the components of U simultaneously. We plot the stability factor as well, and the linear

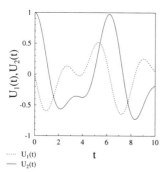
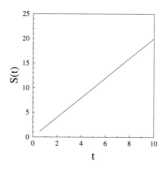

Figure 10.8: Results for the cG(1) approximation of (10.33) computed with error tolerance .2. The figure on the left shows the components of the solution and the figure on the right shows the approximation of the stability factor $S(t)$

growth is clearly demonstrated.

Problem 10.44. Derive error estimates, including stability estimates, for the dG(0) method for

$$\begin{cases} \dot{u}_1 + u_2 = f_1, & t > 0, \\ \dot{u}_2 = f_2, & t > 0, \\ u(0) = u_0. \end{cases}$$

10.8. The motion of a satellite

In the advanced companion volume, we will extend the scope to nonlinear differential equations modeling many-body systems, chemical reactions, fluid flow and wave propagation, among other phenomena. To illustrate some of the complications that can arise due to nonlinearity, we discuss a simple example of an n-body system. A classic problem of celestial mechanics is to determine the long time behavior of our solar system, which consists of the sun, the nine planets together with a variety of smaller bodies like moons, comets and asteroids. This is an example of an n-body system. A set of model equations for the motion of the bodies in an n-body system can be formulated readily using Newton's law of gravitation. However, the resulting nonlinear system of differential equations in general cannot be solved analytically if $n > 2$. Thus, to determine the fate of our solar system, we are left with approximate numerical solutions and the accuracy of computed solutions over long time becomes a major issue.

To illustrate, we consider the motion of a satellite under the influence of the gravitational force of the earth. This is called the restricted two-body problem when the heavy body is considered fixed. This problem was solved by Newton in the *Principia* in the case the motion takes place in a plane, and is one of the few problems of celestial mechanics that can be solved analytically. The analytical solution shows that the light body either moves in an elliptical orbit, if it is captured by the gravitational field of the heavy body, or otherwise along a parabolic or hyperbolic curve (see Chapter 11 below).

Letting $x(t) \in \mathbb{R}^3$ denote the coordinates of the satellite at time t, assuming that the center of the earth is at the origin, Newton's inverse square law of gravitation states that

$$\ddot{x} = -\frac{x}{|x|^3}, \tag{10.43}$$

where we chose the units to make the constant of proportionality one. Here we assumed the system acts as if all of the mass of the earth is located at its center of gravity, which is justifiable if the earth is approximately a homogeneous sphere (cf. Problem 10.8). We note that for nearly circular orbits with $|x| \approx 1$, the system (10.43) resembles the familiar equation $\ddot{x} + x = 0$. Introducing the new variable $u = (x_1, x_2, x_3, \dot{x}_1, \dot{x}_2, \dot{x}_3)$ the system changes into a first order system of the form

$$\begin{cases} \dot{u} + f(u) = 0 & \text{in } (0, T], \\ u(0) = u_0. \end{cases} \tag{10.44}$$

where the initial data u_0 is determined by initial position $x(0)$ and velocity $\dot{x}(0)$.

Problem 10.45. Determine f in (10.44).

We can directly apply the cG(q) and dG(q) methods to this problem. For instance, the dG(0) method takes the form

$$U_n + k_n f(U_n) = U_{n-1},$$

which gives a nonlinear system of equations to solve to determine $U_n \in \mathbb{R}^6$ from U_{n-1}. In Fig. 10.9, we present computations made with dG(1) for (10.44) over the time interval $[0, 1000]$ starting with the initial data $x(0) = (.4, 0, 0)$, $\dot{x}(0) = (0, 2, 0)$ in which case the satellite moves in the plane $x_3 = 0$.

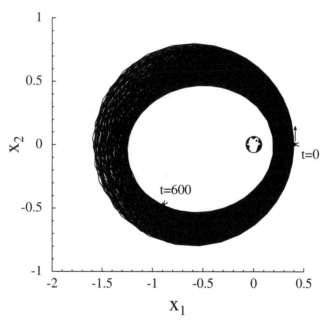

Figure 10.9: The orbit of a dG(1) solution of the two body problem. The heavy planet is located at the origin and the satellite begins at $(.4, 0)$.

Without further knowledge of the solution, from Fig. 10.9, we would conclude that the satellite slowly moves towards the earth and eventually enters the atmosphere and burns up. But to be confident in our prediction, we need to know something about the error of the numerical approximation. In fact, the exact analytical solution in this case is known to be a periodic elliptic orbit and the numerical results are misleading. The dG(1) method adds a kind of energy dissipation to the system that causes the satellite's orbit to decay. In more general n-body systems, we do not have a formula for the solution and this case it is critical to address the issue of error estimation if we want to make accurate predictions based on numerical results.

> More appealing than knowledge itself is the feeling of knowing. (D. Boorstin)

> I'll be here in this cell,
> till my body's just a shell,
> and my hair turns whiter than snow.
> I'll never see that love of mine,
> Lord, I'm in Georgia doin' time.
> And I hear that lonesome whistle blow. (H. Williams Sr.)

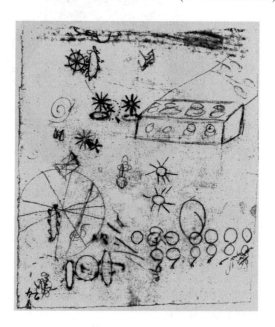

Figure 10.10: Design sketches for a calculator by Leibniz.

11

Calculus of Variations

Dans les modifications des mouvements, l'action devient ordinaire-
ment un Maximum ou un Minimum. (Leibniz)

The *calculus of variations* was created by Euler and Lagrange in
the 18th century to formulate problems in mechanics and physics as
minimization problems, or more generally, variational problems. A basic
minimization problem considered by Euler and Lagrange is to find a
function $u(x)$ that minimizes the integral

$$F(v) = \int_0^1 f(v, v') \, dx \qquad (11.1)$$

over all functions $v(x)$ defined on the interval $I = (0, 1)$ satisfying $v(0) = u_0$ and $v(1) = u_1$, where $f(v, w)$ is a given function of two variables,
and u_0 and u_1 are given boundary conditions. We refer to $F(v)$ as the
Lagrangian of the function v, and thus the problem is to minimize the
Lagrangian over some set of functions. It turns out that the solution
of the minimization problem also solves a differential equation that is
related to the Lagrangian. As an example, consider the model of the
elastic string with elastic modulus one loaded by a transversal weight
$g(x)$ derived in Chapter 8. The corresponding Lagrangian $F(v)$ defined
by (11.1) with

$$f(v, v') = \frac{1}{2}(v')^2 - gv, \qquad (11.2)$$

represents the total energy with contributions from the internal elastic
energy $\int_I \frac{1}{2}(v')^2 \, dx$ and the load potential $- \int_I gv \, dx$ of a deflection $v(x)$

satisfying $v(0) = v(1) = 0$. In Chapter 8, we saw that the minimizing function $u(x)$ satisfies the differential equation $-u'' = g$ in $(0,1)$ and of course the boundary conditions $u(0) = u(1) = 0$, which illustrates the basic connection between the calculus of variations and differential equations. This connection is particularly useful for computing numerical approximations of solutions and in fact, this book may be viewed as an extended application of the calculus of variations. Furthermore, questions about solutions of a differential equation, such as existence, are often approachable after reformulating the problem as a variational problem on which analysis might be easier. We use this approach in Chapter 21.

The calculus of variations was extended in the 19th century by Hamilton among others to more general variational principles. Hamilton's principle describes solutions to basic dynamical problems in mechanics and physics as stationary points (instead of minimum points) of a Lagrangian, or *action integral*, representing a time integral of the difference between the kinetic and potential energies.

We start by considering the minimization problem (11.1) and deriving the corresponding differential equation, referred to as the *Euler-Lagrange equation*, satisfied by a minimizing function. We then give a couple of basic examples. After that we discuss Hamiltonian formulations, including some basic examples.

Note the abuse of our standard notation: In this chapter, f does not represent a load as usual; rather the integrand in the integral to be minimized is denoted by $f(v, v')$, and the load is denoted by $g(x)$.

11.1. Minimization problems

We shall prove that the function $u(x)$ that minimizes the Lagrangian $F(v)$ given by (11.1) over all functions $v(x)$ satisfying $v(0) = u_0$ and $v(1) = u_1$, solves the differential equation

$$f_u(u, u') - \frac{d}{dx} f_{u'}(u, u') = 0 \quad \text{in } (0,1), \tag{11.3}$$

where we denote the partial derivatives of $f(v, w)$ with respect to v and w by f_v and f_w respectively. This is the Euler-Lagrange equation for the minimization problem (11.1). For example, in the case of the elastic string with $f(v, v')$ defined by (11.2), we have $f_v = -g$ and $f_{v'} = v'$ and the Euler-Lagrange equation takes the expected form $-g - u'' = 0$.

To prove the claim, we first observe that the minimization property of u implies that

$$F(u) \leq F(u + \epsilon v) \quad \text{for all } \epsilon \in \mathbb{R}, \tag{11.4}$$

for any function $v(x)$ satisfying $v(0) = v(1) = 0$ (note that $u(x) + \epsilon v(x)$ should satisfy the same boundary conditions as u). We conclude from (11.4) that the function $g(\epsilon) = F(u + \epsilon v)$ has a minimum for $\epsilon = 0$ and therefore $g'(0) = 0$ if the derivative $g'(\epsilon)$ exists for $\epsilon = 0$. We compute the derivative of $g(\epsilon)$ using the chain rule and set $\epsilon = 0$ to get

$$0 = g'(0) = \int_0^1 \left(f_u(u, u')v + f_{u'}(u, u')v' \right) dx \tag{11.5}$$

for all $v(x)$ with $v(0) = v(1) = 0$. Integrating by parts and using the fact the boundary terms vanish, we get

$$\int_0^1 \left(f_u(u, u') - \frac{d}{dx} f_{u'}(u, u') \right) v\, dx = 0,$$

for all $v(x)$ with $v(0) = v(1) = 0$. Under appropriate assumptions that guarantee that the integrand is continuous, by varying $v(x)$ we conclude that $u(x)$ must satisfy (11.3).

We refer to (11.5) as the *weak form* or *variational form* of the Euler-Lagrange equation and the corresponding differential equation (11.3) as the *strong form*. The weak form requires only first derivatives of the unknown u, while the strong form involves second derivatives of u. The two formulations are thus not completely equivalent because of the different regularity requirements; see the discussion in Section 8.1.2.

In the example above, the Lagrangian is *convex*, i.e.

$$F(\theta v + (1 - \theta)w) \leq \theta F(v) + (1 - \theta)F(w) \quad \text{for } 0 \leq \theta \leq 1,$$

because it is the sum of a quadratic and a linear function. A stationary point of a convex function is necessarily a global minimum point, but this does not have to be true for non-convex functions.

When can we expect a unique minimizing function to exist? Based on our experience in solving minimization problems in one-variable calculus, we expect a continuous function $F(x)$ defined on $\mathbb{R}$ to have a unique minimum if F is convex and $F(x) \to \infty$ as $x \to \pm\infty$. A corresponding result holds for the more general minimization problem (11.1). The Euler-Lagrange equation (11.3) results from the equation $\frac{d}{d\epsilon}F(u+\epsilon v)|_{\epsilon=0} = 0$ which can be expressed formally as $F'(u) = 0$ where F' is the "derivative" of the functional $F(v)$.

We now consider some examples. Further examples are given in Chapters 13 and 21.

The elastic bar

The Euler-Lagrange equation for the Lagrangian

$$F(v) = \int_0^1 \left(\frac{1}{2}a(v')^2 - fv\right) dx,$$

which represents the total energy of an elastic bar with modulus of elasticity a including contributions from the internal elastic energy and the load potential defined for functions $v(x)$ satisfying $v(0) = v(1) = 0$, is the differential equation (8.2) for the displacement u under the load f, i.e. $-(au')' = f$ in $(0,1)$, together with the boundary conditions $u(0) = 0$, $u(1) = 0$.

The weak form of the Euler-Lagrange equation reads: find the displacement $u(x)$ with $u(0) = u(1) = 0$ such that

$$\int_0^1 au'v' \, dx = \int_0^1 fv \, dx, \tag{11.6}$$

for all displacements v with $v(0) = v(1) = 0$. This is also referred to as the *principle of virtual work* in mechanics which states that the stress $\sigma = au'$ satisfies

$$\int_0^1 \sigma v' \, dx = \int_0^1 fv \, dx, \tag{11.7}$$

for all displacements v with $v(0) = v(1) = 0$. The principle states that for any displacement v the internal elastic "virtual work" on the left is equal to the external virtual work on the right. In general, we may think of the Euler-Lagrange equation as expressing a balance of external and internal forces.

The shortest distance between two points

The minimization problem of finding the curve $u(x)$ of minimal length connecting the two points $(0, u_0)$ and $(1, u_1)$ in $\mathbb{R}^2$ takes the form: minimize the integral

$$\int_0^1 \sqrt{1 + (v')^2} \, dx$$

over all functions $v(x)$ such that $v(0) = u_0$ and $v(1) = u_1$. The Euler-Lagrange equation is

$$\left(\frac{u'}{\sqrt{1 + (u')^2}}\right)' = 0 \quad \text{in } (0, 1),$$

together with the boundary conditions $u(0) = u_0$, and $u(1) = u_1$. Differentiating, we see that the Euler-Lagrange equation is satisfied if $u'' = 0$, and we conclude that the minimizing curve is a straight line, as expected.

The brachistochrone problem

The *brachistochrone* problem is to find the shape of a piece of wire connecting two points so that a bead that slides down the wire under the force of gravity, neglecting friction, does so in the least amount of time; see Fig. 11.1 for an illustration. It was first formulated by Johann Bernoulli and solved by Leibniz. To formulate the minimization problem, we denote in a (x, u) coordinate system with the u-axis downward, the two points by (x_0, u_0) and (x_1, u_1) with $x_1 > x_0$ and $u_1 > u_0$. We note that the total time of descent is given by

$$\int_0^{s_1} \frac{ds}{v} = \int_{x_0}^{x_1} \frac{\sqrt{1 + (u')^2}}{v} \, dx,$$

where (cf. Chapter 13) the element of curvelength is ds and s_1 is the total length of the wire, $v = v(x)$ is the absolute velocity of the bead at the point $(x, u(x))$, and the set of points $(x, u(x))$ with $x_0 \leq x \leq x_1$ defines the shape of the wire. Conservation of energy states (assuming the gravity constant is one) that $\frac{v^2(x)}{2} = u(x) - u_0$, which leads to the minimization problem: find the function $u(x)$ that minimizes

$$\frac{1}{\sqrt{2}} \int_{x_0}^{x_1} \frac{\sqrt{1 + (w')^2}}{\sqrt{w - u_0}} \, dx,$$

over all functions $w(x)$ such that $w(x_0) = u_0$ and $w(x_1) = u_1$.

The solution curve is given in parameterized form by $x = x_0 + a(\theta - \sin(\theta))$, $u = u_0 + a(1 - \cos(\theta))$, where $0 \leq \theta \leq \theta_1$ is the parameter and the constants a and θ_1 are determined by the condition that $u(x_1) = u_1$. The solution curve is a cycloid generated by the motion of a fixed point on the circumference of a circle that rolls on the positive side of the given line $u = u_0$.

Problem 11.1. Verify that this is the solution.

In Fig. 11.1, we show some copies of sketches made by Leibniz in working out the solution of the brachistochrone problem. His solution can be viewed as the first application of the finite element method with piecewise linear approximation. The objective of Leibniz' construction was to derive the differential equation for the brachistochrone problem through minimization over piecewise linear functions.

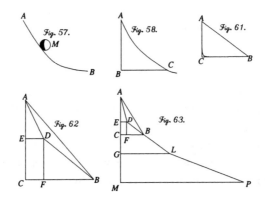

Figure 11.1: Some sketches made by Leibniz in his solution of the brachistochrone problem. Note the successive refinement involving piecewise linear functions.

11.2. Hamilton's principle

The Euler-Lagrange equation, in weak or strong form, states that the Lagrangian is stationary at a minimum point. Above, we considered convex Lagrangian functions for which a stationary point is necessarily the unique minimum point. Lagrangian functions which are not convex also play a fundamental role in modelling in mechanics and physics. *Hamilton's principle* states that the dynamics of certain physical systems may be characterized as stationary points of a Lagrangian function that represents the difference of the kinetic and potential energies. A stationary point of a Lagrangian $F(v)$ is a function u such that for all perturbations w, $\frac{d}{d\epsilon} F(u + \epsilon w) = 0$ at $\epsilon = 0$.

We now consider some basic example of applications of Hamilton's principle.

Mass-spring system

The equation $m\ddot{u} + ku = 0$ describing a system consisting of one mass m connected to a spring with spring constant k, see (10.3), is the Euler-Lagrange equation that results when Hamilton's principle is applied to the Lagrangian or action integral

$$F(v) = \int_{t_1}^{t_2} \left(\frac{m}{2} \dot{v}^2 - \frac{k}{2} v^2 \right) dt,$$

where the integrand is the difference between the kinetic and potential energies related to the displacement $v(x)$, and t_1 and t_2 are two arbitrary times, considering $v(t_1)$ and $v(t_2)$ to be fixed. To see how the equation $m\ddot{u} + ku = 0$ arises, we compute the derivative of $F(u + \epsilon w) = 0$ with respect to ϵ at $\epsilon = 0$, where $w(x)$ is a perturbation satisfying $w(t_1) = w(t_2) = 0$, to get

$$\int_{t_1}^{t_2} \left(m\dot{u}\dot{w} - kuw \right) dt,$$

which gives the desired equation after integration by parts and varying the function w.

The pendulum

We consider a pendulum of mass one attached to a string of unit length under the action of a vertical gravity force normalized to one. The action integral, which again is the difference between kinetic and potential energy, is given by

$$\int_{t_1}^{t_2} \left(\frac{1}{2} \dot{v}^2 - (1 - \cos(v)) \right) dt,$$

where $v(t)$ represents the angle measured from the lowest point of the pendulum range of motion. See Fig. 11.2. The corresponding Euler-

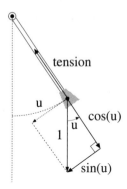

Figure 11.2: The notation for a model of a pendulum.

Lagrange equation is $\ddot{u} + \sin(u) = 0$, which is a nonlinear differential equation. The equation is complemented by initial conditions for the position u and velocity $\dot{u}$.

Problem 11.2. Supply the missing details in the derivation of the equation for the pendulum. If the angle u stays small during the motion, then the simpler *linearized* model $\ddot{u} + u = 0$ may be used. Solve this equation analytically and compare with numerical results for the nonlinear pendulum equation to determine limits of validity of the linear model.

The two-body problem

We return to the two-body problem considered in Chapter 10 of a satellite orbiting around the globe. We assume that the motion takes place in a plane and use polar coordinates (r, θ) to denote the position of the satellite measured from the orbit center. The action integral, representing the difference between kinetic and potential energy, is given by

$$\int_{t_1}^{t_2} \left(\frac{1}{2}\dot{r}^2 + \frac{1}{2}(\dot{\theta}r)^2 + \frac{1}{r} \right) dt \tag{11.8}$$

because the velocity is $(\dot{r}, r\dot{\theta})$ in the radial and angular directions respectively, and the gravity potential is $-r^{-1} = -\int_r^\infty s^{-2}\, ds$ corresponding to the work needed to move a particle of unit mass a distance r from the orbit center to infinity. The corresponding Euler-Lagrange equations are

$$\begin{cases} \ddot{r} - r\dot{\theta}^2 = -\frac{1}{r^2}, & t > 0, \\ \frac{d}{dt}(r^2\dot{\theta}) = 0, & t > 0, \end{cases} \qquad (11.9)$$

which is a second order system to be complemented with initial values for position and velocity.

We construct the analytical solution of this system in the following set of problems, which may be viewed as a short course on Newton's Principia. We hope the reader will take this opportunity of getting on speaking terms with Newton himself.

Problem 11.3. Prove that the stationary point of the action integral (11.8) satisfies (11.9).

Problem 11.4. Prove that the total energy is constant in time.

Problem 11.5. Introducing the change of variables $u = r^{-1}$, show that $\dot{\theta} = cu^2$ for c constant. Use this relation together with the fact that the chain rule implies that

$$\frac{dr}{dt} = \frac{dr}{du}\frac{du}{d\theta}\frac{d\theta}{dt} = -c\frac{du}{d\theta} \quad \text{and} \quad \ddot{r} = -c^2 u^2 \frac{d^2 u}{d\theta^2}$$

to rewrite the system (11.9) as

$$\frac{d^2 u}{d\theta^2} + u = c^{-2}. \qquad (11.10)$$

Show that the general solution of (11.10) is

$$u = \frac{1}{r} = \gamma \cos(\theta - \alpha) + c^{-2},$$

where γ and α are constants.

Problem 11.6. Prove that the solution is either an ellipse, parabola, or hyperbola. Hint: Use the fact that these curves can be described as the loci of points for which the ratio of the distance to a fixed point and to a fixed straight line, is constant. Polar coordinates are suitable for expressing this relation.

Problem 11.7. Prove Kepler's three laws for planetary motion using the experience from the previous problem.

What we know is very slight, what we don't know is immense. (Laplace)

12

Computational Mathematical Modeling

> Quand on envisage les divers poblèmes de Calcul Intégral qui se posent naturellement lorsqu'on veut approfondir les parties les plus différentes de la Physique, il est impossible de n'etre pas frappé par des analogies que tous ce problèmes présentent entre eux. Qu'il s'agisse de l'electricité ou de l'hydrodynamique, on est toujours conduit à des équations différentielles de même famille et les conditions aux limites, quoique différentes, ne sont pas pourtant sans offrir quelques resemblances. (Poincaré)

> Was mann mit Fehlerkontrolle nicht berechnen kann, darüber muss man schweigen. ("Wittgenstein")

In this chapter, we discuss the basic issues of *computational mathematical modeling* based on differential equations and computation. We have already met most of these issues and we now seek to formulate a framework. This book and the advanced companion volume give body to this framework.

We consider a mathematical model of the generic form

$$A(u) = f \tag{12.1}$$

where A describes a mathematical model in the form of a differential operator with certain coefficients and boundary conditions, f represents given data, and u is the unknown solution. As formulated, this represents a typical *forward* problem, where we assume that the model A and the data f are given and we seek the unknown solution u. In an *inverse* problem, we would try to determine the model A, for example the coefficients of the differential operator such as heat conductivity or modulus of elasticity, or the right-hand side f, from data corresponding to solutions u that is obtained from measurements for example. We focus in this volume on forward problems and discuss inverse problems in the advanced companion book.

Sources of error in mathematical modeling

There are three basic sources contributing to the total error in computational mathematical modeling of physical phenomena:

- *modeling* (A)

- *data imprecision* (f)

- *numerical computation* (u).

We refer to the effect on the solution from these sources as the error from modeling or *modeling error*, the error from data or *data error*, and the error from computation or *computational error*. Modeling errors result from approximations in the mathematical description of real phenomena, for example in the coefficients of the differential operator A related to material properties such as heat capacity. Data errors come about because data related to initial or boundary conditions or forcing terms like f often cannot be measured or determined exactly. Finally, the computational error results from the numerical computation of the solution u. The total error thus has contributions from all these three.

Sources of computational error

In the finite element method, the total computational error also has contributions from three sources:

- *the Galerkin discretization of the differential equation*

- *quadrature errors arising from the construction of the discrete equations*

- *numerical solution of the discrete equations.*

The Galerkin discretization error arises from using Galerkin's method to compute an approximation of the true solution in a finite dimensional function space like piecewise polynomials, assuming that the integrals that occur in the Galerkin formulation are evaluated exactly and the resulting discrete equations are solved exactly. The quadrature error comes from evaluating the integrals arising in the Galerkin formulation using numerical quadrature. Finally, the discrete solution error is the error resulting from solving the discrete system of equations only approximately. It is important to distinguish the three sources of error because the different errors propagate and accumulate differently.

For now, we focus on the Galerkin discretization error, leaving quadrature and discrete solution errors mostly to the advanced companion volume. However, we do discuss quadrature error briefly in Chapter 9 and discrete solution error in Chapter 7.

The goal of computational mathematical modeling

A natural goal is to seek to control the total error in some given norm to some given tolerance with minimal total computational work. We may formulate this goal as a combination of

- *reliability*

- *efficiency.*

Reliability means that the total error is controlled in a given norm on a given tolerance level with some degree of security; for instance, the total error is guaranteed to be within 1 percent at every point in space and time with a probability of 95 percent. Efficiency means that the total work to achieve this error control is essentially as small as possible. In this volume, we mainly focus on the computational error, and mostly leave modeling and data errors to the advanced volume. However, we do give an example of control of modeling and data errors in Chapter 8.

Thus we formulate our goal as reliable and efficient control of the computational error. We have seen that realizing this goal in general requires adaptive computational methods. Typically, the adaptivity concerns the underlying mesh in space and time.

Adaptive methods

An *adaptive method* consists of a computational method together with an *adaptive algorithm*. An adaptive algorithm consists of

- a *stopping criterion* guaranteeing control of the computational error on a given tolerance level

- a *modification strategy* in case the stopping criterion is not satisfied.

The goal of the adaptive algorithm is to determine through computation the "right" mesh satisfying the reliability and efficiency criteria. In practice, this usually results from an iterative process, where in each step an approximate solution is computed on a given mesh. If the stopping criterion is satisfied, then the mesh and the corresponding approximate solution are accepted. If the stopping criterion is not satisfied, then a new mesh is determined through the modification strategy and the process is continued. To start the procedure, a first mesh is required. Central in this process is the idea of *feedback* in the computational process. Information concerning the nature of the underlying problem and exact solution is successively revealed through computation, and this information is used to determine the right mesh.

Adaptive methods are based on error estimates. We now turn to error estimates focusing on estimates for the computational error.

Error estimates for Galerkin discretization errors

We sum up so far: our general objective is to control the total error using a reliable and efficient adaptive method. Here we have narrowed down our goal to adaptive control of the Galerkin discretization error and we now turn to error estimates for this error. Analogous error estimates and related adaptive error control for modeling, data, quadrature and discrete solution errors, are presented in detail in the advanced companion volume.

The basic concepts underlying error estimates are

- *stability*

- *accuracy*.

Accuracy refers to the effect of discretization in localized regions in space and time. Stability measures how these effects or perturbations are propagated and accumulated to add up to the Galerkin discretization error.

Different numerical methods generally have different accuracy and stability properties, and a "good" numerical method for a given differential equation is a method which combines "good" local accuracy with "good" stability, which means that the stability properties of the discrete equation appropriately reflect those of the differential equation. We saw an example of this in Chapter 9 in a comparison of the dG(0) and cG(1) methods for a parabolic model problem.

Galerkin discretization error estimates come in two forms:

- *a priori error estimates*

- *a posteriori error estimates.*

An a priori error estimate measures the Galerkin discretization error in terms of the error in direct interpolation of the exact solution and the stability properties of the discretized differential equation or discrete equation. The basic form of an a priori error estimate is

error $\propto$ stability factor for the discrete equation $\times$ interpolation error.

The stability factor for the discrete equation, or the *discrete stability factor*, measures the perturbation growth through a stability estimate for a dual discrete equation. The interpolation error involves derivatives of the exact solution and thus represents quantities that are not known. If we have some information on these derivatives, then we can estimate the interpolation error (a priori) without solving the discrete equations. To give the a priori error error estimate a quantitative meaning, the discrete stability factor has to be estimated. This can sometimes be done analytically. The alternative is to estimate this factor by solving the discrete dual problem.

An a posteriori error estimate measures the Galerkin discretization error in terms of the residual of the computed approximate solution and stability properties of the differential equation. The basic form of an a posteriori error estimate is

error $\propto$ stability factor for the differential equation $\times$ residual error.

The residual error can be evaluated (a posteriori) once the approximate solution has been computed. The stability factor for the differential equation, or the *continuous stability factor*, measures the growth of perturbations through a stability estimate for a dual differential equation. To give the a posteriori error estimate a quantitative meaning, the corresponding continuous stability factor has to be estimated. This may be done analytically in special cases and in general computationally by solving the dual problem numerically. The reliability of the adaptive method is directly related to the reliability in the estimate of the continuous stability factor. In the advanced companion volume, we discuss in considerable detail how to reliably compute continuous stability factors in a variety of contexts.

The stopping criterion of an adaptive method is usually based directly on an a posteriori error estimate. The modification criterion may be based on both a priori and a posteriori error estimates. The interpolation error in the a priori error estimate may then be estimated replacing the required derivatives of the exact solution by computed approximations.

Different errors are connected to different stability factors measuring the accumulation of the specific error. The orthogonality property of Galerkin methods couples naturally to a certain stability concept measuring derivatives of solutions of dual problems in terms of given data. This is referred to as *strong stability*. The orthogonality property of Galerkin perturbations sometimes cause the Galerkin discretization error to accumulate more favorably than, for example, quadrature errors. In general each specific error naturally couples to a specific stability concept and stability factor. We saw examples of this in Chapters 8, 9 and 10.

If stability factors are large, then perturbations are amplified considerably and more computational work, and greater accuracy in modelling and data, is needed to reach a certain tolerance for the total error. The Lorenz system (see the advanced companion volume for a detailed discussion), is an example of a problem where stability factors quickly grow with time, which reflects a strong sensitivity to initial data (the butterfly effect) and makes prediction difficult over long time intervals (limits of weather prediction).

> There is nothing without reason, there is no cause without effect.
> (Leibniz)

> Yesterday, when weary with writing, I was called to supper, and a salad I had asked for was set before me. "It seems then," I said,

"if pewter dishes, leaves of lettuce, grains of salt, drops of water, vinegar, oil and slices of eggs had been flying about in the air from all eternity, it might at last happen by chance that there would come a salad." "Yes," responded my lovely, "but not so nice as this one of mine." (Kepler)

The investigation of nature is an infinite pasture-ground, where all may graze, and where the more bite, the longer the grass grows, the sweeter is its flavor, and the more it nourishes. (T. Huxley)

The discovery of nature, of the ways of the planets, and plants and animals, require first the conquest of common sense. Science would advance, not by authenticating everyday experience but by grasping paradox, adventuring into the unknown. (D. Boorstin)

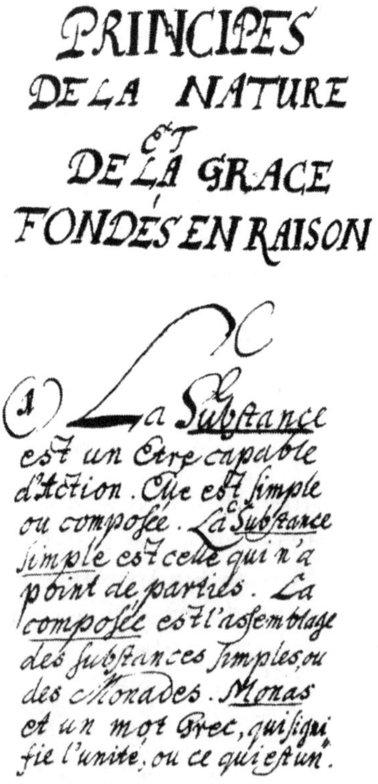
PRINCIPES
DE LA NATURE
ET
DE LA GRACE
FONDÉS EN RAISON

(1) La Subſtance est un Etre capable d'Action. Elle est ſimple ou compoſée. La Subſtance ſimple est celle qui n'a point de parties. La compoſée est l'aſſemblage des ſubſtances ſimples, ou des Monades. Monas est un mot Grec, qui ſignifie l'unité, ou ce qui eſt un.

Figure 12.1: From a summary of Leibniz's philosophy dedicated to Prince Eugen von Savoyen.

Part III

Problems in several dimensions

In the last part of this book, we extend the scope to the real world of two and three dimensions. We start by recalling some basic results from calculus of several variables including some facts about piecewise polynomial approximation. We then consider the basic types of linear partial differential equations including the elliptic Poisson equation, the parabolic heat equation, the hyperbolic wave equation, and the mixed parabolic/elliptic-hyperbolic convection-diffusion equation. We also consider eigenvalue problems and conclude with an abstract development of the finite element method for elliptic problems.

13

Calculus of Several Variables

> Music is the pleasure the human soul experiences from counting
> without being aware that it is counting. (Leibniz)

In this chapter, we review some basic material from the calculus of
several variables or calculus in several dimensions including the basic
differential operators of the gradient, divergence, rotation and Laplacian,
and integration together with the basic theorems of Green and Stokes.
We then illustrate the use of this material by deriving a model for non-
stationary and stationary heat flow, with special cases including the
heat equation and the Poisson equation. Finally, we discuss calculus of
variations in several variables briefly and conclude by extending some of
the one-dimensional models met in the first two parts of the book.

The extension of calculus into two and three dimensions began in
the early 18th century. Partial derivatives were used by James Bernoulli
in his work on isoperimetric problems and by his brother Nicholaus in
a paper on orthogonal trajectories in Acta Eruditorum from 1720, but
the more systematic study of partial derivatives was initiated by Euler, Clairaut and d'Alembert. Multiple integrals in principle were involved already in the *Principia* in discussions on the gravitational fields
of spherical masses, and were used explicitly by Euler for similar purposes.

13.1. Functions

We say that $u : \mathbb{R}^n \to \mathbb{R}^m$ is a *function* from $\mathbb{R}^n$ into $\mathbb{R}^m$, where
$n, m \geq 1$, if for each vector $x \in \mathbb{R}^n$ there is a unique associated vec-

tor $u(x) \in \mathbb{R}^m$. If $m = 1$, then $u(x) \in \mathbb{R}$ is a *scalar function*, and if $m > 1$, then $u(x)$ is a *vector function* or *vector field* with components $u(x) = (u_1(x), ..., u_m(x))^\top \in \mathbb{R}^m$. We use the same notation for functions defined on subsets of $\mathbb{R}^n$. In particular, $u : (a, b) \to \mathbb{R}^m$, where $(a, b) \subset \mathbb{R}$ is an interval, represents a *curve* in $\mathbb{R}^m$ and $u : \Omega \to \mathbb{R}^m$, where $\Omega \subset \mathbb{R}^2$, represents a *surface*, see Fig. 13.1. In particular, in the case $m = 3$, a surface is given by the *graph* of $u : \Omega \to \mathbb{R}$, that is the set of points $(x_1, x_2, u(x_1, x_2)) \in \mathbb{R}^3$ with $(x_1, x_2) \in \Omega$. The notion of limit

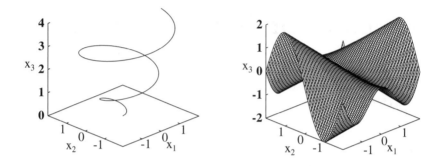

Figure 13.1: On the left, we plot the curve $(t^{1/2} \cos(\pi t),\ t^{1/2} \sin(\pi t),\ t)^\top$, and on the right, we plot the surface $x_3 = x_1 \sin\left((x_1 + x_2)\pi/2\right)$.

and continuity for a function $u : \mathbb{R} \to \mathbb{R}$ directly extend to a vector function $u : \mathbb{R}^n \to \mathbb{R}^m$ after replacing the absolute value by the Euclidean norm. The notions of a sequence of functions and uniform convergence extend similarly. If u is continuous on a closed and bounded subset of $\mathbb{R}^n$, then u attains a maximum and minimum and u is uniformly continuous on the subset. A subset S of $\mathbb{R}^n$ is closed if any point not in S has a positive distance to S. A closed set "contains its boundary". For example, $\{x \in \mathbb{R}^n : |x| \leq 1\}$ is closed.

Problem 13.1. Let $u : \mathbb{R}^2 \to \mathbb{R}$ be defined by $u(x) = x_1 x_2 / |x|^\alpha$ for $x \neq 0$. Prove that $\lim_{x \to 0} u(x)$ exists if and only if $\alpha < 2$.

13.2. Partial derivatives

For a scalar function $u : \mathbb{R}^2 \to \mathbb{R}$, that is a function $u(x) \in \mathbb{R}$ defined for $x = (x_1, x_2) \in \mathbb{R}^2$, the *partial derivative* $\frac{\partial u}{\partial x_1}$ of u with respect to x_1 is the function whose value at $(\bar{x}_1, \bar{x}_2)$ is given by

$$\frac{\partial u}{\partial x_1}(\bar{x}_1, \bar{x}_2) = \lim_{y_1 \to \bar{x}_1} \frac{u(y_1, \bar{x}_2) - u(\bar{x}_1, \bar{x}_2)}{y_1 - \bar{x}_1},$$

where the second coordinate $\bar{x}_2$ is held fixed. The partial derivative $\partial u / \partial x_2$ is defined analogously by holding the first coordinate fixed. Higher order derivatives are defined by repeated first order derivation so that for example

$$\frac{\partial^2 u}{\partial x_1 \partial x_2} = \frac{\partial}{\partial x_1}\left(\frac{\partial u}{\partial x_2}\right) = \frac{\partial}{\partial x_2}\left(\frac{\partial u}{\partial x_1}\right) \tag{13.1}$$

is a second order partial derivative, where the order of application of the first order derivatives is immaterial if the second order partial derivatives turn out to be continuous. This was proved by Euler in 1734. Partial derivatives of vector valued functions are defined componentwise.

Problem 13.2. (a) Compute the partial derivatives up to second order of the function $u : \mathbb{R}^3 \to \mathbb{R}$ defined by $u(x) = |x|^\alpha$ for $x \neq 0$, with $\alpha = 2$ and $\alpha = -1$. (b) Prove (13.1) under the stated condition. Hint: interpret the expression $u(y_1, y_2) - u(x_1, y_2) - u(y_1, x_2) + u(x_1, x_2)$ as the difference of two differences in two different ways, divide by $(y_1 - x_1)(y_2 - x_2)$, and let $y_i \to x_i$.

13.2.1. The gradient, divergence, rotation, and Laplacian

The *gradient* of a function $u : \mathbb{R}^2 \to \mathbb{R}$, denoted grad u or ∇u, is the vector function formed by the set of first order partial derivatives of u, i.e.

$$\nabla u = \left(\frac{\partial u}{\partial x_1}, \frac{\partial u}{\partial x_2}\right)^\mathsf{T}.$$

The *divergence* of a function $u : \mathbb{R}^2 \to \mathbb{R}^2$, denoted div u or $\nabla \cdot u$, is the scalar function defined by

$$\text{div } u = \nabla \cdot u = \frac{\partial u_1}{\partial x_1} + \frac{\partial u_2}{\partial x_2}.$$

Next, we recall the *rotation* of a function $u(x_1, x_2)$, denoted by rot u or curl u, which has a different definition depending on whether u is a scalar function or vector function. If $u = (u_1, u_2)^\top$ is a vector function, then rot u is the scalar function

$$\text{rot } u = \frac{\partial u_2}{\partial x_1} - \frac{\partial u_1}{\partial x_2},$$

while if u is a scalar, then rot u is the vector function

$$\text{rot } u = \left(\frac{\partial u}{\partial x_2}, -\frac{\partial u}{\partial x_1} \right)^\top.$$

The following identities follow directly from these definitions

$$\text{div rot } u = 0 \quad \text{and} \quad \text{rot grad } u = 0. \tag{13.2}$$

Finally, the *Laplacian* Δu of a function $u : \mathbb{R}^2 \to \mathbb{R}$ is defined by

$$\Delta u = \nabla \cdot (\nabla u) = \text{div grad } u = \frac{\partial^2 u}{\partial x_1^2} + \frac{\partial^2 u}{\partial x_2^2},$$

where $\frac{\partial^2 u}{\partial x_i^2} = \frac{\partial}{\partial x_i}\left(\frac{\partial u}{\partial x_i}\right)$.

Problem 13.3. Show that in two dimensions rot rot u $= -\Delta u$ with u scalar.

Problem 13.4. Prove that $\Delta u(x) = 0$ for $x \neq 0$ if $u : \mathbb{R}^2 \to \mathbb{R}$ is given by $u(x) = \log(|x|^{-1})$.

The definitions of the gradient and the Laplacian extend to functions $u(x)$ of $x \in \mathbb{R}^3$. For a vector function $u : \mathbb{R}^3 \to \mathbb{R}^3$, the divergence div $u = \nabla \cdot u$ is a scalar function defined by

$$\text{div } u = \nabla \cdot u = \sum_{i=1}^{3} \frac{\partial u_i}{\partial x_i},$$

and rot $u = \nabla \times u$ is the vector function

$$\text{rot } u = \nabla \times u = \left(\frac{\partial u_3}{\partial x_2} - \frac{\partial u_2}{\partial x_3}, \frac{\partial u_1}{\partial x_3} - \frac{\partial u_3}{\partial x_1}, \frac{\partial u_2}{\partial x_1} - \frac{\partial u_1}{\partial x_2} \right)^\top.$$

Problem 13.5. Prove (13.2) for functions u defined in $\mathbb{R}^3$.

Problem 13.6. Show that the definition of $\text{rot}\,u$ for $u : \mathbb{R}^3 \to \mathbb{R}^3$ in the cases that $u = (u_1, u_2, 0)$ and $u = (0, 0, u_3)$, where the functions are independent of x_3, reduces to the above definitions of the rotation in two dimensions with the obvious identifications.

Problem 13.7. Prove that in three dimensions $\nabla \times (\nabla \times u) = -\Delta u + \nabla(\nabla \cdot u)$, where $\Delta u = (\Delta u_1, \Delta u_2, \Delta u_3)^\top$.

13.2.2. The chain rule and the gradient

Let Γ be a curve in $\mathbb{R}^2$ defined by $s : (a, b) \to \mathbb{R}^2$. The tangent to Γ at the point $s(t)$ is the line through $s(t)$ in the direction of $s'(t) = (s_1'(t), s_2'(t))^\top$, where the prime indicates differentiation with respect to t. The chain rule states that if $s : (a, b) \to \mathbb{R}^2$ and $u : \mathbb{R}^2 \to \mathbb{R}$, where ∇u is continuous in $\mathbb{R}^2$ and s is differentiable on (a, b), then the composite function $v(t) = u(s(t))$ satisfies for $t \in (a, b)$,

$$v'(t) = \frac{d}{dt} u(s(t)) = \nabla u(s(t)) \cdot s'(t). \tag{13.3}$$

In particular, if $s(t) = \bar{x} + tn$, where $n \in \mathbb{R}^2$ is a given vector of length one, then $\partial_n u(\bar{x}) = \nabla u(\bar{x}) \cdot n = v'(0)$ is the derivative of u at $\bar{x}$ in the direction n.

Problem 13.8. Prove the chain rule (13.3). Hint: write

$$u(s_1(\tau), s_2(\tau)) - u(s_1(t), s_2(t)) = u(s_1(\tau), s_2(\tau)) - u(s_1(t), s_2(\tau))$$
$$+ u(s_1(t), s_2(\tau)) - u(s_1(t), s_2(t)),$$

use the mean value theorem, divide by $\tau - t$ and let τ tend to t .

Problem 13.9. Let $u : \mathbb{R}^n \to \mathbb{R}$ have a local minimum at $y \in \mathbb{R}^n$, that is $u(y) \leq u(x)$ for all x satisfying $|x - y| \leq \delta$ for some $\delta > 0$. Prove that if the gradient $\nabla u(x)$ is continuous, then $\nabla u(y) = 0$.

13.2.3. Surface plots and level curves

Plotting the surface given by the graph of a function $u : \mathbb{R}^2 \to \mathbb{R}$ seen from one point of view may not be sufficient to represent the function completely, because the surface itself can hide, or "shadow", features.

This can be seen both in Fig. 13.1 and Fig. 13.2. As an alternative, it is convenient to use a plot of *level curves*, or *isolines*. A level curve of a function $u : \mathbb{R}^2 \to \mathbb{R}$ is a set of points in $\mathbb{R}^2$ on which u takes on a particular value. A contour plot is made by drawing a set of level curves computed by specifying regularly spaced values for the function. We illustrate this in Fig. 13.2. We can measure the change in function

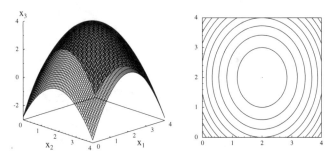

Figure 13.2: Plots of a graph of a function and the corresponding level curves of corresponding to specifying function values every .7 units starting at the maximum height of 4.

value (or height of the graph) between two points in a contour plot by counting the number of level curves intersected by a line that joins the two points. This is useful on a hiking map for example.

More precisely, we define the level curve of a function $u : \mathbb{R}^2 \to \mathbb{R}$ is a curve $s = (s_1, s_2) : (a, b) \to \mathbb{R}^2$ such that

$$u(s_1(t), s_2(t)) = c \quad \text{for } t \in (a, b), \tag{13.4}$$

where c is a constant. The local existence of a level curve is guaranteed by the following *implicit function theorem*.

Theorem 13.1. *Assume $u : \mathbb{R}^2 \to \mathbb{R}$ has continuous partial derivatives and $u(0, 0) = 0$. If $\partial u / \partial x_2(0, 0) \neq 0$, then there is a $\delta > 0$ such that for a given x_1 with $|x_1| < \delta$, the equation $u(x_1, x_2) = 0$ has a unique solution $x_2 = x_2(x_1)$, which defines x_2 as a function of x_1 for $|x_1| < \delta$.*

Problem 13.10. Use theorem 13.1 to prove that if $u(\xi_1, \xi_2) = c$ for some (ξ_1, ξ_2) and $\partial u / \partial x_2(\xi_1, \xi_2) \neq 0$ (or more generally $\nabla u(\xi_1, \xi_2) \neq 0$), then there is a level curve (13.4) near (ξ_1, ξ_2).

Problem 13.11. Prove Theorem 13.1. Hint: for a given x_1, consider the fixed point iteration, see Problem 13.14, $x_{2j} = x_{2j-1} - \alpha u(x_1, x_{2j-1})$, where α is chosen suitably to make the iteration converge.

Differentiating both sides of (13.4), we get by the chain rule

$$\frac{d}{dt}u(s(t)) = \nabla u(s(t)) \cdot s'(t) = 0.$$

Since $s'(t)$ is tangent to the level curve at $s(t)$, this means that level curves of a function are orthogonal to the gradient of the function. Recall that the gradient points in the direction of the steepest change in the function. Therefore, the direction perpendicular to the gradient is the direction in which u does not change value, which is a tangent to the corresponding level curve, see Fig. 13.3.

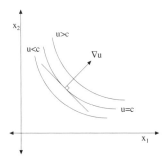

Figure 13.3: The gradient of u at a point is perpendicular to the level curve of u at the point.

Problem 13.12. Verify this observation for $u(x_1, x_2) = x_1^2 + 4x_2^2$.

Problem 13.13. Similarly for a function $u : \mathbb{R}^3 \to \mathbb{R}$, we define a level surface as a set of points x for which $u(x)$ is equal to some constant. (a) Show that if $\partial u/\partial x_3(\xi) \neq 0$, then the level surface through ξ may in neighborhood of ξ be expressed as a function $x_3 = \varphi(x_1, x_2)$ satisfying $u(x_1, x_2, \varphi(x_1, x_2)) = u(\xi)$. (b) Show that the gradient of u at a point is perpendicular to the level surface through the point.

13.2.4. Taylor's theorem

Taylor's theorem for a function $u : \mathbb{R}^2 \to \mathbb{R}$ with second order remainder reads

Theorem 13.2. *Assume u has continuous second derivatives near the point* $a = (a_1, a_2)$. *Then for all* $x = (x_1, x_2)$ *near* a,

$$u(x) = u(a) + \frac{\partial u}{\partial x_1}(a)(x_1 - a_1) + \frac{\partial u}{\partial x_2}(a)(x_2 - a_2)$$

$$+ \frac{1}{2}\left(\frac{\partial^2 u}{\partial x_1^2}(\xi)(x_1 - a_1)^2 + 2\frac{\partial^2 u}{\partial x_1 \partial x_2}(\xi)(x_1 - a_1)(x_2 - a_2)\right.$$

$$\left. + \frac{\partial^2 u}{\partial x_2^2}(\xi)(x_2 - a_2)^2\right),$$

for some point ξ *on the line segment joining* x *and* a.

Problem 13.14. Prove Taylor's theorem for u by applying Taylor's formula for one variable to the function $f(t) = u(a + t(x - a))$ where $t \in \mathbb{R}$, to get $f(1) = f(0) + f'(0) + ...$, that is $u(x) = u(a) + \nabla u(a) \cdot (x - a) + ...$

Problem 13.15. State and prove Taylor's theorem in $\mathbb{R}^n$ with first and second order remainder.

13.2.5. The Jacobian and changing coordinates

The *Jacobian* of a mapping $u : \mathbb{R}^n \to \mathbb{R}^n$ at a point $y \in \mathbb{R}^n$ is the matrix $Du(y)$ with elements $\partial u_i / \partial x_j(y)$. By Taylor's theorem, if the second order partial derivatives of the u_i are bounded then for some constant C,

$$|u(x) - u(y) - Du(y)(x - y)| \leq C|x - y|^2,$$

which shows that for x close to y we may approximate $u(x)$ to second order accuracy by the linear function $u(y) + Du(y)(x - y)$ of x. The *inverse function theorem* states that if $Du(y)$ is nonsingular, then $x \to u(x)$ gives a one-to-one correspondence of points close to y with points close to $u(y)$, see Problem 13.16. In particular we may view $x \to u(x)$ as a local change of coordinates around y. The change of scale (the area if $m = 2$ and volume if $m = 3$) at y is equal to the modulus of the determinant of $Du(y)$, i.e., $|\det Du(y)|$, so that formally $du = |\det Du(y)|dx$.

Problem 13.16. (*Inverse function theorem*) Let $u : \mathbb{R}^n \to \mathbb{R}^n$ and let $y \in \mathbb{R}^n$ be fixed. For given $v \in \mathbb{R}^n$ with $|v|$ small, consider the problem of finding x close to y such that $u(x) = u(y) + v$. To solve this equation, consider the sequence $\{x_j\}$ in $\mathbb{R}^n$ generated by

$$x_j = x_{j-1} - A(u(x_{j-1}) - u(y) - v) \quad \text{for } j = 1, 2, ...,$$

with $x_0 = y$, where A is a constant matrix; see Problem 13.14. Prove that if Du is continuous and $Du(y)$ is nonsingular, then with a suitable choice of the matrix A, the sequence $\{x_j\}$ converges to a limit x close to y satisfying $u(x) = u(y) + v$. Conclude that $x \to u(x)$ gives a one-to-one mapping of points close to y with points close to $u(y)$.

13.2.6. Particle paths and incompressible velocity fields

Let $u(x,t)$ be a velocity field in $\mathbb{R}^2$ depending on $x \in \mathbb{R}^2$ and time $t > 0$. A *particle path* is a curve $x(t) \in \mathbb{R}^2$ parameterized by t satisfying the ordinary differential equation $\dot{x}(t) = u(x(t), t)$ for $t > 0$, where t represents time and as usual $\dot{x} = \frac{dx}{dt}$. We denote the particle path

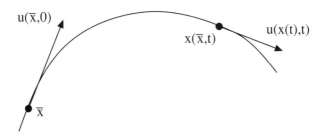

Figure 13.4: A particle path.

starting at $\bar{x} \in \mathbb{R}^2$ at time $t = 0$ by $x(\bar{x}, t)$; see Fig. 13.4. For a fixed t, we can view $\bar{x} \to x(\bar{x}, t)$ as a map from $\mathbb{R}^2$ to $\mathbb{R}^2$, which takes the initial position $\bar{x}$ of a particle to the position $x(\bar{x}, t)$ of the same particle at time t. Denoting the determinant of the Jacobian of this mapping by $A(t)$, it follows by a direct computation based on the chain rule that

$$\frac{dA}{dt} = \nabla \cdot u \, A. \tag{13.5}$$

In the case that $\nabla \cdot u = 0$, it follows that $A(t) = 1$ for $t > 0$, which means that the area occupied by a given set of particles is constant in time. For this reason, a velocity field u satisfying $\nabla \cdot u = 0$ is said to be *incompressible*.

Problem 13.17. Prove (13.5).

13.2.7. Properties of the Laplacian

The Laplacian is one of the fundamental differential operators studied in this book and we describe some of its properties that are useful later on. The Laplace operator typically occurs in *isotropic* models which have the same properties in all directions. In particular, the Laplace operator is invariant under rotations and translations in $\mathbb{R}^2$, i.e., *rigid transformations* of the form

$$\tilde{x}_1 = \cos(\alpha)x_1 + \sin(\alpha)x_2 + a_1$$
$$\tilde{x}_2 = -\sin(\alpha)x_1 + \cos(\alpha)x_2 + a_2,$$

where (x_1, x_2) are the old coordinates and $(\tilde{x}_1, \tilde{x}_2)$ the new ones.

> **Problem 13.18.** Verify that the given transformation corresponds to a counterclockwise rotation of the (x_1, x_2)-coordinate system by an angle α followed by a translation $(a_1, a_2)^\top$.

The invariance of the Laplacian means that the operator is given by the same formula in the two coordinate systems:

$$\frac{\partial^2 u}{\partial x_1^2} + \frac{\partial^2 u}{\partial x_2^2} = \frac{\partial^2 u}{\partial \tilde{x}_1^2} + \frac{\partial^2 u}{\partial \tilde{x}_2^2}.$$

> **Problem 13.19.** Prove this using the chain rule.

The invariance of Δ under rotations is related to the fact that the Laplacian has a simple form in polar coordinates. If we set

$$x_1 = r\cos(\theta) \quad \text{and} \quad x_2 = r\sin(\theta),$$

then

$$\Delta u = \frac{1}{r}\frac{\partial}{\partial r}\left(r\frac{\partial u}{\partial r}\right) + \frac{1}{r^2}\frac{\partial^2 u}{\partial \theta^2}. \tag{13.6}$$

> **Problem 13.20.** Compute the Laplacian in polar coordinates. Hint: show that the Jacobian of the change of variables from (x_1, x_2) to (r, θ) through the mapping $(x_1, x_2) = (r\cos(\theta), r\sin(\theta))$ is
>
> $$\begin{pmatrix} \cos(\theta) & -r\sin(\theta) \\ \sin(\theta) & r\cos(\theta) \end{pmatrix} \tag{13.7}$$
>
> and use this to show that
>
> $$\frac{\partial}{\partial x_1} = \cos(\theta)\frac{\partial}{\partial r} - \frac{\sin(\theta)}{r}\frac{\partial}{\partial \theta} \quad \text{and} \quad \frac{\partial}{\partial x_2} = \sin(\theta)\frac{\partial}{\partial r} + \frac{\cos(\theta)}{r}\frac{\partial}{\partial \theta}.$$

Problem 13.21. Show using (13.6) that the function $u = c_1 \log(r) + c_2$ where the c_i are arbitrary constants, is a solution of the Laplace equation $\Delta u(x) = 0$ in $\mathbb{R}^2$ for $x \neq 0$. Are there other solutions of the Laplace equation in $\mathbb{R}^2$ which are invariant under rotation (i.e. it depends only on $r = |x|$)?

The same invariance property of Δ holds in $\mathbb{R}^3$. In this case, a rotation from x to $\tilde{x}$ is given by $\tilde{x} = Bx$ where B is an *orthogonal* 3×3 matrix, which means that $B^\mathsf{T} B = B B^\mathsf{T} = I$. This invariance property is related to the formula for the Laplacian in *spherical coordinates*.

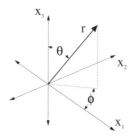

Figure 13.5: The spherical coordinate system (r, θ, φ).

With $r = \sqrt{x^2 + y^2 + z^2}$ and θ and φ defined as illustrated in Fig. 13.5, the spherical coordinates are

$$x_1 = r \sin(\theta)\cos(\varphi), \quad x_2 = r\sin(\theta)\sin(\varphi), \quad \text{and} \quad x_3 = r\cos(\theta),$$

and the Laplacian in spherical coordinates is given by

$$\Delta u = \frac{1}{r^2}\frac{\partial}{\partial r}\left(r^2\frac{\partial u}{\partial r}\right) + \frac{1}{r^2 \sin(\theta)}\frac{\partial}{\partial \theta}\left(\sin(\theta)\frac{\partial u}{\partial \theta}\right) + \frac{1}{r^2 \sin^2(\theta)}\frac{\partial^2 u}{\partial \varphi^2}.$$

Problem 13.22. Prove that the function $u = c_1/r + c_2$, where the c_i are constants, is a solution of Laplace's equation $\Delta u(x) = 0$ in $\mathbb{R}^3$ with $x \neq 0$. Are there other solutions invariant under rotation?

13.3. Integration

In this section we discuss basic aspects of integration along curves, over two-dimensional domains and surfaces, and over three-dimensional domains. We present the basic theorems of Gauss and Stokes together with Green's formula.

13.3.1.   Integrals over curves

Let Γ be a curve in $\mathbb{R}^2$ given by the function $s : (a, b) \to \mathbb{R}^2$ and let $u : \Gamma \to \mathbb{R}$ be a function defined on Γ. We define the integral of u over Γ by

$$\int_\Gamma u \, ds = \int_a^b u(s(t))|s'(t)| \, dt, \qquad (13.8)$$

where $s'(t)$ is the tangent vector to the curve at $a < t < b$. The value the integral is independent of the parameterization.

Problem 13.23.   Prove this. Hint: consider a different parametrization $\sigma : (c, d) \to \mathbb{R}^2$ and associate to each $\tau \in (c, d)$ the unique value $t \in (a, b)$ such that $s(t) = \sigma(\tau)$, which defines $t = t(\tau)$ as a function of τ (assuming that the curve does not cross itself). Now use the formula for change of variables in (13.8) and the fact that $\frac{d\sigma}{d\tau} = \frac{ds}{dt}\frac{dt}{d\tau}$.

If $u(x) = 1$, then the integral of u over Γ gives the length of Γ. In particular,

$$\sigma(\bar{t}) = \int_a^{\bar{t}} |s'| \, dt$$

is the *arclength* of the part of the curve from $s(a)$ to $s(\bar{t})$. By the fundamental theorem of calculus, $\frac{d\sigma}{dt} = |s'|$. Choosing the arc length σ as the parameter corresponds to a parametrization with $|s'| = 1$, see Fig. 13.6.

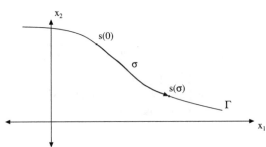

Figure 13.6: A curve Γ parametrized by arclength σ.

13.3.2. The curvature of a plane curve

The *curvature* of a curve $s(t) = (s_1(t), s_2(t))$ in $\mathbb{R}^2$, which measures of how quickly the curve bends as we move along the curve, is defined by

$$\kappa = \frac{d\theta}{d\sigma},$$

where θ is the polar angle of the tangent vector $s' = (s'_1, s'_2)$ defined by $\theta(t) = \tan^{-1}\big(x'_2(t)/x'_1(t)\big)$, and σ is the arclength. In the case of a straight line, the polar angle θ is constant, and thus the curvature is zero, see Fig. 13.7.

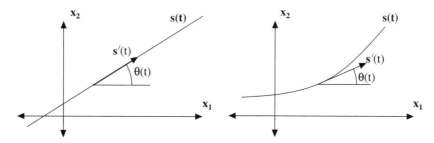

Figure 13.7: The polar angle θ of the tangent vector of a straight line is constant as shown on the right. The tangent vector of a curve that bends, like the example on the left, has a different polar angle at each point.

Problem 13.24. Find the curvature of the plane curve $\big(R\cos(t), R\sin(t)\big)$ where R is constant. What is the curvature of a circle of radius R?

Since $\frac{d\sigma}{dt} = |s'|$, so that $\frac{dt}{d\sigma} = |s'|^{-1}$, we have by the chain rule

$$\kappa(t) = \frac{d\theta}{dt}\frac{dt}{d\sigma} = \frac{\theta'(t)}{|s'(t)|}.$$

Thus, computing $\theta'(t)$ we find that

$$\kappa(t) = \frac{x'_1(t)x''_2(t) - x''_1(t)x'_2(t)}{\big(x'_1(t)^2 + x'_2(t)^2\big)^{3/2}}.$$

In particular, if the curve is parameterized by $s(t) = (t, f(t))$, where $f : \mathbb{R} \to \mathbb{R}$ has two continuous derivatives, then the curvature at the point $(t, f(t))$ is given by

$$\kappa(t) = \frac{f''(t)}{\left(1 + (f'(t))^2\right)^{3/2}}.$$

Problem 13.25. Verify the two formulas for the curvature.

Problem 13.26. (a) Compute the curvature of the curve (t, t^2). (b) Do the same for (t, t^3), and then discuss what happens at the inflection point.

We define the *circle of curvature* at a point $P = s(t)$ on a curve $s : (a, b) \to \mathbb{R}^2$, as the circle of radius $|\kappa(t)|^{-1}$ (assuming $\kappa \neq 0$) that shares the same tangent line as Γ at P and points to the left of Γ if $\kappa > 0$ and to the right if $\kappa < 0$, see Fig. 13.8. The *radius of curvature* at P is $|\kappa(t)|^{-1}$.

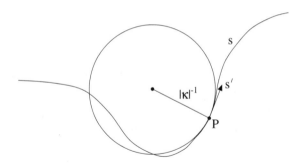

Figure 13.8: The circle of curvature of Γ at P.

Problem 13.27. Find the circle of curvature of the curve $x_2 = x_1^2$ at $x_1 = 0$.

Problem 13.28. Compute the integral $\int_\Gamma u \, ds$, where $u(x) = x_1 x_2$ and Γ is the boundary of the unit square $[0, 1] \times [0, 1]$.

13.3.3. Work and line integrals

Let $F : \mathbb{R}^2 \to \mathbb{R}^2$ be a vector function representing a variable force, or a *force field*, defined in $\mathbb{R}^2$, and let Γ be a curve in $\mathbb{R}^2$ from a point A

to B. It follows from the definition of work as "force in the direction of the displacement $\times$ distance" that the *work* performed, when a particle acted upon by the force F moves along Γ from A to B, is

$$\int_\Gamma F_\tau \, ds,$$

where F_τ is the component of F in the direction of the tangent vector of Γ. Often the notation $F \cdot ds = F_1 dx_1 + F_2 dx_2$, with $ds = (dx_1, dx_2)^\top$ defined as the vector $\frac{ds}{dt} dt = s'(t) dt$ is used instead of $F_\tau \, ds$, where $s : (a, b) \to \mathbb{R}^2$ represents Γ and the integral is called a *line integral*. This is motivated from the computation

$$\int_\Gamma F_\tau \, ds = \int_a^b F \cdot \frac{s'}{|s'|} |s'| \, dt = \int_a^b F \cdot s' \, dt = \int_\Gamma F \cdot ds.$$

If $F = \nabla\varphi$, then the chain rule implies

$$\int_\Gamma F_\tau \, ds = \int_a^b \nabla\varphi(s(t)) \cdot s'(t) \, dt = \int_a^b \frac{d}{dt}\varphi(s(t)) \, dt = \varphi(s(b)) - \varphi(s(a)).$$

We conclude that if the force field F is given as a gradient $F = \nabla\varphi$, then the work by F along a curve is equal to the difference of the values of φ at the end and starting points. In particular, if the curve is closed so that $s(b) = s(a)$, then the work is zero.

The definitions of integrals over curves and line integrals directly extend to curves in $\mathbb{R}^3$ represented by $s : (a, b) \to \mathbb{R}^3$.

13.3.4. Integration in two dimensions

In the case of a two-dimensional domain Ω whose boundary Γ is described by two curves $x_2 = \gamma_1(x_1)$ and $x_2 = \gamma_2(x_1)$ for $a \le x_1 \le b$ as shown in Fig. 13.9, we define the integral of a function $u(x)$ over Ω in terms of iterated integration in one dimension as follows:

$$\begin{cases} \int_\Omega u(x) \, dx &= \int_\Omega u(x_1, x_2) \, dx_1 \, dx_2 \\ &= \int_a^b \left(\int_{\gamma_2(x_1)}^{\gamma_1(x_1)} u(x_1, x_2) \, dx_2 \right) dx_1. \end{cases} \tag{13.9}$$

The role of x_1 and x_2 may be interchanged and the integral is independent of the particular representation of Γ. To handle a more general domain Ω, we split Ω into sub-domains Ω_j of this form and define

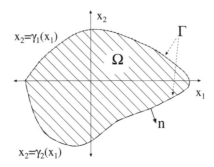

Figure 13.9: A domain in two dimensions.

$\int_\Omega u\,dx = \sum_j \int_{\Omega_j} u\,dx$. By this definition, it follows that the integral of u over Ω represents the volume of the domain under the graph of u over Ω.

Evaluation of an integral over a two-dimensional domain by repeated integration was used by Euler in 1738, when he computed the gravitational attraction of an elliptic lamina.

If $g : \tilde\Omega \to \Omega$ is a one-to-one mapping of $\tilde\Omega \subset \mathbb{R}^2$ onto $\Omega \subset \mathbb{R}^2$, so that $x = g(y) \in \Omega$ for $y \in \tilde\Omega$, then

$$\int_\Omega u(x)\,dx = \int_{\tilde\Omega} u(g(y))|\det Dg(y)|\,dy, \qquad (13.10)$$

where $|\det Dg(y)|$ is the modulus of the determinant of the Jacobian $Dg(y)$ of $g(y)$. This is the formula for "change of coordinates" in a two-dimensional integral. Formally we have $dx = |\det Dg|dy$. For example, using polar coordinates (r, θ) we have formally $dx = r\,drd\theta$ since the determinant of the Jacobian (13.7) is equal to r.

Problem 13.29. (a) Let $u : [0,1] \times [0,1] \to \mathbb{R}$ be continuous, and for $N > 1$, let $(x_{1i}, x_{2j}) = (ih, jh)$ where $i, j = 0, 1, ..., 2^N$, and $h = 2^{-N}$, be a uniform partition of $\Omega = [0,1] \times [0,1]$. Prove that

$$\lim_{N \to \infty} \sum_{i,j=1}^{2^N} u(x_{1i}, x_{2j})h^2$$

$$= \int_0^1 \left(\int_0^1 u(x_1, x_2)dx_1 \right) dx_2 = \int_0^1 \left(\int_0^1 u(x_1, x_2)dx_2 \right) dx_1.$$

By the definition (13.9) the limit is equal to $\int_\Omega u\,dx$. Hint: follow relevant parts of the proof of the Fundamental Theorem of Calculus, and in the discrete sum, carry out the summation in different orders. (b) Estimate the quadrature error in terms of h and ∇u. (c) Extend to integration over domains Ω in $\mathbb{R}^2$ of the form indicated in Fig. 13.9.

13.3.5. Integration by parts, Gauss' theorem, and Green's formula

We recall the following important variant in $\mathbb{R}^2$ of the Fundamental Theorem of Calculus:

$$\int_\Omega \frac{\partial u}{\partial x_2}\,dx = \int_\Gamma u\,n_2\,ds, \tag{13.11}$$

where $n = (n_1, n_2)$, is the unit outward normal to the boundary Γ of Ω. This follows from the definition (13.9) by using the Fundamental Theorem with respect to x_2 and noting that

$$\int_\Gamma un_2\,ds = \int_a^b u(x_1, \gamma_1(x_1))\,dx_1 - \int_a^b u(x_1, \gamma_2(x_1))\,dx_1.$$

since Γ is composed of the two curves $s_1(x_1) = (x_1, \gamma_1(x_1))$ and $s_2(x_1) = (x_1, \gamma_2(x_1))$ with $a \leq x_1 \leq b$, and

$$\left|\frac{ds_1}{dx_1}\right| = \sqrt{1 + (\gamma_1')^2}, \quad n_2 = \frac{1}{\sqrt{1 + (\gamma_1')^2}},$$

$$\left|\frac{ds_2}{dx_1}\right| = \sqrt{1 + (\gamma_2')^2}, \quad n_2 = -\frac{1}{\sqrt{1 + (\gamma_2')^2}}.$$

Note that formally $n_2ds_1 = dx_1$ and $n_2ds_2 = -dx_1$.

Applying (13.11) to the product vw of two functions v and w, we obtain the analog of integration by parts in two dimensions,

$$\int_\Omega \frac{\partial v}{\partial x_i}w\,dx = \int_\Gamma vw\,n_i\,ds - \int_\Omega v\frac{\partial w}{\partial x_i}\,dx, \quad i = 1, 2. \tag{13.12}$$

Applying (13.12) to the components v_i of a vector-valued function $v = (v_1, v_2)$ with $w = 1$ and summing over i, we obtain the important *divergence theorem*, or *Gauss' theorem*:

$$\int_\Omega \nabla \cdot v\,dx = \int_\Gamma v \cdot n\,ds, \tag{13.13}$$

where $v \cdot n = v_1 n_1 + v_2 n_2$. Another consequence of (13.12) is *Green's formula*:

$$\int_\Omega \nabla v \cdot \nabla w \, dx = \int_\Gamma v \partial_n w \, ds - \int_\Omega v \Delta w \, dx, \qquad (13.14)$$

where $\partial_n v = \nabla v \cdot n = \frac{\partial v}{\partial x_1} n_1 + \frac{\partial v}{\partial x_2} n_2$ is the outward normal derivative of v on Γ. We often use Green's formula in the form

$$\int_\Omega v \Delta w \, dx - \int_\Omega \Delta v \, w \, dx = \int_\Gamma v \, \partial_n w \, ds - \int_\Gamma \partial_n v \, w \, ds, \qquad (13.15)$$

which results after applying (13.14) twice.

The above directly generalizes to three dimensions. In particular, using spherical coordinates we have that $dx = r^2 \sin(\theta) d\varphi \, d\theta \, dr$ since the determinant of the corresponding Jacobian is equal to $r^2 \sin(\theta)$. This was used by Lagrange in his work in 1773 on the gravitational attraction of ellipsoids of revolution.

Problem 13.30. (a) Prove Green's formula (13.14) using (13.12). (b) Prove (13.15).

13.3.6. Surface integrals and Stokes' theorem

If S is a surface in $\mathbb{R}^3$ represented by $s : \Omega \to \mathbb{R}^3$, where $\Omega \subset \mathbb{R}^2$, and $u(x)$ is a function defined on the surface, then the surface integral of u over S is defined by

$$\int_S u \, ds = \int_\Omega u(s(\xi)) \, |s'(\xi)| \, d\xi,$$

where $s'(\xi)$ measures the local stretching of the parameter domain defined as the vector product $s' = s'_{\xi_1} \times s'_{\xi_2}$, where $s'_{\xi_i} = (\frac{\partial s_1}{\partial \xi_i}, \frac{\partial s_2}{\partial \xi_i}, \frac{\partial s_3}{\partial \xi_i})$, $i = 1, 2$. Thus $ds = ds(x) = |s'(\xi)| \, d\xi$ is the local element of surface area at $x = \varphi(\xi)$. Again, the integral is independent of the parametrization used. Note that the surface $s : \Omega \to \mathbb{R}^3$ is the two-dimensional analog of the line $s : I \to \mathbb{R}^3$ considered above. In the case S is given by the graph of a function $f : \Omega \to \mathbb{R}$ so that $s(x_1, x_2) = (x_1, x_2, f(x_1, x_2))$, then

$$\int_S u \, ds = \int_\Omega u(x_1, x_2, f(x_1, x_2)) \sqrt{1 + f_{x_1}^2 + f_{x_2}^2} \, dx_1 dx_2, \qquad (13.16)$$

where f_{x_i} denotes the partial derivative of f with respect to x_i.

We recall *Stokes' theorem*

$$\int_S (\text{rot } u)_n \, ds = \int_\Gamma u_\tau \, ds \, ,$$

where Γ is the boundary of S, and $(\text{rot } u)_n$ is the component of $\text{rot } u$ in the normal direction to S, and u_τ is the component of u in the tangent direction to Γ (with a clockwise orientation of the tangent looking the direction of the normal). Sometimes the notation $\text{rot } u \cdot ds = (\text{rot } u)_n \, ds$ is used, where the first ds is the *vector* $s'(\xi)d\xi$. The integral on the right-hand side is the *circulation* of u around Γ. The special case of a plane surface S in the plane $\{x \in \mathbb{R}^3 : x_3 = 0\}$ with boundary Γ, takes the form

$$\int_S \left(\frac{\partial u_2}{\partial x_1} - \frac{\partial u_1}{\partial x_2} \right) dx_1 \, dx_2 = \int_\Gamma (u_2 n_1 - u_1 n_2) \, ds = \int_\Gamma (u_1 dx_1 + u_2 dx_2).$$
$$(13.17)$$

This also follows from (13.12) after identifying the plane $x_3 = 0$ with $\mathbb{R}^2$ and is often referred to as Green's formula in two dimensions. Note that the unit tangent direction is given by $\tau = (-n_2, n_1)$, where (n_1, n_2) is the outward normal direction to Γ in the plane $\{x : x_3 = 0\}$ corresponding to a counter clockwise orientation. An important consequence of Stokes' theorem is that if $\int_\Gamma F_\tau ds = 0$ for all closed curves Γ then $\text{rot} F = 0$.

The divergence theorem (13.13) and Green's formula (13.14) directly extend to a volume Ω in $\mathbb{R}^3$ with boundary surface Γ.

Problem 13.31. (a) Prove the analog of (13.11) in three dimensions. (b) Prove the analogs of (13.12)-(13.15) in three dimensions.

Problem 13.32. Prove (13.16).

Problem 13.33. Prove that $\int_\Omega \text{rot } u \, dx = \int_\Gamma n \times u \, ds$, where Ω is a subset of $\mathbb{R}^3$ with boundary Γ with outward normal n. Hint: apply the divergence theorem to $u \times a$ with a an arbitrary constant vector.

Problem 13.34. Study the relation between Green's formula (13.17) and the divergence theorem applied to the two-dimensional domain S with boundary Γ:

$$\int_S \left(\frac{\partial v_1}{\partial x_1} + \frac{\partial v_2}{\partial x_2} \right) dx_1 \, dx_2 = \int_\Gamma (v_1 n_1 + v_1 n_2) \, ds \, ,$$

with the identification $(u_1, u_2) = (-v_2, v_1)$ corresponding to counter clockwise rotation of the vector (v_1, v_2) by $\pi/2$. Explain how the clockwise direction in Stokes' theorem becomes a counter clockwise direction in (13.17).

Problem 13.35. (a) Prove the *Principle of Archimedes* which states that the total lifting force on a body immersed in a fluid resulting from the pressure of the fluid on the boundary of the body is equal to the weight of the displaced fluid. Hint: use the fact that the pressure in the fluid at the depth z is equal to ρz, where ρ is the (constant) density of the fluid, assuming the pressure is zero at the fluid surface, together with the divergence theorem. (b) Prove that the resultant of the fluid pressure on the boundary is directed vertically and passes through the center of gravity of the displaced fluid volume. Hint: use Problem 13.33. (c) Study the stability of different floating bodies by analyzing how the center of gravity of the displaced fluid volume and the center of gravity of the body changes when the body is tilted slightly. Consider in particular the cases when the displaced fluid is a spherical cap and a very flat rectangular box.

13.4. Representation by potentials

Under general assumptions, a vector function $u : \mathbb{R}^2 \to \mathbb{R}^2$ may be represented as the sum of the gradient and the rotation, respectively, of two scalar functions φ and ψ:

$$u = \operatorname{grad} \varphi + \operatorname{rot} \psi.$$

To determine φ and ψ we apply the divergence and the rotation to both sides and obtain the equations $\Delta\varphi = \operatorname{div} u$ and $-\Delta\psi = \operatorname{rot} u$. This is Poisson's equation for φ and ψ, which we will study in Chapter 15. Two special cases stand out: if div $u = 0$ in $\mathbb{R}^2$, then we may choose $\varphi = 0$ and if rot $u = 0$ in $\mathbb{R}^2$, then we may choose $\psi = 0$. The first case was settled by Clairaut 1739.

We have to be a little careful when restricting the above statements to functions defined on a subset $\Omega \subset \mathbb{R}^2$. Consider for example the function $u(x) = (-x_2, x_1)/|x|^2$ defined for $x \neq 0$. This function satisfies rot $u(x) = 0$ for $x \neq 0$, but u cannot be written as a gradient for $x \neq 0$. This follows by noting that the integral $\int_\Gamma u \cdot ds \neq 0$ if Γ is a circle around $x = 0$, while the integral would be zero if u was a gradient. The reason that u is not a gradient is that the set $\{x \in \mathbb{R}^2 : x \neq 0\}$ where rot $u = 0$, is not simply connected, that is the set has a "hole" namely the point $x = 0$. Thus u is a gradient in $\Omega \subset \mathbb{R}^2$ if rot $u = 0$ in Ω and Ω is simply connected. A special case of a simply connected domain is a *convex domain* with the property that if two points x and y belong to Ω, then the whole line segment between x and y belongs to Ω.

Problem 13.36. Prove that if $u : \mathbb{R}^2 \to \mathbb{R}^2$ satisfies rot $u = 0$ in a convex domain $\Omega \subset \mathbb{R}^2$, then there is a scalar function φ defined in Ω such that $u = \nabla\varphi$ in Ω. Hint: define $\varphi(x) = \int_{\Gamma_x} u \cdot ds$, where Γ_x is a curve joining a fixed point $x_0 \in \Omega$ to $x \in \Omega$. Show that $\varphi(x)$ is independent of the choice of curve from x_0 to x, and conclude by letting Γ_x join x along the x_1 and x_2 axis, that $u(x) = \nabla\varphi(x)$. What if Ω is not convex or not simply connected?

The above representation extends to vector functions $u = (u_1, u_2, u_3)^\top$ in $\mathbb{R}^3$, in which case the potential ψ related to the rotation is a vector. We may prescribe that div $\psi = 0$.

13.5. A model of heat conduction

As an example of application of calculus in several dimensions, we model heat conduction in a heat-conducting material occupying the volume $\Omega \subset \mathbb{R}^3$ with boundary Γ, over a time interval $I = [0, T]$. We let $u(x, t)$ denote the *temperature* and $q(x, t)$ the *heat flux* at (x, t). The heat flux is a vector $q = (q_1, q_2, q_3)$, where q_i is the heat flux in the direction of increasing x_i. The intensity of the heat source is given by $f(x, t)$, while we use $a(x, t)$ to denote the *heat conductivity coefficient* and $\lambda(x, t)$ the *heat capacity coefficient*, both of which are positive.

We derive the model by formulating the basic conservation law expressing that for each $t \in I$, the amount of heat accumulated per unit time in an arbitrary but fixed subdomain $V \subset \Omega$ plus the total heat flux through the boundary of V in the outward direction is equal to the heat produced in V per unit time. With S denoting the boundary of V and n denoting the outward unit normal to S, see Fig. 13.10, the conservation law is

$$\frac{\partial}{\partial t} \int_V \lambda u \, dx + \int_S q \cdot n \, ds = \int_V f \, dx, \qquad (13.18)$$

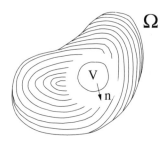

Figure 13.10: An arbitrary subset of the heat conducting body Ω.

where $\frac{\partial}{\partial t}$ denotes differentiation with respect to t and all functions are evaluated at a specific time $t \in I$. By the divergence theorem,

$$\int_S q \cdot n \, ds = \int_V \nabla \cdot q \, dx,$$

and combined with (13.18), this implies that

$$\int_V \left(\frac{\partial}{\partial t}(\lambda u) + \nabla \cdot q \right) dx = \int_V f \, dx,$$

where the time derivative could be moved under the integral sign because V does not depend on time t. Since V is arbitrary, assuming the integrands are continuous, it follows that

$$\frac{\partial}{\partial t}(\lambda u)(x,t) + \nabla \cdot q(x,t) = f(x,t) \quad \text{for all } x \in \Omega, 0 < t \leq T.$$
$$(13.19)$$

The other component of a mathematical model of heat flow is a *constitutive equation* that couples the heat flux q to the temperature gradient ∇u. An example is *Fourier's law* which states that heat flows from warm to cold regions with the heat flux proportional to the temperature gradient, where the factor of proportionality is the *heat conductivity* $a(x,t)$:

$$q(x,t) = -a(x,t)\nabla u(x,t) \quad \text{for } x \in \Omega, \, 0 < t \leq T.$$
$$(13.20)$$

Combining (13.19) and (13.20), we obtain the basic differential equation describing heat conduction:

$$\frac{\partial}{\partial t}(\lambda u) - \nabla \cdot (a\nabla u) = f \quad \text{in } \Omega \times (0,T].$$
$$(13.21)$$

To define the solution uniquely, the differential equation is complemented by initial and boundary conditions. The general heat equation
with Dirichlet boundary conditions reads

$$\begin{cases} \frac{\partial}{\partial t}(\lambda u) - \nabla \cdot (a\nabla u) = f & \text{in } \Omega \times (0, T], \\ u = u_1 & \text{on } \Gamma, \\ u(x, 0) = u_0(x) & \text{for } x \in \Omega, \end{cases} \tag{13.22}$$

where u_0 is the initial temperature and u_1 is the boundary temperature. Other commonly encountered boundary conditions are Neumann
and Robin boundary conditions. The Dirichlet boundary condition corresponds to immersing the body Ω in a large reservoir with a specified
temperature u_1 and assuming that the boundary acts as a perfect ther-
mal conductor. A Neumann boundary condition corresponds to prescribing the heat flow on the boundary. The homogeneous Neumann
case corresponds to a perfectly insulating boundary across which there
is no heat flow. The Robin boundary condition is intermediate with the
boundary neither perfectly conducting nor perfectly insulated, with the
heat flow proportional to the temperature difference across the bound-
ary. Partitioning the boundary as $\Gamma = \Gamma_1 \cup \Gamma_2 \cup \Gamma_3$ according to the
boundary conditions that are posed, the general initial boundary value
problem for the heat equation has the form,

$$\begin{cases} \frac{\partial}{\partial t}(\lambda u) - \nabla \cdot (a\nabla u) = f & \text{in } \Omega \times (0, T], \\ u = u_1 & \text{on } \Gamma_1 \times (0, T], \\ a\partial_n u = g_2 & \text{on } \Gamma_2 \times (0, T], \\ a\partial_n u + \kappa(u - u_3) = g_3 & \text{on } \Gamma_3 \times (0, T], \\ u(x, 0) = u_0(x) & \text{for } x \in \Omega, \end{cases} \tag{13.23}$$

where the coefficient κ represents the heat conductivity of the boundary,
u_1 and u_3 represent "exterior" boundary temperatures, and g_2 and g_3
represent boundary heat fluxes.

Problem 13.37. Derive the heat equation describing the heat conduction in a thin piece of wire of length one whose ends are kept at a fixed
temperature (i.e., derive the heat equation in one dimension):

$$\begin{cases} \dot{u} - u'' = f & \text{in } (0, 1) \times (0, T], \\ u(0, t) = u(1, t) = 0 & \text{for } t \in (0, T], \\ u(x, 0) = u_0(x) & \text{for } x \in (0, 1). \end{cases} \tag{13.24}$$

We now consider some special cases.

13.5.1. The heat equation

We refer to the special case when $\lambda = a = 1$ as the *heat equation*. In the case with homogeneous Dirichlet conditions on the boundary, we get the model

$$\begin{cases} \dot{u} - \Delta u = f & \text{in } \Omega \times (0, T], \\ u = 0 & \text{on } \Gamma \times (0, T], \\ u(x, 0) = u_0(x) & \text{for } x \in \Omega, \end{cases}$$

where u_0 is the initial temperature, and $\Delta u = \nabla \cdot (\nabla u)$ is the Laplacian. The heat equation serves as a basic prototype of a parabolic problem.

13.5.2. Stationary heat conduction: Poisson's equation

The stationary analog of (13.23) reads

$$\begin{cases} -\nabla \cdot (a\nabla u) = f & \text{in } \Omega, \\ u = u_1 & \text{on } \Gamma_1, \\ a\partial_n u = g_2 & \text{on } \Gamma_2, \\ a\partial_n u + \kappa(u - u_3) = g_3 & \text{on } \Gamma_3. \end{cases} \tag{13.25}$$

Choosing again $a = 1$ leads to the *Poisson equation*

$$\begin{cases} -\Delta u = f & \text{in } \Omega, \\ u = u_1 & \text{on } \Gamma_1, \\ a\partial_n u = g_2 & \text{on } \Gamma_2, \\ a\partial_n u + \kappa(u - u_3) = g_3 & \text{on } \Gamma_3. \end{cases} \tag{13.26}$$

In the case of homogeneous Dirichlet boundary conditions on the whole of the boundary, the Poisson equation reads

$$\begin{cases} -\Delta u = f & \text{in } \Omega, \\ u = 0 & \text{on } \Gamma. \end{cases}$$

Poisson's equation serves as a basic model of an elliptic problem. The equation $-\Delta u = 0$ is referred to as *Laplace's equation*, and a function satisfying this equation is said to be a *harmonic function*.

Problem 13.38. The *Euler equations* for an incompressible inviscid fluid of density one take the form

$$\frac{\partial u}{\partial t} + (u \cdot \nabla)u + \nabla p = f, \quad \nabla \cdot u = 0,$$

where $u(x,t)$ is the velocity and $p(x,t)$ the pressure of the fluid at the point x at time t, and f is an applied volume force like a gravitational force. The second equation $\nabla \cdot u = 0$ expresses the incompressibility, cf. Section 13.2.6. Prove that the first equation follows using the chain rule from Newton's law stating that the acceleration $\frac{d}{dt}u(x(t),t)$, where $x(t)$ is the trajectory followed by a fluid particle satisfying $\frac{dx}{dt} = u(x(t),t)$, is equal to the force $-\nabla p + f$. Also write out the first equation in component form. The *Navier-Stokes equations* are modifications of the Euler equations with an additional viscous force term of the form $\epsilon \Delta u$ in the simplest case, where $\epsilon > 0$ is a constant viscosity. See also Chapters 19 and 21.

13.6. Calculus of variations in several dimensions

We comment briefly on the extension of the calculus of variations to problems in several dimensions. The basic minimization problem takes the form: minimize

$$\int_\Omega f(v, \nabla v)\, dx$$

over all functions $v(x)$ such that $v = u_0$ on Γ, where Ω is a domain in R^d with boundary Γ, $f(v, w_1, ..., w_d)$ is a given function from $\mathbb{R}^{d+1}$ to $\mathbb{R}$, and u_0 is given boundary data. The corresponding Euler-Lagrange equation satisfied by a minimizing function reads

$$-\nabla \cdot f_w(u, \nabla u) + f_u(u, \nabla u) = 0 \quad \text{in} \quad \Omega,$$

where $f_w = (f_{w_1}, ..., f_{w_d})$ and $f_{w_i} = \partial f / \partial w_i$.

A fundamental example is the minimization problem for the function

$$f(v, \nabla v) = \frac{1}{2}|\nabla v|^2 - fv,$$

for which the Euler-Lagrange equation in weak form reads (assuming $u_0 = 0$): find $u(x)$ with $u = 0$ on Γ such that

$$\int_\Omega \nabla u \cdot \nabla v\, dx = \int_\Omega fv\, dx,$$

for all v such that $v = 0$ on Γ. In Chapters 15 and 21, we will see that this is the variational formulation of Poisson's equation $-\Delta u = f$ in Ω, $u = 0$ on Γ.

Another example is given by the *minimal surface problem* of finding the function $u(x)$ minimizing the surface area

$$\int_\Omega \sqrt{1 + |\nabla v|^2} \, dx$$

over all functions $v(x)$ taking on prescribed boundary values $v = u_0$ on Γ.

13.7. Mechanical models

We conclude by extending some of the basic models of mechanical systems consisting of masses, springs, and dashpots presented in Chapter 10 to one-dimensionalcontinuous analogs modeling elastic-viscous materials. For simplicity, we consider systems in one space dimension with constitutive laws modelling an elastic-viscous material that behaves locally as a mass-spring-dashpot system. The continuous analogs are obtained from the ordinary differential equation models by replacing the displacement u by the strain u' in the constitutive equation and σ by $-\sigma'$ in the equilibrium equation. In the presence of viscous damping, these diffrential equations are not Euler-Lagrange equations expressing the stationarity of a Lagrangian.

An elastic-viscous material that locally acts like a mass-spring-dashpot with the spring and dashpot coupled in parallel (Kelvin material) may be modeled by

$$\sigma = ku' + \mu\dot{u}' \quad \text{and} \quad m\ddot{u} - \sigma' = f.$$

In this case, the total stress σ is the sum of an elastic stress proportional to the strain u' and a viscous stress proportional to the strain rate $\dot{u}'$. Eliminating σ, we obtain the model

$$m\ddot{u} - \mu\dot{u}'' - ku'' = f.$$

With $\mu = 0$, we obtain the equation for longitudinal vibrations in an elastic string, see Chapter 17.

If the material acts like a spring-dashpot in series (Maxwell material), the model takes the form

$$\dot{u}' = \frac{1}{\mu}\sigma + \frac{1}{k}\dot{\sigma} \quad \text{and} \quad m\ddot{u} - \sigma' = f.$$

Here the total strain rate $\dot{u}'$ is the sum of the strain rates related to the elastic and viscous stresses. In this case, elimination of the stress σ is less direct.

More generally, we may model more complex materials including the components of mass, springs and dashpots in various configurations involving couplings in series or parallel.

Problem 13.39. Motivate the above basic elastic-viscous models, and construct systems building on the basic models.

Problem 13.40. Consider the following model of a "shock absorber" in one dimension consisting of an elastic bar occupying the interval $(0, 1)$ coupled in series with a liquid in a porous medium occupying the interval $(1, 2)$:

$$\begin{cases} \ddot{u} - u'' = 0, & 0 < x < 1, 0 < t \leq T, \\ \dot{v} - v'' = 0, & 1 < x < 2, 0 < t \leq T, \\ u(1, t) = v(1, t), u'(1, t) = v'(1, t), & 0 < t \leq T, \end{cases} \quad (13.27)$$

together with suitable initial and boundary data. The compatibility conditions at the interface $x = 1$ express continuity of velocity and stress. (a) Study the action of the above shock absorber. (b) Propose a better model.

It's like a book, I think, this bloomin' world,
Which you can read and care for just so long,
But presently you feel that you will die
Unless you get the page you're readin' done,
An' turn another – likely not so good;
But what you're after is to turn 'em all. (R. Kipling)

14

Piecewise Polynomials in Several Dimensions

> The universal mathematics is, so to speak, the logic of the imagination. (Leibniz)

In this chapter, we prepare for the application of the finite element method to partial differential equations by discussing approximation of functions by piecewise polynomial functions in several dimensions. We consider three main topics. The first is the construction of a mesh, or triangulation, for a domain in $\mathbb{R}^2$ or $\mathbb{R}^3$, the second is the construction of spaces of piecewise polynomials on a triangulation, and the third topic is the estimation of the interpolation errors.

Recall that in Chapter 8, we used mesh adaptivity to compute an approximation of a two-point boundary value problem of a desired accuracy using a minimum of computational work. We will use mesh adaptivity for similar purposes in higher dimensions. In addition, in higher dimensions, there are geometric considerations that did not arise in one dimension, and in particular the mesh will also be adapted to resolve the features of the domain of the problem. We plot some examples in Fig. 14.1. The construction of a mesh in higher dimensions is more complicated than in one dimension and we give only a brief description of the issues here. We will discuss the details in the advanced companion volume.

We further discuss the construction of vector spaces of piecewise polynomial functions, concentrating mainly on piecewise linear and quadratic functions on triangulations in two dimensions. We conclude by analyzing

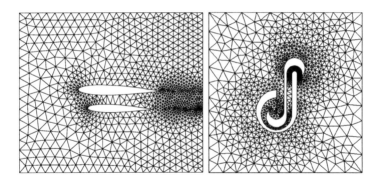

Figure 14.1: The mesh on the left was used in a computation of the flow of air around two airfoils. The mesh on the right, provided through the courtesy of Roy Williams, Center for Advanced Computing Research at Caltech, see http://www.ccsf.caltech.edu/~roy, was used to discretize a piece of metal that has been punched with a fancy character. In both cases, the meshes are adapted to allow accurate computation, taking into account both the behavior of the solution and the shape of the domain.

the error in piecewise linear interpolation on triangles noting in particual that the error in the gradient is affected by the shape of the triangles. We also briefly discuss the L_2 projection and quadrature in two space dimensions.

14.1. Meshes in several dimensions

We start by considering a two-dimensional domain Ω with a polygonal boundary Γ. A *triangulation* $\mathcal{T}_h = \{K\}$ is a sub-division of Ω into a non-overlapping set of triangles, or *elements*, K constructed so that no vertex of one triangle lies on the edge of another triangle. We use $\mathcal{N}_h = \{N\}$ to denote the set of *nodes* N, or corners of the triangles, and $\mathcal{S}_h = \{S\}$ to denote the set of *edges* S of the triangles. Depending on the problem, we may or may not distinguish *internal* nodes and edges, which do not lie on Γ, from the boundary nodes and edges that lie on Γ. We illustrate the notation in Fig. 14.2.

We measure the size of a triangle $K \in \mathcal{T}_h$, by the length h_K of its

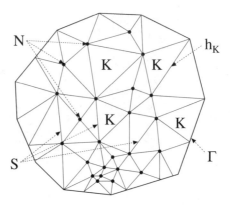

Figure 14.2: A triangulation of a domain Ω.

largest side, which is called the *diameter* of the triangle. The *mesh function* $h(x)$ associated to a triangulation $\mathcal{T}_h$ is the piecewise constant function defined so $h(x) = h_K$ for $x \in K$ for each $K \in \mathcal{T}_h$. We measure the degree of *isotropy* of an element $K \in \mathcal{T}_h$ by its smallest angle τ_K. If $\tau_K \approx \pi/3$ then K is almost isosceles, while if τ_K is small then K is thin, see Fig. 14.3. For the error analysis we present in this book, we need to limit the degree of anisotropy of the mesh. We use the smallest angle among the triangles in $\mathcal{T}_h$, i.e.

$$\tau = \min_{K \in \mathcal{T}_h} \tau_K$$

as a measure of the degree of anistropy of the triangulation $\mathcal{T}_h$, and we control the degree by assuming that τ is greater than a fixed value.

Problem 14.1. For a given triangle K, determine the relation between the smallest angle τ_K, the triangle diameter h_K and the diameter ρ_K of the largest inscribed circle.

The basic problem of mesh generation is to generate a triangulation of a given domain with mesh size given (approximately) by a prescribed mesh function $h(x)$. This problem arises in each step of an adaptive algorithm, where a new mesh function is computed from an approximate solution on a given mesh, and a new mesh is constructed with mesh size given by the new mesh function. The process is then repeated until a stopping criterion is satisfied. The new mesh may be constructed

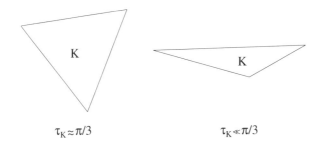

Figure 14.3: Measuring the isotropy of the triangle.

from scratch or by modification of the previous mesh including local refinement or coarsening.

In the *advancing front* strategy a mesh with given mesh size is constructed beginning at some point (often on the boundary) by successively adding one triangle after another, each with a mesh size determined by the mesh function. The curve dividing the domain into a part already triangulated and the remaining part is called the *front*. The front sweeps through the domain during the triangulation process. An alternative is to use a *h-refinement* strategy, where a mesh with a specified local mesh size is constructed by successively dividing elements of an initial coarse triangulation with the elements referred to as *parents*, into smaller elements, called the *children*. We illustrate the refinement and advancing front strategies in Fig. 14.4. It is often useful to combine the two strategies using the advancing front strategy to construct an initial mesh that represents the geometry of the domain with adequate accuracy, and use adaptive *h*-refinement.

There are various strategies for performing the division in an *h*-refinement aimed at limiting the degree of anisotropy of the elements. After the refinements are completed, the resulting mesh is fixed up by the addition of edges aimed at avoiding nodes that are located in the middle of element sides. This causes a mild "spreading" of the adapted region. We illustrate one technique for *h*-refinement in Fig. 14.5. In general, refining a mesh tends to introduce elements with small angles, as can be seen in Fig. 14.5 and it is an interesting problem to construct algorithms for mesh refinement that avoid this tendency in situations where the degree of anisotropy has to be limited. On the other hand, in certain circumstances, it is important to use "stretched" meshes that

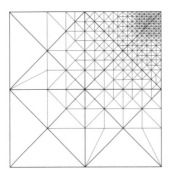 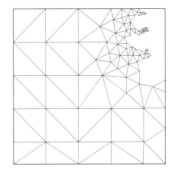

Figure 14.4: The mesh on the left is being constructed by successive *h* refinement starting from the coarse parent mesh drawn with thick lines. The mesh on the right is being constructed by an advancing front strategy. In both cases, high resolution is required near the upper right-hand corner.

have regions of thin elements aligned together to give a high degree of refinement in one direction. In these cases, we also introduce mesh functions that give the local stretching, or degree of anisotropy, and the orientation of the elements. We discuss the construction and use of such meshes in the advanced companion volume.

> **Problem 14.2.** Draw the refined mesh that results from sub-dividing the smallest two triangles in the mesh on the right in Fig. 14.5.

Mesh generation in three dimensions is similar to that in two dimensions with the triangles being replaced by tetrahedra. In practice, the geometric constraints involved become more complicated and the number of elements also increases drastically. We show some examples in Fig. 14.6 and Fig. 14.7.

> **Problem 14.3.** Identify the tetrahedrons shown in the figure on the left in Fig. 14.6.

14.2. Vector spaces of piecewise polynomials

We focus on the case of a triangulation $\mathcal{T}_h = \{K\}$ of a two-dimensional domain Ω with piecewise polynomial boundary Γ. We begin by dis-

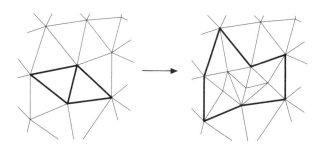

Figure 14.5: On the left, two elements in the mesh have been marked for refinement. The refinement uses the Rivara algorithm in which an element is divided into two pieces by inserting a side connecting the node opposite the longest side to the midpoint of the longest side. Additional sides are added to avoid having a node of one element on the side of another element. The refinement is shown in the mesh on the right along with the boundary of all the elements that had to be refined in addition to those originally marked for refinement.

cussing the finite dimensional vector space V_h consisting of the continuous piecewise linear functions on $\mathcal{T}_h$ defined by

$$V_h = \{v \;:\; v \text{ is continuous on } \Omega, v|_K \in \mathcal{P}^1(K) \text{ for } K \in \mathcal{T}_h\},$$

where $\mathcal{P}^1(K)$ denotes the set of linear functions on K, i.e., the set of functions v of the form $v(x) = c_0 + c_1 x_1 + c_2 x_2$ for some constants c_i. We can describe functions in V_h by their nodal values $N \in \mathcal{N}_h$ because of two facts. The first is that a linear function is uniquely determined by its values at three points, as long as they don't lie on a straight line. To prove this claim, let $K \in \mathcal{T}_h$ have vertices $a^i = (a_1^i, a_2^i)$, $i = 1, 2, 3$, see Fig. 14.8. We want to show that $v \in \mathcal{P}^1(K)$ is determined uniquely by $\{v(a^1), v(a^2), v(a^3)\} = \{v_1, v_2, v_3\}$. A linear function v can be written $v(x_1, x_2) = c_0 + c_1 x_1 + c_2 x_2$ for some constants c_0, c_1, c_2. Substituting the nodal values of v into this expression yields a linear system of equations:

$$\begin{pmatrix} 1 & a_1^1 & a_2^1 \\ 1 & a_1^2 & a_2^2 \\ 1 & a_1^3 & a_2^3 \end{pmatrix} \begin{pmatrix} c_0 \\ c_1 \\ c_2 \end{pmatrix} = \begin{pmatrix} v_1 \\ v_2 \\ v_3 \end{pmatrix}.$$

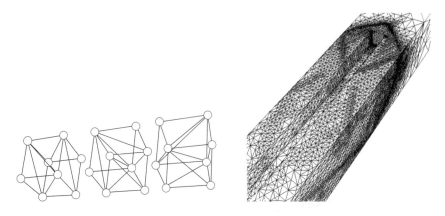

Figure 14.6: On the left, we plot an "exploded" sequence of cubes divided into tetrahedral elements. The nodes of the mesh are marked with circles. On the right, we plot a tetrahedral mesh constructed for a model of the high speed flow of fluid inside a three dimensional channel with an indent in the top. The flow is from the upper right to the lower left. Notice the refinement in the wake caused by the obstacle. These figures provided by the courtesy of Roy Williams, Center for Advanced Computing Research at Caltech, see http://www.ccsf.caltech.edu/~roy.

It is straightforward to check that the matrix is invertible as long as the points $\{a^i\}$ do not fall on a line.

Problem 14.4. Prove this claim. Hint: relate the area of K to the determinant of the coefficient matrix.

The second fact is that if a function is linear in each of two neighboring triangles and its nodal values on the two common nodes of the triangles are equal, then the function is continuous across the common edge. To see this, let K_1 and K_2 be adjoining triangles with common boundary $\partial K_1 = \partial K_2$; see the figure on the left in Fig. 14.9. Parametrizing v along this boundary, we see that v is a linear function of one variable there. Such functions are determined uniquely by the value at two points, and therefore since the values of v on K_1 and K_2 at the common nodes agree, the values of v on the common boundary between K_1 and K_2 agree, and v is indeed continuous across the boundary.

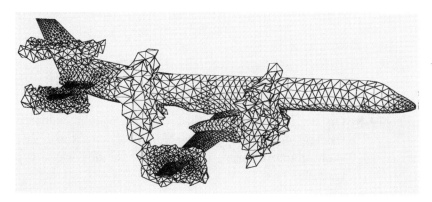

Figure 14.7: The surface mesh on the body, and parts of a tetrahedral mesh around a Saab 2000. Mesh generated by an advancing front mesh generator developed by P. Hansbo and P. Möller in a project supported by the Swedish Institute of Applied Mathematics.

To construct a set of basis functions for V_h, we begin by describing a set of *element basis functions* for triangles. Once again, assuming that a triangle K has nodes at $\{a^1, a^2, a^3\}$, the element nodal basis is the set of functions $\lambda_i \in \mathcal{P}^1(K)$, $i = 1, 2, 3$, such that

$$\lambda_i(a^j) = \begin{cases} 1, & i = j, \\ 0, & i \neq j. \end{cases}$$

We show these functions in Fig. 14.10.

Problem 14.5. Compute explicit formulas for the λ_i.

We construct the *global* basis functions for V_h by piecing together the element basis functions on neighboring elements using the continuity requirement, i.e. by matching element basis functions on neighboring triangles that have the same nodal values on the common edge. The resulting set of basis functions $\{\varphi_j\}_{j=1}^M$, where N_1, N_2,..., N_M is an enumeration of the nodes $N \in \mathcal{N}_h$, is called the set of *tent* functions. The tent functions can also be defined by specifying that $\varphi_j \in V_h$ satisfy

$$\varphi_j(N_i) = \begin{cases} 1, & i = j, \\ 0, & i \neq j, \end{cases}$$

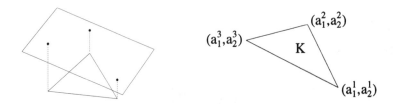

Figure 14.8: On the left, we show that the three nodal values on a triangle determine a linear function. On the right, we show the notation used to describe the nodes of a typical triangle.

Figure 14.9: On the left, we show that a function that is piecewise linear on triangles reduces to a linear function of one variable on triangle edges. On the right, we plot a function that is piecewise linear on triangles whose values at the common nodes on two neighboring triangles do not agree.

for $i, j = 1, ..., M$. We illustrate a typical tent function in Fig. 14.10. We see in particular that the *support* of φ_i is the set of triangles that share the common node N_i.

The tent functions are a nodal basis for V_h because if $v \in V_h$ then

$$v(x) = \sum_{i=1}^{M} v(N_i)\varphi_i(x).$$

Problem 14.6. Prove this.

Problem 14.7. Let K be a triangle with nodes $\{a^i\}$ and let the midpoints of the edges be denoted $\{a^{ij}, 1 \leq i < j \leq 3\}$ (see Fig. 14.11). (a) Show that a function $v \in \mathcal{P}^1(K)$ is uniquely determined by the degrees of freedom

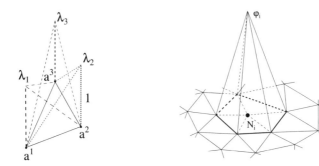

Figure 14.10: On the left, we show the three element nodal basis functions for the linear functions on K. On the right, we show a typical global basis "tent" function.

$\{v(a^{ij}), 1 \leq i < j \leq 3\}$. (b) Are functions continuous in the corresponding finite element space of piecewise linear functions?

Problem 14.8. Let K be a tetrahedron with vertices $\{a^i, i = 1, ..., 4\}$. Show that a function $v \in \mathcal{P}^1(K)$ is uniquely determined by the degrees of freedom $\{v(a^i), i = 1, ..., 4\}$. Show that the corresponding finite element space V_h consists of continuous functions.

Problem 14.9. Explain in what sense the tent functions are "nearly orthogonal" with respect to the L_2 inner product.

14.2.1. Continuous piecewise quadratic functions

The finite element method may be used with piecewise polynomial approximation of degree $q > 1$. We refer to the finite element method in this general form as the (h, q)-method, where both the mesh size h and the degree q may vary from element to element. We have seen that taking q large may lead to technical numerical difficulties, but using moderately large q such as $q = 2$ and $q = 3$ can lead to substantial increase in overall computational efficiency. We now consider the case of piecewise quadratic polynomial functions on triangles.

We let V_h denote the space of continuous piecewise quadratic polynomials on a triangulation $\mathcal{T}_h$:

$$V_h = \{v : v \text{ is continuous in } \Omega, v \in \mathcal{P}^2(K), K \in \mathcal{T}_h\},$$

where $\mathcal{P}^2(K)$ is the set of polynomials on K of degree at most 2, that is

$$\mathcal{P}^2(K) = \left\{ v : v(x) = \sum_{0 \leq i+j \leq 2} c_{ij} x_1^i x_2^j \text{ for } x = (x_1, x_2) \in K, \ c_{ij} \in \mathbb{R} \right\}.$$

Each polynomial $v \in \mathcal{P}^2(K)$ is determined by the six coefficients $c_{00}, c_{10}, c_{01}, c_{20}, c_{11}$ and c_{02}, and the dimension of $\mathcal{P}^2(K)$ is thus six. We seek suitable degrees of freedom to describe continuous piecewise quadratic functions. These turn out to be the vertex values together with the values at midpoints of the edges. Let K be a triangle with vertices a^1, a^2 and a^3 and let a^{12}, a^{23} and a^{13} denote the midpoints of the triangle sides, see Fig. 14.11. We call these points the *element nodes*. The claim is

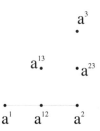

Figure 14.11: Nodes marking the degrees of freedom for quadratic functions.

that a function $v \in \mathcal{P}^2(K)$ is uniquely determined by its values at the 6 element nodes. Since the dimension of $\mathcal{P}^2(K)$ is 6, it suffices to show uniqueness, that is to show that if $v \in \mathcal{P}^2(K)$ satisfies $v(a^i) = 0$ and $v(a^{ij}) = 0$, for $i < j$, $i = 1, 2, 3$, $j = 2, 3$, then $v \equiv 0$ in K.

Problem 14.10. Assuming this fact, show that if v and w are functions in $\mathcal{P}^2(K)$ and agree in value at a^i and a^{ij}, for $i < j$, $i = 1, 2, 3$, $j = 2, 3$, then $v \equiv w$ in K.

Along the side $a^2 a^3$, the function $v|_{a^2 a^3}$ varies like a quadratic polynomial of one variable. Such a function is determined uniquely by the values at three points. Hence, $v|_{a^2 a^3} = 0$ at a^2, a^{23}, and a^3 implies that $v|_{a^2 a^3} \equiv 0$ on all of $a^2 a^3$. This implies that v can be factored

$$v(x) = \lambda_1(x) v_1(x),$$

where λ_1 is the linear element basis function for $\mathcal{P}^1(K)$ with the property that $\lambda_1(a^1) = 1$ and $\lambda_1(a^j) = 0$ for $1 \neq j$, and $v_1 \in \mathcal{P}^1(K)$.

Problem 14.11. Prove this claim. Hint: this is similar to "factoring" out a root of a polynomial in one dimension. Note that $\lambda_1 \equiv 0$ on $a^2 a^3$.

Since a similar argument shows that $v \equiv 0$ on $a^1 a^3$ as well, we must also be able to factor out λ_2 and write

$$v(x) = \lambda_1(x)\lambda_2(x)v_2,$$

where v_2 is a constant. So far we have used the information at a^1, a^{13}, a^2, a^{23}, and a^3. At a^{12} we know that $\lambda_1(a^{12}) \neq 0$ and $\lambda_2(a^{12}) \neq 0$.

Problem 14.12. Prove this.

Since $v(a^{12}) = 0$, we conclude that $v_2 = 0$ and therefore $v \equiv 0$.

To prove that the indicated degrees of freedom may be used to describe V_h, we have to show that this choice leads to continuous functions. In other words, we have to show that a piecewise quadratic polynomial, whose values agree on the three nodes on the common side between neighboring triangles, is continuous. We leave this as a problem.

Problem 14.13. Prove that if K_1 and K_2 are neighboring triangles and $w_1 \in \mathcal{P}^2(K_1)$ and $w_2 \in \mathcal{P}^2(K_2)$ agree at the three nodes on the common boundary, then $w_1 \equiv w_2$ on the common boundary.

To determine the element basis, we note that $v \in \mathcal{P}^2(K)$ may be represented as follows

$$v(x) = \sum_{i=1}^{3} v(a^i)\lambda_i(x)(2\lambda_i(x) - 1) + \sum_{1 \leq i < j \leq 3} v(a^{ij})4\lambda_i(x)\lambda_j(x).$$

Problem 14.14. Verify this claim. Hint: it suffices to check that the expansion on the right agrees with v at the six nodal points.

It follows that the element nodal basis functions $\{\psi_i\}_{i=1}^{6}$ for $\mathcal{P}^2(K_1)$ are given by

$$\psi_1 = \lambda_1(2\lambda_1 - 1), \psi_2 = \lambda_2(2\lambda_2 - 1), \psi_3 = \lambda_3(2\lambda_3 - 1),$$
$$\psi_4 = 4\lambda_1\lambda_2, \psi_5 = 4\lambda_1\lambda_3, \psi_6 = 4\lambda_2\lambda_3.$$

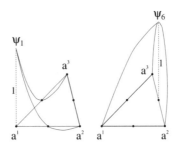

Figure 14.12: Two of six element basis functions for quadratics.

We plot ψ_1 and ψ_6 in Fig. 14.12.

In the same way as for piecewise linear functions, we construct the corresponding global basis functions by piecing together the element basis functions using the continuity requirement.

Problem 14.15. Construct the global basis functions for the space of piecewise quadratic functions on a uniform triangulation of a square into right-angled triangles. Plot a sample of the basis functions and determine the total number.

Problem 14.16. Let K be a tetrahedron with vertices $\{a^i, i = 1, ..., 4\}$. Let a^{ij} denote the midpoint of the segment $a^i a^j$ for $i < j$. (a) Show that a function $v \in \mathcal{P}^2(K)$ is uniquely determined by the degrees of freedom $\{v(a^i), v(a^{ij}), i, j = 1, ..., 4, i < j\}$. (b) Show that the corresponding finite element space V_h has continuous functions.

14.2.2. Examples with polynomials of higher degree

As the degree q of the piecewise polynomial functions increases, there are more choices for the degrees of freedom for a given element K and space $\mathcal{P}^q(K)$ of polynomials on K of degree at most q. As an example, we consider the space of continuous piecewise cubic polynomials

$$V_h = \{v : v \text{ is continuous and } v \in \mathcal{P}^3(K), K \in \mathcal{T}_h, v = 0 \text{ on } \Gamma\}.$$

As degrees of freedom we may choose the function values at the nodal points

$$a^{iij} = \frac{1}{3}(2a^i + a^j), \ i, j = 1, 2, 3, i \neq j \text{ and } a^{123} = \frac{1}{3}(a^1 + a^2 + a^3),$$

Figure 14.13: One choice of nodes for cubic polynomials on triangles.

where $\{a^i\}$ are the vertices of K as usual; see Fig. 14.13. Another choice of degrees of freedom is the set of values

$$\left\{ v(a^i), \frac{\partial v}{\partial x_j}(a^i), v(a^{123}), \, i = 1, 2, 3, \, j = 1, 2 \right\}.$$

In both cases the corresponding finite elements space of piecewise cubics consists of continuous functions.

Problem 14.17. Prove the above claims.

We conclude by describing a finite element space of functions with continuous gradients. Such spaces would be needed in direct applications of the finite element method to partial differential of fourth order, such as the two-dimensional analogs of the beam equation modeling thin plates. The requirement of continuity of the gradient forces the use of higher degree polynomials; with low degree polynomials it is not known how to conveniently choose the degrees of freedom. One possibility is to use the following space of piecewise quintics: $V_h = \{v : v \text{ and } \nabla v \text{ are continuous}, v \in \mathcal{P}^5(K), \, K \in \mathcal{T}_h\}$ on a triangulation of the domain. Using the node notation defined above for quadratics, the degrees of freedom for $v \in V_h$ may be chosen as the values of the partial derivatives of v of total order less than or equal to two at a^i, $i = 1, 2, 3$, together with $\partial_{n_{ij}} v(a^{ij})$, $i, j = 1, 2, 3$, $i < j$, where n_{ij} is the outward normal to the side $a^i a^j$.

Problem 14.18. It is possible to use other basic geometric shapes as elements. One example is to use rectangles when $\Omega \subset \mathbb{R}^2$ is a rectangular domain. Assume the sides of Ω are parallel to the coordinate axes and let K be a smaller rectangle in a "triangulation" of Ω with vertices $\{a^i, i =$

$1, ..., 4\}$ and sides parallel to the coordinate axes as well. Define the space $\mathcal{Q}^1(K)$ to be the set of *bilinear functions* on K, i.e., $v \in \mathcal{Q}^1(K)$ implies that

$$v = c_0 + c_1 x_1 + c_2 x_2 + c_{12} x_1 x_2,$$

for some constants c_0, c_1, c_2, c_{12}. (a) Prove that a function in $\mathcal{Q}^1(K)$ is uniquely determined by the element degrees of freedom $\{v(a^i)\}$. (b) Show that it is possible to define V_h to be the space of continuous functions that are in $\mathcal{Q}^1(K)$ on K. (c) Define an appropriate "triangulation" of Ω into rectangles. (d) Assuming that Ω and the elements K are squares, describe the element and global basis functions.

14.3. Error estimates for piecewise linear interpolation

In this section we prove the basic pointwise maximum norm error estimate for linear interpolation on a triangle, which states that the interpolation error depends on the second order partial derivatives of the function being interpolated, i.e. on the "curvature" of the function, the mesh size and the shape of the triangle. Analogous results hold for other norms. The results also extend directly to more than two space dimensions.

Let K be a triangle with vertices $a^i, i = 1, 2, 3$. Given a continuous function v defined on K, let the linear interpolant $\pi_h v \in \mathcal{P}^1(K)$ be defined by

$$\pi_h v(a^i) = v(a^i), \quad i = 1, 2, 3.$$

We illustrate this in Fig. 14.14.

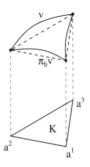

Figure 14.14: The nodal interpolant of v.

Theorem 14.1. *If v has continuous second derivatives, then*

$$\|v - \pi_h v\|_{L_\infty(K)} \le \frac{1}{2} h_K^2 \|D^2 v\|_{L_\infty(K)},\tag{14.1}$$

$$\|\nabla(v - \pi_h v)\|_{L_\infty(K)} \le \frac{3}{\sin(\tau_K)} h_K \|D^2 v\|_{L_\infty(K)},\tag{14.2}$$

where h_K is the largest side of K, τ_K is the smallest angle of K, and

$$D^2 v = \left(\sum_{i,j=1}^{2} \left(\frac{\partial^2 v}{\partial x_i \partial x_j} \right)^2 \right)^{1/2}.$$

Remark 14.3.1. Note that the gradient estimate depends on the reciprocal of the sine of the smallest angle of K, and therefore this error bound deteriorates as the the triangle gets thinner.

Proof. The proof follows the same general outline as the proofs of Theorem 5.1 and Theorem 5.2. Let λ_i, $i = 1, 2, 3$, be the element basis functions for $\mathcal{P}^1(K)$ defined by $\lambda_i(a^j) = 1$ if $i = j$, and $\lambda_i(a^j) = 0$ otherwise. Recall that a function $w \in \mathcal{P}^1(K)$ has the representation

$$w(x) = \sum_{i=1}^{3} w(a^i) \lambda_i(x) \quad \text{for } x \in K,$$

so that

$$\pi_h v(x) = \sum_{i=1}^{3} v(a^i) \lambda_i(x) \quad \text{for } x \in K,\tag{14.3}$$

since $\pi_h v(a^i) = v(a^i)$. We shall derive representation formulas for the interpolation errors $v - \pi_h v$ and $\nabla(v - \pi_h v)$, using a Taylor expansion at $x \in K$:

$$v(y) = v(x) + \nabla v(x) \cdot (y - x) + R(x, y),$$

where

$$R(x, y) = \frac{1}{2} \sum_{i,j=1}^{2} \frac{\partial^2 v}{\partial x_i \partial x_j}(\xi)(y_i - x_i)(y_j - x_j),$$

is the remainder term of order 2 and ξ is a point on the line segment between x and y. In particular choosing $y = a^i = (a_1^i, a_2^i)$, we have

$$v(a^i) = v(x) + \nabla v(x) \cdot (a^i - x) + R_i(x),\tag{14.4}$$

where $R_i(x) = R(x, a^i)$. Inserting (14.4) into (14.3) gives for $x \in K$

$$\pi_h v(x) = v(x) \sum_{i=1}^{3} \lambda_i(x) + \nabla v(x) \cdot \sum_{i=1}^{3} (a^i - x)\lambda_i(x) + \sum_{i=1}^{3} R_i(x)\lambda_i(x).$$

We shall use the following identities that hold for $j, k = 1, 2$, and $x \in K$,

$$\sum_{i=1}^{3} \lambda_i(x) = 1, \quad \sum_{i=1}^{3} (a_j^i - x_j)\lambda_i(x) = 0, \tag{14.5}$$

$$\sum_{i=1}^{3} \frac{\partial}{\partial x_k} \lambda_i(x) = 0, \quad \sum_{i=1}^{3} (a_j^i - x_j)\frac{\partial \lambda_i}{\partial x_k} = \delta_{jk}, \tag{14.6}$$

where $\delta_{jk} = 1$ if $j = k$ and $\delta_{jk} = 0$ otherwise.

Problem 14.19. Prove these identities. Hint: use the fact that $\pi_h v = v$ if $v \in \mathcal{P}^1(K)$. For example, choosing $v(x) \equiv 1$ shows the first of the identities in (14.5). The second follows by choosing $v(x) = d_1 x_1 + d_2 x_2$ with $d_i \in \mathbb{R}$. Also, show the identities by direct computation for the reference triangle with corners at $(0,0)$, $(1,0)$ and $(0,1)$. Finally, (14.6) follows by differentiating (14.5).

Using (14.5), we obtain the following representation of the interpolation error,

$$v(x) - \pi_h v(x) = -\sum_{i=1}^{3} R_i(x)\lambda_i(x).$$

Since $|a^i - x| \le h_K$, we can estimate the remainder term $R_i(x)$ as

$$|R_i(x)| \le \frac{1}{2}h_K^2 \|D^2 v\|_{L_\infty(K)}, \quad i = 1, 2, 3.$$

Problem 14.20. Prove this using Cauchy's inequality twice to estimate an expression of the form $\sum_{ij} x_i c_{ij} x_j = \sum_i x_i \sum_j c_{ij} x_j$.

Now, using the fact that $0 \le \lambda_i(x) \le 1$ if $x \in K$, for $i = 1, 2, 3$, we obtain

$$|v(x) - \pi_h v(x)| \le \max_i |R_i(x)| \sum_{i=1}^{3} \lambda_i(x) \le \frac{1}{2}h_K^2 \|D^2 v\|_{L_\infty(K)} \quad \text{for } x \in K,$$

which proves (14.1).

To prove (14.2), we differentiate (14.3) with respect to $x_k, k = 1, 2$ to get

$$\nabla(\pi_h v)(x) = \sum_{i=1}^{3} v(a^i) \nabla \lambda_i(x),$$

which together with (14.4) and (14.6) gives the following error representation:

$$\nabla(v - \pi_h v)(x) = -\sum_{i=1}^{3} R_i(x) \nabla \lambda_i(x) \quad \text{for } x \in K.$$

We now note that

$$\max_{x \in K} |\nabla \lambda_i(x)| \leq \frac{2}{h_K \sin(\tau_K)},$$

which follows by an easy estimate of the shortest height (distance from a vertex to the opposite side) of K. We now obtain (14.2) as above and the proof is complete. ∎

Problem 14.21. Complete the proof by proving the last claims.

We also use π_h to denote the continuous piecewise linear interpolation operator into V_h.

Problem 14.22. Using these element error results, state and derive error bounds for the error of a continuous piecewise linear interpolant $\pi_h v$ of a function v on a domain Ω triangulated by $\mathcal{T}_h$.

14.3.1. A general result in the L_2 norm

The following theorem summarizes the interpolation error estimates that we need in the analysis of the finite element approximation for Poisson's equation. This result estimates the error of the continuous piecewise linear interpolant of a function over the entire domain in the L_2 norm.

Theorem 14.2. *There are interpolation error constants C_i, depending only on the minimum angle in the mesh τ and the order of the estimate m, such that the piecewise linear nodal interpolant $\pi_h w \in V_h$ of a function w satisfies for $m = 0$ and 1,*

$$\|D^m(w - \pi_h w)\| \leq C_i \|h^{2-m} D^2 w\|, \tag{14.7}$$

where $D^1 w = Dw = \nabla w$, and

$$\|h^{-2+m} D^m (w - \pi_h w)\| + \Big(\sum_{K \in \mathcal{T}_h} h_K^{-3} \|w - \pi_h w\|_{\partial K}^2 \Big)^{1/2} \le C_i \|D^2 w\|. \tag{14.8}$$

Furthermore, there is an interpolant $\tilde{\pi}_h w_h$ in V_h of w such that for $m = 0$ and 1,

$$\|h^{-1+m} D^m (w - \tilde{\pi}_h w)\| + \Big(\sum_{K \in \mathcal{T}_h} h_K^{-1} \|w - \tilde{\pi}_h w\|_{\partial K}^2 \Big)^{1/2} \le C_i \|Dw\|. \tag{14.9}$$

The interpolant $\tilde{\pi}_h w$ is defined using suitable averages of w around the nodal points, because $\|Dw\| < \infty$ does not necessarily guarantee that the nodal values of w are well defined (while $\|D^2 w\| < \infty$ does). Note the presence of an additional square root of h in the second term on the left that acts to balance the integrals over ∂K compared to the integrals over K in the other two terms. The boundary integral term will be used in some a posteriori error estimates we derive below.

The proof is similar to the proof of the maximum norm result presented above. A full proof is given in the advanced companion volume, see also Brenner and Scott ([4]).

14.3.2. The error of a piecewise quadratic interpolant

The interpolation properties of the piecewise quadratics are similar to those of the piecewise linears with an increase of one power of the mesh size h and one derivative of the function being interpolated. For example, we have the following L_2 norm interpolation error estimate for the piecewise quadratic interpolant $\pi_h \in V_h$ taking the same values as a given function u at the nodes:

$$\|u - \pi_h u\| \le C_i \|h^3 D^3 u\|,$$

where C_i is an interpolation constant and $D^3 u$ is the square-root of the sum of squares of all the third order derivatives of u.

14.3.3. The L_2 projection

The L_2 projection $P_h u \in V_h$ of a function $u \in L_2(\Omega)$ into the space of continuous piecewise linear functions V_h on a triangulation $\mathcal{T}_h = \{K\}$ of a domain Ω is defined by

$$(u - P_h u, v) = 0 \quad \text{for all } v \in V_h. \tag{14.10}$$

In other words, the error $u - P_h u$ is orthogonal to V_h. (14.10) yields a linear system of equations for the coefficients of $P_h u$ with respect to the nodal basis of V_h. We discuss the computation of the system and its solution in Chapter 15.

The L_2 projection arises as a natural alternative to nodal interpolation in some situations. For example, the L_2 projection of a function $u \in L_2$ is well defined, while the nodal interpolation in general requires u to be continuous, which is a more stringent requirement. Further, the L_2 projection conserves the total "mass" in the sense that

$$\int_\Omega P_h u \, dx = \int_\Omega u \, dx,$$

which follows by choosing $v(x) \equiv 1$ in (14.10), while the nodal interpolation operator does not preserve total mass in general. The L_2 projection also gives the best approximation of a function u in V_h with respect to the L_2 norm. Using (14.10) for $v \in V_h$ and Cauchy's inequality, we estimate

$$\begin{aligned}
\|u - P_h u\|^2 &= (u - P_h u, u - P_h u) \\
&= (u - P_h u, u - v) + (u - P_h u, v - P_h u) = (u - P_h u, u - v) \\
&\leq \|u - P_h u\| \, \|u - v\|,
\end{aligned}$$

or

$$\|u - P_h u\| \leq \|u - v\| \quad \text{for all } v \in V_h. \tag{14.11}$$

Choosing $v = \pi_h u$ and recalling Theorem 14.2, we conclude

Theorem 14.3. *If u has square integrable second derivatives, then the L_2 projection P_h satisfies*

$$\|u - P_h u\| \leq C_i \|h^2 D^2 u\|.$$

14.4. Quadrature in several dimensions

In the process of computing finite element approximations, we have to compute integrals of the form $\int_K g(x) \, dx$, where K is a finite element and g a given function. Sometimes we may evaluate these integrals exactly, but usually it is either impossible or inefficient. In this case, we have to evaluate the integrals approximately using quadrature formulas. We briefly present some quadrature formulas for integrals over triangles.

In general, we would like to use quadrature formulas that do not affect the accuracy of the underlying finite element method, which of course requires an estimate of the error due to quadrature. A quadrature formula for an integral over an element K has the form

$$\int_K g(x)\, dx \approx \sum_{i=1}^{q} g(y^i)\omega_i, \qquad (14.12)$$

for a specified choice of *nodes* $\{y^i\}$ in K and *weights* $\{\omega_i\}$. We now list some possibilities using the notation a_K^i to denote the vertices of a triangle K, a_K^{ij} to denote the midpoint of the side connecting a_K^i to a_K^j, and a_K^{123} to denote the center of mass of K, see Fig. 14.11 and Fig. 14.13, and denote by $|K|$ the area of K:

$$\int_K g\, dx \approx g\big(a_K^{123}\big)|K|, \qquad (14.13)$$

$$\int_K g(x)\, dx \approx \sum_{j=1}^{3} g(a_K^j)\frac{|K|}{3}, \qquad (14.14)$$

$$\int_K g\, dx \approx \sum_{1\le i<j\le 3} g(a_K^{ij})\frac{|K|}{3}, \qquad (14.15)$$

$$\int_K g\, dx \approx \sum_{j=1}^{3} g(a_K^j)\frac{|K|}{20} + \sum_{1\le i<j\le 3} g(a_K^{ij})\frac{2|K|}{15} + g\big(a_K^{123}\big)\frac{9|K|}{20}. \qquad (14.16)$$

We refer to (14.13) as the center of gravity quadrature, to (14.14) as the vertex quadrature, and to (14.15) as the midpoint quadrature. Recall that the accuracy of a quadrature formula is related to the *precision* of the formula. A quadrature formula has precision r if the formula gives the exact value of the integral if the integrand is a polynomial of degree at most $r - 1$, but there is some polynomial of degree r such that the formula is not exact. The quadrature error for a quadrature rule of precision r is proportional to h^r, where h is the mesh size. More precisely, the error of a quadrature rule of the form (14.12) satisfies

$$\left| \int_K g\, dx - \sum_{i=1}^{q} g(y^i)\omega_i \right| \le Ch_K^r \sum_{|\alpha|=r} \int_K |D^\alpha g|\, dx,$$

where C is a constant. Vertex and center of gravity quadrature have precision 2, midpoint quadrature has precision 3, while (14.16) has precision 4.

In finite element methods based on continuous piecewise linear functions, we often use nodal quadrature, often also referred to as *lumped mass* quadrature, because the mass matrix computed this way becomes diagonal.

Problem 14.23. Prove the above quadrature formulas have the indicated precision.

Problem 14.24. Prove that using nodal quadrature to compute a mass matrix for piecewise linears, gives a diagonal mass matrix where the diagonal term is the sum of the terms in the corresponding exactly computed mass matrix. Motivate the term "lumped".

Problem 14.25. In reference to Problem 14.18, construct a quadrature formula for integrals over rectangles by using the two point Gauss rule for integration over an interval in each direction. Check the precision of the resulting formula.

I walked in the restaurant
for something to do.
The waitress yelled at me,
and so did the food.
And the water tastes funny
when you're far from your home.
But it's only the thirsty
that hunger to roam. (J. Prine)

Figure 14.15: Leibniz choosing between the old and new philosophy at the age of 15.

15

The Poisson Equation

> Nature resolves everything to its component atoms and never reduces anything to nothing. (Lucretius)

In this chapter, we extend the material of Chapter 8 to Poisson's equation $-\Delta u = f$ in a domain $\Omega \subset \mathbb{R}^d$, where $d = 2$ or $d = 3$, together with various boundary conditions. We begin by presenting some models from physics and mechanics that are modeled by Poisson's equation and describing some of the properties of its solutions. We then discuss the finite element method for the Poisson equation: constructing the discrete system of linear equations determining the approximation, deriving a priori and a posteriori error estimates, formulating an adaptive error control algorithm, and briefly addressing some implementation issues. The material directly extends e.g. to problems with variable coefficients of the form (13.25) and to three space dimensions using piecewise linear approximation based on tetrahedral meshes.

15.0.1. Applications of Poisson's equation

We derived Poisson's equation in Chapter 13 as a model of stationary heat conduction. Poisson's equation is the prototype of the class of elliptic equations and has numerous applications in physics and mechanics. These include

- *Elasticity.* The model (8.1) of the deflection of an elastic string discussed in Chapter 8 can be extended to describe the transversal deflection due to a transverasal load of a horizontal elastic membrane of uniform tension stretched over a plane curve Γ enclosing a region Ω in $\mathbb{R}^2$; see Fig. 15.1. The equation takes the form of the Poisson equation $-\Delta u = f$ in Ω together with the boundary condition $u = 0$ on Γ, where $f(x)$ is the transversal load.

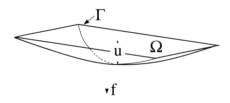

Figure 15.1: An elastic membrane under the load f supported at Γ.

- *Electrostatics.* A basic problem in *electrostatics* is to describe the *electric field* $E(x)$ in a volume Ω containing charges of density $\rho(x)$ and enclosed by a perfectly conducting surface Γ. *Coulomb's law*, one of the famous *Maxwell equations* describing electromagnetic phenomena, can be written

$$\nabla \cdot E = \rho \quad \text{in } \Omega. \tag{15.1}$$

It follows from Faraday's law $\nabla \times E = 0$ (see Chapter 16 below), that the electric field E is the gradient of a scalar *electric potential* φ, i.e. $E = \nabla\varphi$. This leads to the Poisson equation $\Delta\varphi = \rho$ with a Dirichlet boundary condition $\varphi = c$ on Γ, where c is a constant.

- *Fluid mechanics.* The velocity field u of rotation-free fluid flow satisfies $\nabla \times u = 0$, from which it follows that $u = \nabla\varphi$ where φ is a (scalar) velocity potential. If the fluid is incompressible, then $\nabla \cdot u = 0$, and we obtain the Laplace equation $\Delta\varphi = 0$ for the potential of rotation-free incompressible flow. At a solid boundary, the normal velocity is zero, which translates to a homogeneous Neumann boundary condition for the potential. Note that fluid flow is rarely rotation-free in the whole region occupied by the fluid. In particular, if the fluid is viscous, then rotation is generated at solid boundaries.

- *Statistical physics.* The problem is to describe the motion of particles inside a container Ω that move at random until they hit the

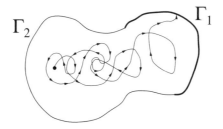

Figure 15.2: An illustration of Brownian motion.

boundary where they stop. We illustrate this in Fig. 15.2. Suppose the boundary Γ of Ω is divided into two pieces $\Gamma = \Gamma_1 \cup \Gamma_2$. Let $u(x)$ be the probability that a particle starting at $x \in \Omega$ winds up stopping at some point on Γ_1, so that $u(x) = 1$ means that it is certain and $u(x) = 0$ means it never happens. It turns out that u solves $\Delta u = 0$ in Ω together with $u = 1$ on Γ_1 and $u = 0$ on Γ_2. Note that the solution of this problem is not continuous on the boundary.

15.0.2. Solution by Fourier series

One time I was sitting visiting the show at the Old Copley Theatre, an idea came into my mind which simply distracted all my attention from the performance. It was the notion of an optical machine for harmonic analysis. I had already learned not to disregard these stray ideas, no matter when they came to my attention, and I promptly left the theatre to work out some of the details of my new plan....The projected machine will solve boundary value problems in the field of partial differential equations. (Wiener)

For special domains, it is possible to write down a formula for the solution of Poisson's equation using Fourier series. For example in Cartesian coordinates, this is possible if the domain is a square or cube. Using polar, cylindrical or spherical coordinates, the set of domains for which Fourier's method may be used includes discs, cylinders, and spheres.

As an illustration, we use Fourier series to solve Poisson's equation $-\Delta u = f$ in a cube $\Omega = (0, \pi) \times (0, \pi) \times (0, \pi)$ with homogeneous Dirichlet boundary conditions. Because the sides of the cube are parallel to the coordinate axes, we can use *separation of variables* to reduce the problem to finding a Fourier series in each variable independently. We start by seeking a solution of the eigenvalue problem $-\Delta v = \lambda v$ in Ω, with $v = 0$ on the boundary of Ω, of the form

$$v(x_1, x_2, x_3) = V_1(x_1)V_2(x_2)V_3(x_3),$$

where each factor satisfies an independent boundary condition $V_i(0) = V_i(\pi) = 0$, $i = 1, 2, 3$. Substituting this into the differential equation yields

$$\frac{V_1''}{V_1} + \frac{V_2''}{V_2} + \frac{V_3''}{V_3} = -\lambda.$$

Because x_1, x_2, and x_3 vary independently, each term V_i''/V_i must be constant. Denoting this constant by λ_i we find that each V_i must solve

$$V_i'' + \lambda_i V_i = 0 \quad \text{in } (0, \pi), \quad V_i(0) = V_i(\pi) = 0.$$

This is the one-dimensional eigenvalue problem considered in Section 6.3 with solution $V_i(x_i) = \sin(jx_i)$ and $\lambda_i = j^2$, where j is an arbitrary integer. It follows that

$$\lambda = \lambda_{jkl} = j^2 + k^2 + l^2, \tag{15.2}$$

for integers j, k, and l with the corresponding eigenfunction

$$v = v_{jkl} = \sin(jx_1)\sin(kx_2)\sin(lx_3).$$

Using the orthogonality of the eigenfunctions, the solution u can be expressed as a Fourier series

$$u(x) = \sum_{j,k,l} A_{jkl} \sin(jx_1)\sin(kx_2)\sin(lx_3),$$

with Fourier coefficients

$$A_{jkl} = \lambda_{jkl}^{-1}\left(\frac{2}{\pi}\right)^3 \int_\Omega f(x)\sin(jx_1)\sin(kx_2)\sin(lx_3)\,dx.$$

The discussion about convergence is nearly the same as in one dimension. In particular, if $f \in L_2(\Omega)$ then the Fourier series of u converges and defines a solution of the given Poisson equation.

Problem 15.1. Prove the formula for A_{jkl}.

Problem 15.2. Prove that the set of eigenfunctions $\{v_{jkl}\}$ are pairwise orthogonal.

Problem 15.3. (a) Compute the Fourier series for the solution of $-\Delta u = 1$ in the square $(0, \pi) \times (0, \pi)$ with homogeneous Dirichlet boundary conditions. (b) Do the same with the Dirichlet condition replaced by a Neumann condition on one side of the square.

Note that there can be several different eigenfunctions for a specific eigenvalue. The *multiplicity* of an eigenvalue is the number of linearly independent eigenvectors that share that eigenvalue. Computing the multiplicity of an eigenvalue λ given by (15.2) is equivalent to determining the number of ways λ be written as a sum of the squares of three integers counting order. For example, $\lambda = 6$ has multiplicity three because $6 = 2^2 + 1 + 1 = 1 + 2^2 + 1 = 1 + 1 + 2^2$.

Problem 15.4. Show that $\lambda = 17$ is an eigenvalue of the Poisson equation posed on $(0, \pi)^3$ with Dirichlet boundary conditions and compute its multiplicity.

15.0.3. Gravitational fields and fundamental solutions

> ... on aura donc $\Delta u = 0$; cette équation remarquable nous sera de la plus grande utilité dans la theorie de la figure des corps célestes. (Laplace)

In his famous treatise *Mécanique Céleste* in five volumes published 1799-1825, Laplace extended Newton's theory of gravitation and in particular developed a theory for describing gravitational fields based on using gravitational potentials that satisfy Laplace's equation, or more generally Poisson's equation.

We consider a gravitational field in $\mathbb{R}^3$ with gravitational force $F(x)$ at position x, generated by a distribution of mass of density $\rho(x)$. We recall that the work of a unit mass, moving along a curve Γ joining a point A to a point B, is given by

$$\int_\Gamma F_\tau \, ds,$$

where F_τ is the component of F in the direction of the tangent to the curve. We illustrate this in Fig. 15.3. If the path Γ is closed, then

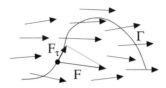

Figure 15.3: The motion of a particle in a field F along a curve Γ.

the total work performed is zero. By Stokes' theorem, it follows that a gravitational field F satisfies $\nabla \times F = 0$ and using the results in Chapter 13, we conclude that F is the gradient of a scalar potential u, i.e.

$$F = \nabla u. \tag{15.3}$$

Laplace proposed the following relation between the gravitational field F and the mass distribution ρ:

$$- \nabla \cdot F = g\rho, \tag{15.4}$$

where g is a gravitational constant. This is analogous to Coulomb's law $\nabla \cdot E = \rho$ in electrostatics, see (15.1), and also to the energy balance equation $\nabla \cdot q = f$ for stationary heat conduction, where q is the heat flux and f a heat source, which we derived in Chapter 13. A corresponding "derivation" of (15.4) does not appear to be available, reflecting that the nature of gravitation is not yet understood. In particular, (15.4) suggests that $\nabla \cdot F(x) = 0$ at points x where there is no mass so that $\rho(x) = 0$. Combining (15.3) and (15.4), we obtain Poisson's equation $-\Delta u = g\rho$ for the gravitational potential u. In particular, the potential satisfies Laplace's equation $\Delta u = 0$ in empty space.

Newton considered gravitational fields generated by *point masses*. We recall that a unit point mass at a point $z \in \mathbb{R}^3$ is represented mathematically by the *delta function* δ_z at z, which is defined by the property that for any smooth function v,

$$\int_{\mathbb{R}^3} \delta_z \, v \, dx = v(z),$$

where the integration is to be interpreted in a generalized sense. Actually, δ_z is a *distribution*, not a proper function, and there is no conventional "formula" for it; instead we define the delta function by its action inside an average of a smooth function.

Formally, the gravitational potential $E(x)$ (avoid confusion with the notation for an electric field used above) corresponding to a unit point mass at the origin should satisfy

$$-\Delta E = \delta_0 \quad \text{in } \mathbb{R}^3, \tag{15.5}$$

where we assumed that the gravitational constant is equal to one. To give a precise meaning to this equation, we first formally multiply by a smooth test function v vanishing outside a bounded set, to get

$$-\int_{\mathbb{R}^3} \Delta E(x)v(x)\, dx = v(0). \tag{15.6}$$

Next, we rewrite the left-hand side formally integrating by parts using Green's formula (13.15) to move the Laplacian from E to v, noting that the boundary terms dissappear since v vanishes outside a bounded set. We may thus reformulate (15.5) as seeking a potential $E(x)$ satisfying

$$-\int_{\mathbb{R}^3} E(x)\Delta v(x)\, dx = v(0), \tag{15.7}$$

for all smooth functions $v(x)$ vanishing outside a bounded set. This is a weak formulation of (15.5), which is perfectly well defined since now the Laplacian acts on the smooth function $v(x)$ and the potential E is assumed to be integrable. We also require the potential $E(x)$ to decay to zero as $|x|$ tends to infinity, which corresponds to a "zero Dirichlet boundary condition at infinity".

In Chapter 13, we showed that the function $1/|x|$ satisfies Laplace's equation $\Delta u(x) = 0$ for $0 \neq x \in \mathbb{R}^3$, while it is singular at $x = 0$. We shall prove that the following scaled version of this function satisfies (15.7):

$$E(x) = \frac{1}{4\pi}\frac{1}{|x|}. \tag{15.8}$$

We refer to this function as the *fundamental solution* of $-\Delta$ in $\mathbb{R}^3$. We conclude in particular that the gravitational field in $\mathbb{R}^3$ created by a unit point mass at the origin is proportional to

$$F(x) = \nabla E(x) = -\frac{1}{4\pi}\frac{x}{|x|^3},$$

which is precisely Newton's inverse square law of gravitation. Laplace thus gives a motivation why the exponent should be two, which Newton did not (and therefore was criticized by Leibniz). Of course, it still remains to motivate (15.4). In the context of heat conduction, the fundamental solution $E(x)$ represents the stationary temperature in a homogeneous body with heat conductivity equal to one filling the whole of $\mathbb{R}^3$, subject to a concentrated heat source of strength one at the origin and with the temperature tending to zero as $|x|$ tends to infinity.

We now prove that the function $E(x)$ defined by (15.8) satisfies (15.7). We then first note that since Δv is smooth and vanishes outside a bounded set, and $E(x)$ is integrable over bounded sets, we have

$$\int_{\mathbb{R}^3} E\Delta v\, dx = \lim_{a\to 0^+} \int_{D_a} E\Delta v\, dx, \tag{15.9}$$

where $D_a = \{x \in \mathbb{R}^3 : a < |x| < a^{-1}\}$, with a small, is a bounded region obtained from $\mathbb{R}^3$ by removing a little sphere of radius a with boundary surface S_a and also points further away from the origin than a^{-1}, see Fig. 15.4. We now use Green's formula (13.15) on D_a with $w = E$.

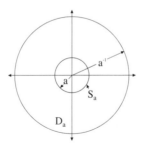

Figure 15.4: A cross-section of the domain D_a.

Since v is zero for $|x|$ large, the integrals over the outside boundary vanish when a is sufficiently small. Using the fact that $\Delta E = 0$ in D_a, $E = 1/(4\pi a)$ on S_a and $\partial E/\partial n = 1/(4\pi a^2)$ on S_a with the normal pointing in the direction of the origin, we obtain

$$-\int_{D_a} E\Delta v\, dx = \int_{S_a} \frac{1}{4\pi a^2} v\, ds - \int_{S_a} \frac{1}{4\pi a} \frac{\partial v}{\partial n}\, ds = I_1(a) + I_2(a),$$

with the obvious definitions of $I_1(a)$ and $I_2(a)$. Now, $\lim_{a\to 0} I_1(a) = v(0)$ because $v(x)$ is continuous at $x = 0$ and the surface area of S_a is equal to $4\pi a^2$, while $\lim_{a\to 0} I_2(a) = 0$. The desired equality (15.7) now follows recalling (15.9).

The corresponding fundamental solution of $-\Delta$ in $\mathbb{R}^2$ is given by

$$E(x) = \frac{1}{2\pi} \log(\frac{1}{|x|}). \qquad (15.10)$$

In this case the fundamental solution is not zero at infinity.

Problem 15.5. Prove that (15.10) is a fundamental solution of $-\Delta$ in $\mathbb{R}^2$.

Problem 15.6. Because the presented mathematical models of heat flow and gravitation, namely Poisson's equation, are the same, it opens the possibility of thinking of a gravitational potential as "temperature" and a gravitational field as "heat flux". Can you "understand" something about gravitation using this analogy?

Replacing 0 by an arbitrary point $z \in \mathbb{R}^3$, (15.7) becomes

$$- \int_{\mathbb{R}^3} E(z - x)\Delta v(x)\, dx = v(z), \qquad (15.11)$$

which leads to a solution formula for Poisson's equation in $\mathbb{R}^3$. For example, if u satisfies the Poisson equation $-\Delta u = f$ in $\mathbb{R}^3$ and $|u(x)| = O(|x|^{-1})$ as $|x| \to \infty$, then u may be represented in terms of the fundamental solution E and the right-hand side f as follows:

$$u(z) = \int_{\mathbb{R}^3} E(z - x)f(x)\, dx = \frac{1}{4\pi} \int_{\mathbb{R}^3} \frac{f(x)}{|z - x|}\, dx.$$
$$\qquad (15.12)$$

We see that $u(z)$ is a mean value of f centered around z weighted so that the influence of the values of $f(x)$ is inversely proportional to the distance from z.

Problem 15.7. Present a corresponding solution formula in the case $d = 2$.

Similarly, the potential u resulting from a distribution of mass of density $\rho(x)$ on a (bounded) surface Γ in $\mathbb{R}^3$ is given by

$$u(z) = \frac{1}{4\pi} \int_\Gamma \frac{\rho(\cdot)}{|z - \cdot|}\, ds, \qquad (15.13)$$

where the dot indicates the integration variable. Formally we obtain this formula by simply adding the potentials from all the different pieces of mass on Γ. One can show that the potential u defined by (15.13) is continuous in $\mathbb{R}^3$ if ρ is bounded on Γ, and of course u satisfies Laplace's equation away from Γ. Suppose now that we would like to determine the distribution of mass ρ on Γ so that the corresponding potential u defined by (15.13) is equal to a given potential u_0 on Γ, that is we seek in particular a function u solving the boundary value problem $\Delta u = 0$ in Ω and $u = u_0$ on Γ, where Ω is the volume enclosed by Γ. This leads to the following *integral equation*: given u_0 on Γ find the function ρ on Γ, such that

$$\frac{1}{4\pi} \int_\Gamma \frac{\rho(y)}{|x - y|}\, ds = u_0(x) \quad \text{for } x \in \Gamma. \tag{15.14}$$

This is a *Fredholm integral equation of the first kind*, named after the Swedish mathematician Ivar Fredholm (1866-1927). In the beginning of the 20th century, Fredholm and Hilbert were competing to prove the existence of solutions of the basic boundary value problems of mechanics and physics using integral equation methods. The integral equation (15.14) is an alternative way of formulating the boundary value problem of finding u such that $\Delta u = 0$ in Ω, and $u = u_0$ on Γ. Integral equations may also be solved using Galerkin methods. We return to the topic of integral equations and their numerical solution in the advanced volume.

Problem 15.8. Show that the potential from a uniform distribution of mass on the surface of a sphere is given as follows: (a) outside the sphere the potential is the same as the potential from a point mass at the origin of the sphere with the same mass as the total surface mass. (b) inside the sphere the potential is constant. Hint: rewrite the surface integral in spherical coordinates and consult a calculus book to evaluate the resulting standard integral.

15.0.4. Green's functions

There is an analog of the formula (15.12) for the solution of Poisson's equation in a bounded domain Ω based on using a *Green's function*, which is the analog of the fundamental solution on a domain different from $\mathbb{R}^d$. The Green's function $G_z(x)$ for the Laplace operator with homogeneous Dirichlet boundary conditions on a bounded domain Ω with boundary Γ satisfies:

$$\begin{cases} -\Delta G_z(x) = \delta_z(x) & \text{for } x \in \Omega, \\ G_z(x) = 0 & \text{for } x \in \Gamma. \end{cases}$$

$G_z(x)$ has a singularity at z corresponding to that of the fundamental solution and in this sense, it is a modified fundamental solution that satisfies the Dirichlet boundary condition. In heat conduction, the Green's function $G_z(x)$ represents the stationary temperature in a homogeneous heat conducting body occupying Ω with zero temperature at its boundary subjected to a concentrated heat source at $z \in \Omega$. It is possible to compute G_z for special domains. For example if $\Omega = \{x : |x| < a\}$ is the ball of radius a in $\mathbb{R}^3$ centered at the origin, then

$$G_z(x) = \frac{1}{4\pi|x - z|} - \frac{1}{4\pi| |z|x/a - az/|z| |}. \tag{15.15}$$

Problem 15.9. Verify (15.15).

Problem 15.10. Determine the Green's function for a "half space" defined as a part of $\mathbb{R}^3$ that has a given plane as a boundary. Hint: consider the function $(|x - z|^{-1} - |x - z^*|^{-1})/(4\pi)$, where z^* is obtained from z by reflection in the plane defining the half space.

If u satisfies $-\Delta u = f$ in Ω and $u = g$ on Γ, then using Green's formula as above we find that the solution u can be represented as

$$u(z) = -\int_\Gamma g \, \partial_n G_z \, ds + \int_\Omega f \, G_z \, dx. \tag{15.16}$$

In the case Ω is the ball of radius a and $f = 0$, so that

$$u(z) = \frac{a^2 - |z|^2}{2^{d-1}\pi a} \int_{S_a} g K_z \, ds, \tag{15.17}$$

with $S_a = \{x : |x| = a\}$ and $K_z(x) = |x - z|^{-d}$, the representation (15.16) is called *Poisson's formula* for harmonic functions. We note in particular that the value at the center of the sphere S_a is equal to the mean value of u on the surface of the sphere, i.e.

$$u(0) = \frac{1}{(2a)^{d-1}\pi} \int_{S_a} u \, ds.$$

Thus a harmonic function has the property that the value at a point is equal to its spherical mean values.

Problem 15.11. Verify (15.16) and (15.17).

In general it is difficult to use (15.16) to compute a solution of Poisson's equation, since finding a formula for the Green function for a general domain is difficult. Moreover, integrals over the entire domain and its boundary have to be evaluated for each value $u(z)$ desired.

15.0.5. The differentiability of solutions

The Poisson formula may be used to show that a bounded function u satisfying $\Delta u = 0$ in a domain Ω has derivatives of any order inside Ω. Thus a harmonic function is *smooth* inside the domain where it is harmonic. This is because the function $|z - x|^{-d}$ is differentiable with respect to z any number of times as long as $x \neq z$, and if x is strictly inside Ω then the sphere $|z - x| = a$ is contained in Ω for a sufficiently small, so that the Poisson representation formula may be used. Thus a bounded solution u of $\Delta u = 0$ in Ω is smooth away from the boundary of Ω. On the other hand, it may very well have singularities on the boundary; we discuss this below. These results carry over to solutions of the Poisson equation $-\Delta u = f$ in Ω: if f is smooth inside Ω then so is u.

15.1. The finite element method for the Poisson equation

In this section, we develop the finite element method with piecewise linear approximation for the Poisson equation with homogeneous Dirichlet boundary conditions

$$\begin{cases} -\Delta u(x) = f(x) & \text{for } x \in \Omega, \\ u(x) = 0 & \text{for } x \in \Gamma, \end{cases} \tag{15.18}$$

where Ω is a bounded domain in $\mathbb{R}^2$ with polygonal boundary Γ,

15.1.1. The variational formulation

Generalizing the procedure used in one dimension from Chapter 8, we first give (15.18) the following variational formulation: find $u \in V$ such that

$$(\nabla u, \nabla v) = (f, v) \quad \text{for all } v \in V, \tag{15.19}$$

where

$$(w, v) = \int_\Omega wv \, dx, \quad (\nabla w, \nabla v) = \int_\Omega \nabla w \cdot \nabla v \, dx,$$

and

$$V = \left\{ v : \int_\Omega (|\nabla v|^2 + v^2) dx < \infty \text{ and } v = 0 \text{ on } \Gamma \right\}. \tag{15.20}$$

A detailed motivation for the choice of V is given in Chapter 21. Here we note that if v and w belong to V, then $(\nabla v, \nabla w)$ is well defined, and if $v \in V$ and $f \in L_2(\Omega)$, then (f, v) is well defined. This follows from Cauchy's inequality. Thus, (15.19) makes sense. In fact, we may think of V as the largest space with this property.

As in the one-dimensional case, we now seek to show that (15.18) and (15.19) have the same solution if f is smooth. First, to see that a solution u of (15.18) with continuous second derivatives (requiring f to be continuous) also is a solution of the variational problem (15.19), we multiply $-\Delta u = f$ by $v \in V$ and use Green's formula to get

$$\int_\Omega fv \, dx = - \int_\Omega \Delta u \, v \, dx = - \int_\Gamma \partial_n u v \, ds + \int_\Omega \nabla u \cdot \nabla v \, dx = \int_\Omega \nabla u \cdot \nabla v \, dx,$$

where the boundary condition $v = 0$ on Γ was used to eliminate the boundary integral over Γ. Conversely, assuming that the solution of (15.19) has continuous second derivatives, we can use Green's formula in (15.19) to put two derivatives back on u, again using the boundary conditions on v, to get

$$\int_\Omega (-\Delta u - f) v \, dx = 0 \quad \text{for all } v \in V. \tag{15.21}$$

Now suppose that $-\Delta u - f$ is non-zero, say positive, at some point $x \in \Omega$. Since $-\Delta u - f$ is continuous, it is therefore positive in some small neighborhood of x contained in Ω. We choose v to be a smooth "hill" that is zero outside the neighborhood and positive inside. It follows that $(-\Delta u - f)v$ is positive in the small neighborhood and zero outside, which gives a contradiction in (15.21). It remains to show that the solution u of (15.19) in fact has continuous second order derivatives if f is continuous; we prove such a regularity result in Chapter 21. We conclude that the differential equation (15.18) and the variational problem (15.19) have the same solution if the data f is continuous. As in the one-dimensional case, the variational problem (15.19) is meaningful for a wider set of data including $f \in L_2(\Omega)$.

Problem 15.12. Prove that the set of functions that are continuous and piecewise differentiable on Ω and vanish on Γ, is a subspace of V.

Problem 15.13. Assuming that a solution of (15.19) is continuous on $\Omega \cup \Gamma$, show that it is unique. Hint: choose $v = u$ and use the continuity of u.

Problem 15.14. Provide the details of the equivalence of (15.18) and (15.19).

The variational problem (15.19) is equivalent to the following quadratic *minimization problem*: find $u \in V$ such that

$$F(u) \le F(v) \quad \text{for all } v \in V, \tag{15.22}$$

where

$$F(v) = \frac{1}{2} \int_\Omega |\nabla v|^2 \, dx - \int_\Omega fv \, dx.$$

The quantity $F(v)$ may be interpreted as the *total energy* of the function $v \in V$ composed of the *internal energy* $\frac{1}{2} \int_\Omega |\nabla v|^2 \, dx$ and the *load potential* $- \int_\Omega fv \, dx$. Thus, the solution u minimizes the total energy $F(v)$ over V. In Chapter 21 we prove existence of a unique solution to the minimization problem (15.22) and thus existence of a unique solution to the variational problem (15.19) and consequently to (15.18).

Problem 15.15. Prove the equivalence of (15.22) and (15.19).

15.1.2.   The finite element method

Let $\mathcal{T}_h = \{K\}$ be a triangulation of Ω with mesh function $h(x)$ and let V_h be the corresponding finite element space of continuous piecewise linear functions vanishing on Γ. The finite element space V_h is a subspace of the space V defined by (15.20). Let $\mathcal{N}_h = \{N\}$ denote the set of *internal nodes* N and $\mathcal{S}_h = \{S\}$ the set of *internal edges* S of $\mathcal{T}_h$. We exclude the nodes and edges on the boundary because of the homogeneous Dirichlet boundary condition. Let $\{N_1, ..., N_M\}$ be an enumeration of the internal nodes $\mathcal{N}_h$, and $\{\varphi_1, ..., \varphi_M\}$ the corresponding nodal basis for V_h.

The finite element method for (15.18) reads: find $U \in V_h$ such that

$$(\nabla U, \nabla v) = (f, v) \quad \text{for all } v \in V_h. \tag{15.23}$$

As in one dimension, we can interpret this as demanding that U solve the Poisson equation in an "average" sense corresponding to the residual of U being "orthogonal" in a certain sense to V_h. More precisely, using the fact that $(\nabla u, \nabla v) = (f, v)$ for $v \in V_h$ because $V_h \subset V$, (15.23) is equivalent to

$$(\nabla u - \nabla U, \nabla v) = 0 \quad \text{for all } v \in V_h, \tag{15.24}$$

which expresses the Galerkin orthogonality of the finite element approximation.

Problem 15.16. Prove that if (15.23) holds with v equal to each of the nodal basis functions V_h, then (15.23) holds for all $v \in V_h$.

15.1.3.   The discrete system of equations

Expanding U in terms of the basis functions $\{\varphi_i\}$ as

$$U = \sum_{j=1}^{M} \xi_j \varphi_j, \quad \text{where } \xi_j = U(N_j),$$

substituting this into (15.23) and choosing $v = \varphi_i$, gives

$$\sum_{j=1}^{M} (\nabla \varphi_j, \nabla \varphi_i) \xi_j = (f, \varphi_i), \quad i = 1, ..., M.$$

This is equivalent to the linear system of equations

$$A\xi = b, \tag{15.25}$$

where $\xi = (\xi_i)$ is the vector of nodal values, $A = (a_{ij})$ is the *stiffness matrix* with elements $a_{ij} = (\nabla\varphi_j, \nabla\varphi_i)$ and $b = (b_i) = (f, \varphi_i)$ is the *load vector*. The stiffness matrix A is obviously symmetric and it is also positive-definite since for any $v = \sum_i \eta_i\varphi_i$ in V_h,

$$\sum_{i,j=1}^{M} \eta_i a_{ij} \eta_j = \sum_{i,j=1}^{M} \eta_i(\nabla\varphi_i, \nabla\varphi_j)\eta_j$$

$$= \left(\nabla\sum_{i=1}^{M}\eta_i\varphi_i, \nabla\sum_{j}^{M}\eta_j\varphi_j\right) = (\nabla v, \nabla v) > 0,$$

unless $\eta_i = 0$ for all i. This means in particular that (15.25) has a unique solution ξ.

Similarly, we determine the linear system determining the L_2 projection $P_h v$ of a function $v \in L_2(\Omega)$ into V_h defined by

$$(P_h v, w) = (v, w) \quad \text{for all } w \in V_h.$$

Substituting $P_h v = \sum_j \eta_j\varphi_j$ and choosing $w = \varphi_i$, $i = 1, ..., M$, we obtain the linear system

$$M\eta = b, \tag{15.26}$$

where the *mass matrix* M has coefficients (φ_j, φ_i) and the data vector b has coefficients (v, φ_i).

Problem 15.17. Prove that the mass matrix is symmetric and positive definite.

15.1.4.   The discrete Laplacian

It will be convenient below to use a discrete analog Δ_h of the Laplacian Δ defined as follows: For a given $w \in V$, let $\Delta_h w$ be the unique function in V_h that satisfies

$$-(\Delta_h w, v) = (\nabla w, \nabla v) \quad \text{for all } v \in V_h. \tag{15.27}$$

In particular, if $w \in V_h$, denoting the nodal values of w by the vector η and those of $\Delta_h w$ by ζ, we find that (15.27) is equivalent to the system of equations $-M\zeta = A\eta$, where M is the mass matrix and A the Poisson stiffness matrix. In other words, the nodal values of the *discrete Laplacian* $\Delta_h w$ of the function $w \in V_h$ with nodal values η, are given by $-M^{-1}A\eta$. We may think of Δ_h as a linear operator on V_h corresponding to multiplication of nodal values by the matrix $-M^{-1}A$. Using Δ_h, we may express the finite element problem (15.23) as finding $U \in V_h$ such that

$$- \Delta_h U = P_h f, \tag{15.28}$$

where P_h is the L_2 projection onto V_h. If w is smooth we may write (15.27) also as

$$(\Delta_h w, v) = (\Delta w, v) \quad \text{for all } v \in V_h, \tag{15.29}$$

which is the same as to say that $\Delta_h w = P_h \Delta w$. Usually, we don't actually compute $\Delta_h w$, but we shall see that the notation is handy.

Problem 15.18. Verify (15.28).

15.1.5. An example: uniform triangulation of a square

> A man who was famous as a tree climber was guiding someone in climbing a tall tree. He ordered the man to cut the top branches, and, during this time, when the man seemed in great danger, the expert said nothing. Only when the man was coming down and had reached the height of the eaves did the expert call out, "Be careful! Watch your step coming down!" I asked him, "Why did you say that? At that height he could jump the rest of the way if he chose."
>
> "That's the point," said the expert. "As long as the man was up at a dizzy height and the branches were threatening to break, he himself was so afraid I said nothing. Mistakes are always made when people get to easy places." (Kenko, translated by D. Keene)

We compute the stiffness matrix and load vector explicitly on the uniform triangulation of the square $\Omega = [0,1] \times [0,1]$ pictured in Fig. 15.5. We choose an integer $m \geq 1$ and set $h = 1/(m+1)$, then construct the triangles as shown. The diameter of the triangles in $\mathcal{T}_h$ is $\sqrt{2}h$ and there

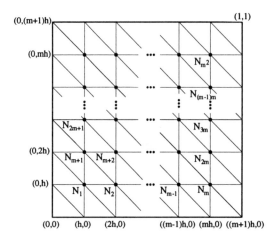

Figure 15.5: The standard triangulation of the unit square.

are $M = m^2$ internal nodes. We number the nodes starting from the lower left and moving right, then working up across the rows.

In Fig. 15.6, we show the support of the basis function corresponding to the node N_i along with parts of the basis functions for the neighboring nodes. As in one dimension, the basis functions are "almost" orthogonal

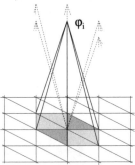

Figure 15.6: The support of the basis function φ_i together with parts of the neighboring basis functions.

in the sense that only basis functions φ_i and φ_j sharing a common triangle in their supports yield a non-zero value in $(\nabla\varphi_i, \nabla\varphi_j)$. We show the nodes neighboring N_i in Fig. 15.7. The support of any two

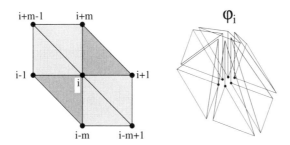

Figure 15.7: The indices of the nodes neighboring N_i and an "exploded" view of φ_i.

neighboring basis functions overlap on just two triangles, while a basis function "overlaps itself" on six triangles.

We first compute

$$(\nabla\varphi_i, \nabla\varphi_i) = \int_\Omega |\nabla\varphi_i|^2\, dx = \int_{\text{support of } \varphi_i} |\nabla\varphi_i|^2\, dx,$$

for $i = 1, ..., m^2$. As noted, we only have to consider the integral over the domain pictured in Fig. 15.7, which is written as a sum of integrals over the six triangles making up the domain. Examining φ_i on these triangles, see Fig. 15.7, we see that there are only two different integrals to be computed since φ_i looks the same, except for orientation, on two of the six triangles and similarly the same on the other four triangles. We shade the corresponding triangles in Fig. 15.6. The orientation affects the direction of $\nabla\varphi_i$ of course, but does not affect $|\nabla\varphi_i|^2$.

We compute $(\nabla\varphi_i, \nabla\varphi_i)$ on the triangle shown in Fig. 15.8. In this case, φ_i is one at the node located at the right angle in the triangle and zero at the other two nodes. We change coordinates to compute $(\nabla\varphi_i, \nabla\varphi_i)$ on the *reference triangle* shown in Fig. 15.8. Again, changing to these coordinates does not affect the value of $(\nabla\varphi_i, \nabla\varphi_i)$ since $\nabla\varphi_i$ is constant on the triangle. On the triangle, φ_i can be written $\varphi_i = ax_1 + bx_2 + c$ for some constants a, b, c. Since $\varphi_i(0,0) = 1$, we get $c = 1$. Similarly, we compute a and b to find that $\varphi_i = 1 - x_1/h - x_2/h$ on this triangle. Therefore, $\nabla\varphi_i = \left(-h^{-1}, -h^{-1}\right)$ and the integral is

$$\int_\triangleright |\nabla\varphi_i|^2\, dx = \int_0^h \int_0^{h-x_1} \frac{2}{h^2}\, dx_2\, dx_1 = 1.$$

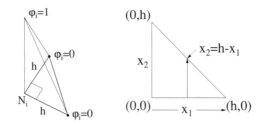

Figure 15.8: First case showing φ_i on the left together with the variables used in the reference triangle.

In the second case, φ_i is one at a node located at an acute angle of the triangle and is zero at the other nodes. We illustrate this in Fig. 15.9. We use the coordinate system shown in Fig. 15.9 to write $\varphi_i = 1 - x_1/h$.

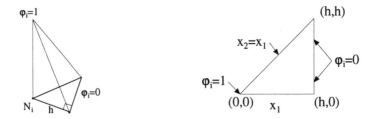

Figure 15.9: Second case showing φ_i and the reference triangle.

When we integrate over the triangle, we get $1/2$.

Problem 15.19. Verify this.

Summing the contributions from all the triangles gives

$$(\nabla \varphi_i, \nabla \varphi_i) = 1 + 1 + \frac{1}{2} + \frac{1}{2} + \frac{1}{2} + \frac{1}{2} = 4.$$

Next, we compute $(\nabla \varphi_i, \nabla \varphi_j)$ for indices corresponding to neighboring nodes. For a general node N_i, there are two cases of inner products (see Fig. 15.6 and Fig. 15.7):

$$(\nabla \varphi_i, \nabla \varphi_{i-1}) = (\nabla \varphi_i, \nabla \varphi_{i+1}) = (\nabla \varphi_i, \nabla \varphi_{i-m}) = (\nabla \varphi_i, \nabla \varphi_{i+m}),$$

and
$$(\nabla\varphi_i, \nabla\varphi_{i-m+1}) = (\nabla\varphi_i, \nabla\varphi_{i+m-1}).$$

The orientation of the triangles in each of the two cases are different, but the inner product of the gradients of the respective basis functions is not affected by the orientation. Note that the the equations corresponding to nodes next to the boundary are special, because the nodal values on the boundary are zero, see Fig. 15.5. For example, the equation corresponding to N_1 only involves N_1, N_2 and N_{m+1}.

For the first case, we next compute $(\nabla\varphi_i, \nabla\varphi_{i+1})$. Plotting the intersection of the respective supports shown in Fig. 15.10, we conclude

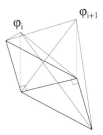

Figure 15.10: The overlap of φ_i and φ_{i+1}.

that there are equal contributions from each of the two triangles in the intersection. We choose one of the triangles and construct a reference triangle as above. Choosing suitable variables, we find that

$$\nabla\varphi_i \cdot \nabla\varphi_{i+1} = \left(-\frac{1}{h}, -\frac{1}{h}\right) \cdot \left(\frac{1}{h}, 0\right) = -\frac{1}{h^2},$$

and integrating over the triangle gives $-1/2$.

Problem 15.20. Carry out this computation in detail.

Since there are two such triangles, we conclude that $(\nabla\varphi_i, \nabla\varphi_{i+1}) = -1$.

Problem 15.21. Prove that $(\nabla\varphi_i, \nabla\varphi_{i-m+1}) = (\nabla\varphi_i, \nabla\varphi_{i+m-1}) = 0$.

We can now determine the stiffness matrix A using the information above. We start by considering the first row. The first entry is $(\nabla\varphi_1, \nabla\varphi_1) = 4$ since N_1 has no neighbors to the left or below. The next entry is $(\nabla\varphi_1, \nabla\varphi_2) = -1$. The next entry after that is zero, because the supports of φ_1 and φ_3 do not overlap. This is true in fact of all the entries up to and including φ_m. However, $(\nabla\varphi_1, \nabla\varphi_{m+1}) = -1$, since these neighboring basis functions do share two supporting triangles. Finally, all the rest of the entries in that row are zero because the supports of the corresponding basis functions do not overlap. We continue in this fashion working row by row. The result is pictured in Fig. 15.11. We see that A has a *block structure* consisting of banded

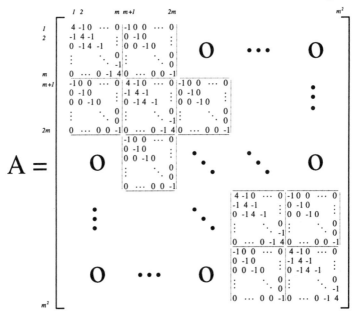

Figure 15.11: The stiffness matrix.

$m \times m$ submatrices, most of which consit only of zeros. Note the pattern of entries around corners of the diagonal block matrices; it is a common mistake to program these values incorrectly.

Problem 15.22. Compute the stiffness matrix for the Poisson equation with homogeneous Dirichlet boundary conditions for (a) the *union jack* triangulation of a square shown in Fig. 15.12 and (b) the triangulation of triangular domain shown in Fig. 15.12.

Problem 15.23. Compute the coefficients of the mass matrix M on the standard triangulation of the square of mesh size h. Hint: it is possible to

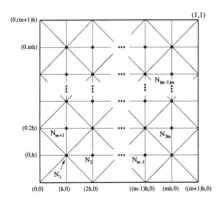
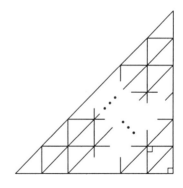

Figure 15.12: The "union jack" triangulation of the unit square and a uniform triangulation of a right triangle.

use quadrature based on the midpoints of the sides of the triangle because this is exact for quadratic functions. The diagonal terms are $h^2/2$ and the off-diagonal terms are all equal to $h^2/12$. The sum of the elements in a row is equal to h^2.

Problem 15.24. Compute the stiffness matrix A for the continuous piecewise quadratic finite element method for the Poisson equation with homogeneous boundary conditions on the unit square using the standard triangulation.

Problem 15.25. Compute the matrix $-\hat{M}^{-1}A$ on the standard triangulation, where $\hat{M}$ is the lumped mass matrix obtained computing the mass matrix using nodal quadrature. Give an interpretation of $-\hat{M}^{-1}A$ related to Δ_h.

The storage of a sparse matrix and the solution of a sparse system are both affected by the *structure* or *sparsity pattern* of the matrix. The sparsity pattern is affected in turn by the enumeration scheme used to mark the nodes.

Problem 15.26. Describe the sparsity pattern of the stiffness matrices A for the Poisson equation with homogeneous Dirichlet data on the unit square corresponding to the continuous piecewise linear finite element method on the standard triangulation using the three numbering schemes pictured in Fig. 15.13.

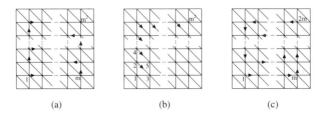

(a)   (b)   (c)

Figure 15.13: Three node numbering schemes for the standard triangulation of the unit square.

There are several algorithms for reordering the coefficients of a sparse
matrix to form a matrix with a smaller bandwidth. Reordering the
coefficients is equivalent to computing a new basis for the vector space.

The load vector b is computed in the same fashion, separating each
integral

$$\int_\Omega f\varphi_i \, dx = \int_{\text{support of } \varphi_i} f(x)\varphi_i(x) \, dx$$

into integrals over the triangles making up the support of φ_i. To compute
the elements (f, φ_i) of the load vector, we often use one of the quadrature
formulas presented in Chapter 14.

Problem 15.27. Compute the load vector b for $f(x) = x_1 + x_2^2$ on the
standard triangulation of the unit square using exact integration and the
lumped mass (trapezoidal rule) quadrature.

15.1.6. General remarks on computing the stiffness matrix and load vector

To compute the finite element approximation U, we have to compute
the coefficients of the stiffness matrix A and load vector b and solve the
linear system of equations (15.25). This is relatively easy to do on a
uniform mesh, but it is a considerable programming problem in general
because of the complexity of the geometry involved.

The first task is to compute the non-zero elements $a_{ij} = (\nabla\varphi_j, \nabla\varphi_i)$
of the stiffness matrix A. As we saw above, $a_{ij} = 0$ unless both N_i and
N_j are nodes of the same triangle K because this is the only way that
the support of different basis functions overlap. The common support

corresponding to a non-zero coefficient is equal to the support of φ_j if $i = j$ and equal to the two triangles with the common edge connecting N_j and N_i if $i \neq j$. In each case a_{ij} is the sum of contributions

$$a_{ij}^K = \int_K \nabla\varphi_j \cdot \nabla\varphi_i \, dx \qquad (15.30)$$

over the triangles K in the common support. The process of adding up the contributions a_{ij}^K from the relevant triangles K to get the a_{ij}, is called *assembling* the stiffness matrix A. Arranging for a given triangle K the numbers a_{ij}^K, where N_i and N_j are nodes of K, into a 3×3 matrix (renumbering locally the nodes 1, 2 and 3 in some order), we obtain the *element stiffness matrix* for the element K. We refer to the assembled matrix A as the *global stiffness matrix*. The element stiffness matrices were originally introduced as a way to organize the computation of A. They are also useful in iterative methods where the assembly (and storage) of A may be avoided completely, and the coefficients a_{ij} are assembled as they are required for the computation of discrete residuals.

Problem 15.28. (a) Show that the element stiffness matrix (15.30) for the linear polynomials on a triangle K with vertices at $(0,0)$, $(h,0)$, and $(0,h)$ numbered 1, 2 and 3, is given by

$$\begin{pmatrix} 1 & -1/2 & -1/2 \\ -1/2 & 1/2 & 0 \\ -1/2 & 0 & 1/2 \end{pmatrix}.$$

(b) Use this result to verify the formula computed for the stiffness matrix A for the continuous piecewise linear finite element method for the Poisson equation with homogeneous boundary conditions on the unit square using the standard triangulation. (c) Compute the element stiffness matrix for a triangle K with nodes $\{a^i\}$.

Problem 15.29. (a) Compute the element stiffness matrix for Poisson's equation for the quadratic polynomials on the reference triangle with vertices at $(0,0)$, $(h,0)$ and $(0,h)$. (b) Use the result to compute the corresponding global stiffness matrix for the standard triangulation of the unit square assuming homogeneous boundary conditions; cf. Problem 15.24.

Problem 15.30. (a) Compute the element stiffness matrix A^K for the continuous bilinear finite element method for the Poisson equation with homogeneous boundary conditions on the unit square using a triangulation into small squares. (b) Use the result to compute the global stiffness matrix.

Problem 15.31. There are speculations that the coupling of two nodes N_i and N_j corresponding to a non-zero coefficient a_{ij} in the stiffness matrix $A = (a_{ij})$, is established through the exchange of particles referred to as *femions*. The nature of these hypothetical particles is unknown. It is conjectured that a femion has zero mass and charge but nevertheless a certain "stiffness". Give your opinion on this question.

15.1.7. Basic data structures

To compute the element stiffness matrices a_{ij}^K, we need the physical coordinates of the nodes of K, and to perform the assembly of A we need the global numbering of the nodes. Similar information is needed to compute the load vector. This information is arranged in a *data structure*, or data base, containing a list of the coordinates of the nodes and a list of the global numbers of the nodes of each triangles. Additional information such as a list of the neighboring elements of each element and a list of the edges, may also be needed for example in adaptive algorithms. It is desirable to organize the data structure so that mesh modification can be handled easily. We discuss this further in the advanced companion book.

15.1.8. Solving the discrete system

Once we have assembled the stiffness matrix, we solve the linear system $A\xi = b$ to obtain the finite element approximation. We discuss this briefly based on the material presented in Chapter 7. The stiffness matrix resulting from discretizing the Laplacian is symmetric and positive-definite and therefore invertible. These properties also mean that there is a wide choice in the methods used to solve the linear system for ξ, which take advantage of the fact that A is sparse.

In the case of the standard uniform discretization of a square, we saw that A is a banded matrix with five non-zero diagonals and bandwidth $m + 1$, where m is the number of nodes on a side. The dimension of A is m^2 and the asymptotic operations count for using a direct method to solve the system is $O(m^4) = O(h^{-4})$. Note that even though A has mostly zero diagonals inside the band, fill-in occurs as the elimination is performed, so we may as well treat A as if it has non-zero diagonals throughout the band. Clever rearrangement of A to reduce the amount of fill-in leads to a solution algorithm with an operations count on the order of $O(m^3) = O(h^{-3})$. In contrast, if we treat A as a full matrix, we get an asymptotic operations count of $O(h^{-6})$, which is considerably larger for a large number of elements.

Problem 15.32. Compute the asymptotic operations count for the direct solution of the system $A\xi = b$ using the three A computed in Problem 15.26.

Problem 15.33. Write a code to solve the system $A\xi = b$ that uses the band structure of A.

In general, we get a sparse stiffness matrix, though there may not be a band structure. If we want to use direct methods efficiently in general, then it is necessary to first reorder the system to bring the matrix into banded form.

We can also apply both the Jacobi and Gauss-Seidel methods to solve the linear system arising from discretizing the Poisson equation. In the case of the uniform standard discretization of a square for example, the operations count is $O(5M)$ per iteration for both methods if we make use of the sparsity of A. Therefore a single step of either method is much cheaper than a direct solve. The question is: How many iterations do we need to compute in order to obtain an accurate solution?

It is not to difficult to show that the spectral radius of the iteration matrix of the Jacobi method M_J is $\rho(M_J) = 1 - h^2\pi^2/2 + O(h^4)$, which means that the convergence rate is $R_J = h^2\pi^2/2 + O(h^4)$. The Gauss-Seidel method is more difficult to analyze, see Isaacson and Keller ([9]), but it can be shown that $\rho(M_{GS}) = 1 - h^2\pi^2 + O(h^4)$ yielding a convergence rate of $R_{GS} = h^2\pi^2 + O(h^4)$, which is twice the rate of the Jacobi method. Therefore, the Gauss-Seidel method is preferable to the Jacobi method. On the other hand, the convergence rate of either method decreases like h^2 so as we refine the mesh, both methods become very slow. The number of operations to achieve an error of $10^{-\sigma}$ is of order $5\sigma/(\pi^2 h^4)$. This is the same order as using a direct banded solver.

There has been a lot of activity in developing iterative methods that converge more quickly. For example, a classic approach is based on modifying M_{GS} in order to decrease the spectral radius, and the resulting method is called an accelerated or over-relaxed Gauss-Seidel iteration. In recent years, very efficient *multi-grid methods* have been developed and are now becoming a standard tool. A multi-grid method is based on a sequence of Gauss-Seidel or Jacobi steps performed on a hierarchy of successively coarser meshes and are optimal in the sense that the solution work is proportional to the total number of unknowns. We discuss multigrid methods in detail in the advanced companion volume.

Problem 15.34. Program codes to solve $A\xi = b$ using both the Jacobi and Gauss-Seidel iteration methods, making use of the sparsity of A in storage and operations. Compare the convergence rate of the two methods using the result from a direct solver as a reference value.

15.2.   Energy norm error estimates

In this section, we derive a priori and a posteriori error bounds in the *energy norm* for the finite element method for Poisson's equation with homogeneous Dirichlet boundary conditions. The energy norm, which is the L_2 norm of the gradient of a function in this problem, arises naturally in the error analysis of the finite element method because it is closely tied to the variational problem. The gradient of the solution, representing heat flow, electric field, flow velocity, or stress for example, can be a variable of physical interest as much as the solution itself, representing temperature, potential or displacement for example, and in this case, the energy norm is the relevant error measure. We also prove optimal order error estimates in the L_2 norm of the solution itself. We discuss analysis in other norms in the advanced companion book.

15.2.1.   A priori error estimate

We first prove that the Galerkin finite element approximation is the best approximation of the true solution in V_h with respect to the energy norm.

Theorem 15.1. *Assume that u satisfies the Poisson equation (15.18) and U is the Galerkin finite element approximation satisfying (15.23). Then*

$$\|\nabla(u - U)\| \leq \|\nabla(u - v)\| \quad \textit{for all } v \in V_h. \tag{15.31}$$

Proof. Using the Galerkin orthogonality (15.24) with $U - v \in V_h$, we can write

$$\|\nabla e\|^2 = (\nabla e, \nabla(u - U)) = (\nabla e, \nabla(u - U)) + (\nabla e, \nabla(U - v)).$$

Adding the terms involving U on the right, whereby U drops out, and using Cauchy's inequality, we get

$$\|\nabla e\|^2 = (\nabla e, \nabla(u - v)) \leq \|\nabla e\|\,\|\nabla(u - v)\|,$$

which proves the theorem after dividing by $\|\nabla e\|$. ∎

Using the interpolation results of Theorem 14.2 choosing $v = \pi_h u$, we get the following concrete quantitative a priori error estimate:

Corollary 15.2. *There exists a constant C_i depending only on the minimal angle τ in $\mathcal{T}_h$, such that*

$$\|\nabla(u - U)\| \leq C_i \|h D^2 u\|. \tag{15.32}$$

15.2.2. A posteriori error estimate

We now prove an a posteriori error estimate following the strategy used for the two-point boundary value problem in Chapter 8. A new feature occurring in higher dimensions is the appearance of integrals over the internal edges S in $\mathcal{S}_h$. We start by writing an equation for the error $e = u - U$ using (15.19) and (15.23) to get

$$\begin{aligned}
\|\nabla e\|^2 &= (\nabla(u - U), \nabla e) = (\nabla u, \nabla e) - (\nabla U, \nabla e) \\
&= (f, e) - (\nabla U, \nabla e) = (f, e - \tilde{\pi}_h e) - (\nabla U, \nabla(e - \tilde{\pi}_h e)),
\end{aligned}$$

where $\tilde{\pi}_h e \in V_h$ is an interpolant of e chosen as in (14.9). We may think of $\tilde{\pi}_h e$ as the usual nodal interpolant of e, although from a technical mathematical point of view, $\tilde{\pi}_h e$ will have to be defined slightly differently. We now break up the integrals over Ω into sums of integrals over the triangles K in $\mathcal{T}_h$ and integrate by parts over each triangle in the last term to get

$$\|\nabla e\|^2 = \sum_K \int_K (f + \Delta U)(e - \tilde{\pi}_h e)\, dx - \sum_K \int_{\partial K} \frac{\partial U}{\partial n_K}(e - \tilde{\pi}_h e)\, ds, \tag{15.33}$$

where $\partial U / \partial n_K$ denotes the derivative of U in the outward normal direction n_K of the boundary ∂K of K. In the boundary integral sum in (15.33), each internal edge $S \in \mathcal{S}_h$ occurs twice as a part of each of the boundaries ∂K of the two triangles K that have S as a common side.

Of course the outward normals n_K from each of the two triangles K sharing S point in opposite directions. For each side S, we choose one of these normal directions and denote by $\partial_S v$ the derivative of a function v in that direction on S. We note that if $v \in V_h$, then in general $\partial_S v$ is different on the two triangles sharing S; see Fig. 14.9, which indicates the "kink" over S in the graph of v. We can express the sum of the boundary integrals in (15.33) as a sum of integrals over edges of the form

$$\int_S [\partial_S U](e - \tilde{\pi}_h e)\, ds,$$

where $[\partial_S U]$ is the difference, or jump, in the derivative $\partial_S U$ computed from the two triangles sharing S. The jump appears because the outward normal directions of the two triangles sharing S are opposite. We further note that $e - \tilde{\pi}_h e$ is continuous across S, but in general does not vanish on S, even if it does so at the end-points of S if $\tilde{\pi}_h$ is the nodal interpolant. This makes a difference with the one-dimensional case, where the corresponding sum over nodes does indeed vanish, because $e - \pi_h e$ vanishes at the nodes. We may thus rewrite (15.33) as follows with the second sum replaced by a sum over internal edges S:

$$\|\nabla e\|^2 = \sum_K \int_K (f + \Delta U)(e - \tilde{\pi}_h e)\, dx + \sum_{S \in \mathcal{S}_h} \int_S [\partial_S U](e - \tilde{\pi}_h e)\, ds.$$

Next, we return to a sum over element edges ∂K by just distributing each jump equally to the two triangles sharing it, to obtain an *error representation* of the energy norm of the error in terms of the residual error:

$$\|\nabla e\|^2 = \sum_K \int_K (f + \Delta U)(e - \tilde{\pi}_h e)\, dx$$

$$+ \sum_K \frac{1}{2} \int_{\partial K} h_K^{-1}[\partial_S U](e - \tilde{\pi}_h e) h_K\, ds,$$

where we prepared to estimate the second sum by inserting a factor h_K and compensating. In crude terms, the residual error results from substituting U into the differential equation $-\Delta u - f = 0$, but in reality, straightforward substitution is not possible because U is not twice differentiable in Ω. The integral on the right over K is the remainder from substituting U into the differential equation inside each triangle

K, while the integral over ∂K arises because $\partial_S U$ in general is different when computed from the two triangles sharing S.

We estimate the first term in the error representation by inserting a factor h, compensating and using the estimate $\|h^{-1}(e - \tilde{\pi}_h e)\| \leq C_i \|\nabla e\|$ of Theorem 14.2, to obtain

$$\left| \sum_K \int_K h(f + \Delta U) h^{-1}(e - \tilde{\pi}_h e) \, dx \right|$$

$$\leq \|h R_1(U)\| \|h^{-1}(e - \tilde{\pi}_h e)\| \leq C_i \|h R_1(U)\| \|\nabla e\|,$$

where $R_1(U)$ is the function defined on Ω by setting $R_1(U) = |f + \Delta U|$ on each triangle $K \in \mathcal{T}_h$. We estimate the contribution from the jumps on the edges similarly. Formally, the estimate results from replacing $h_K \, ds$ by dx corresponding to replacing the integrals over element boundaries ∂K by integrals over elements K. Dividing by $\|\nabla e\|$, we obtain the following a posteriori error estimate:

Theorem 15.3. *There is an interpolation constant C_i only depending on the minimal angle τ such that the error of the Galerkin finite element approximation U of the solution u of the Poisson equation satisfies*

$$\|\nabla u - \nabla U\| \leq C_i \|h R(U)\|, \qquad (15.34)$$

where $R(U) = R_1(U) + R_2(U)$ with

$$R_1(U) = |f + \Delta U| \quad on \ K \in \mathcal{T}_h,$$

$$R_2(U) = \frac{1}{2} \max_{S \subset \partial K} h_K^{-1} |[\partial_S U]| \quad on \ K \in \mathcal{T}_h.$$

As we mentioned, $R_1(U)$ is the contribution to the total residual from the interior of the elements K. Note that in the case of piecewise linear approximation, $R_1(U) = |f|$. Further, $R_2(U)$ is the contribution to the residual from the jump of the normal derivative of U across edges. In the one dimensional problem considered in Chapter 8, this contribution does not appear because the interpolation error may be chosen to be zero at the node points. We observe that the presence of the factor of h in front of the residual error $R(U)$ in (15.34) originates from the Galerkin orthogonality and the estimate $\|h^{-1}(e - \tilde{\pi}_h e)\| \leq C_i \|\nabla e\|$.

Problem 15.35. Derive a priori and a posteriori error bound in the energy norm for the finite element approximation of the solution of the Poisson equation in which the integrals involving the data f are approximated using the one point Gauss quadrature on each triangle or the "lumped mass" nodal quadrature. Hint: recall the modeling error estimate in Chapter 8.

Problem 15.36. Give a more precise proof of the estimate for the jump term in Theorem 15.3 using Theorem 14.2 starting from the error representation.

Problem 15.37. Implement an "error estimation" routine for a code that approximates the Poisson problem using the continuous piecewise linear finite element method. Construct a test problem with a known solution by choosing a function $u(x)$ that is zero on the boundary of the unit square and setting $f = -\Delta u$, then compare the error estimate to the true error on a few meshes.

15.3.   Adaptive error control

An immediate use of an a posteriori error bound is to estimate the error of a computed solution which gives important information to the user. We may also base an adaptive algorithm on the a posteriori error estimate seeking to optimize the computational work needed to reach a certain accuracy.

More precisely, we formulate the basic goal of adaptive error control as: for a given tolerance TOL, find a triangulation $\mathcal{T}_h$ that requires the least amount of computational work to achieve

$$\|\nabla u - \nabla U\| \leq \text{TOL}, \qquad (15.35)$$

where $U \in V_h$ is the finite element approximation corresponding to $\mathcal{T}_h$. Measuring the computational work in terms of the number of nodes of the triangulation $\mathcal{T}_h$ and estimating the unknown error by the computable a posteriori error bound, we are led to the problem of finding the triangulation $\mathcal{T}_h$ with the least number of nodes such that the corresponding finite element approximation U satisfies the stopping criterion

$$C_i\|hR(U)\| \leq \text{TOL}. \qquad (15.36)$$

This is a nonlinear constrained minimization problem with U depending on $\mathcal{T}_h$. If (15.34) is a reasonably sharp estimate of the error, then a solution of this optimization problem will meet our original goal.

We cannot expect to be able to solve this minimization problem analytically. Instead, a solution has to be sought by an iterative process in which we start with a coarse initial mesh and then successively modify the mesh by seeking to satisfy the stopping criterion (15.36) with a minimal number of elements. More precisely, we follow the following *adaptive algorithm*:

1. Choose an initial triangulation $\mathcal{T}_h^{(0)}$.

2. Given the jth triangulation $\mathcal{T}_{h^{(j)}}$ with mesh function $h^{(j)}$, compute the corresponding finite element approximation $U^{(j)}$.

3. Compute the corresponding residuals $R_1(U^{(j)})$ and $R_2(U^{(j)})$ and check whether or not (15.36) holds. If it does, stop.

4. Find a new triangulation $\mathcal{T}_{h^{(j+1)}}$ with mesh function $h^{(j+1)}$ and with a minimal number of nodes such that $C_i \| h^{(j+1)} R(U^{(j)}) \| \leq$ TOL, and then proceed to #2.

The success of this iteration hinges on the mesh modification strategy used to perform step #4. A natural strategy for error control based on the L_2 norm uses the *principle of equidistribution* of the error in which we try to equalize the contribution from each element to the integral defining the L_2 norm. The rationale is that refining an element with large contribution to the error norm gives a large pay-off in terms of error reduction per new degree of freedom.

In other words, the approximation computed on the optimal mesh $\mathcal{T}_h$ in terms of computational work satisfies

$$\| \nabla e \|_{L_2(K)}^2 \approx \frac{\text{TOL}^2}{M} \quad \text{for all } K \in \mathcal{T}_h,$$

where M is the number of elements in $\mathcal{T}_h$. Based on (15.34), we would therefore like to compute the triangulation at step #4 so that

$$C_i^2 \left(\| h^{(j+1)} R(U^{(j+1)}) \|_{L_2(K)}^2 \right) \approx \frac{\text{TOL}^2}{M^{(j+1)}} \quad \text{for all } K \in \mathcal{T}_{h^{(j+1)}}, \tag{15.37}$$

where $M^{(j+1)}$ is the number of elements in $\mathcal{T}_{h^{(j+1)}}$. However, (15.37) is a nonlinear equation, since we don't know $M^{(j+1)}$ and $U^{(j+1)}$ until we have chosen the triangulation. Hence, we replace (15.37) by

$$C_i^2 \left(\left\| h^{(j+1)} R(U^{(j)}) \right\|_{L_2(K)}^2 \approx \frac{\mathrm{TOL}^2}{M^{(j)}} \quad \text{for all } K \in \mathcal{T}_h^{(j+1)}, \tag{15.38}$$

and use this formula to compute the new mesh size $h^{(j+1)}$.

There are several questions that need to be answered about the process described here, including: how much efficiency is lost by replacing (15.35) by (15.36)? In other words, how much bigger is the right-hand side of (15.34) than the left-hand? Does the iterative process #1–#4 converge to a solution of the minimization problem? How should the initial triangulation $\mathcal{T}_{h^{(0)}}$ be chosen and how does this affect the convergence of the adaptive procedure? Is the approximation (15.38) justified? We address these issues in the advanced companion volume.

We conclude this section with an example that illustrates the behavior of this adaptive algorithm in a situation in which the forcing function is highly localized. We use Femlab to approximate the solution

$$u(x) = \frac{a}{\pi} \exp\left(-a(x_1^2 + x_2^2)\right), \quad a = 400,$$

of Poisson's equation $-\Delta u = f$ on the square $(-.5, .5) \times (-.5, .5)$ with $f(x)$ the following "approximate delta function":

$$f(x) = \frac{4}{\pi} a^2 \left(1 - ax_1^2 - ax_2^2\right) \exp\left(-a(x_1^2 + x_2^2)\right),$$

We plot f in Fig. 15.14 (note the vertical scale), together with the initial mesh with 224 elements. The adaptive algorithm took 5 steps to achieve an estimated .5% relative error. We plot the final mesh together with the associated finite element approximation in Fig. 15.15. The algorithm produced meshes with 224, 256, 336, 564, 992, and 3000 elements respectively.

Problem 15.38. Let $\omega(x)$ be a positive weight function defined on the domain $\Omega \subset \mathbb{R}^2$. Assume that the mesh function $h(x)$ minimizes the integral $\int_\Omega h^2(x)\omega(x)\,dx$ under the constraint $\int_\Omega h^{-1}(x)\,dx = N$, where N is a given positive integer. Prove that $h^3(x)\omega(x)$ is constant. Interpret the result as equidistribution in the context of error control. Hint: use the Lagrange multiplier method with the Lagrange function $L(h, \lambda) = \int_\Omega h^2(x)\omega\,dx + \lambda(\int_\Omega h^{-1}(x)\,dx - N)$.

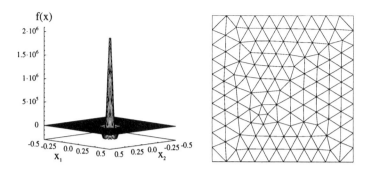

Figure 15.14: The approximate delta forcing function f and the initial mesh used for the finite element approximation.

15.4.   Dealing with different boundary conditions

The variational problem has to be modified when the boundary conditions are changed from homogeneous Dirichlet conditions.

15.4.1.   Non-homogeneous Dirichlet boundary conditions

We first discuss the Poisson's equation with non-homogeneous Dirichlet boundary conditions:

$$\begin{cases} -\Delta u = f & \text{in } \Omega, \\ u = g & \text{on } \Gamma, \end{cases} \tag{15.39}$$

where g is the given boundary data. The variational formulation takes the following form: find $u \in V_g$, where

$$V_g = \left\{ v : \ v = g \text{ on } \Gamma \text{ and } \int_\Omega (|\nabla v|^2 + v^2)dx < \infty \right\},$$

such that

$$(\nabla u, \nabla v) = (f, v) \quad \text{for all } v \in V_0, \tag{15.40}$$

with

$$V_0 = \left\{ v : \ v = 0 \text{ on } \Gamma \text{ and } \int_\Omega (|\nabla v|^2 + v^2)dx < \infty \right\}.$$

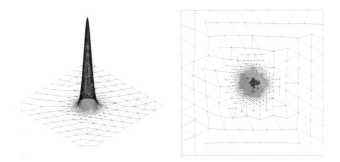

Figure 15.15: The finite element approximation with a relative error of .5% and the final mesh used to compute the approximation. The approximation has a maximum height of roughly 5.

Recall that V_g, where we look for u, is called the *trial space*, while V_0, from which we choose test functions, is called the *test space*. In this case, the trial and test spaces satisfy different boundary conditions, namely, the trial functions satisfy the given non-homogeneous Dirichlet condition $u = g$ on Γ while the test functions satisfy the homogeneous Dirichlet boundary condition. This is important in the construction of the variational formulation (15.40) because when we multiply the differential equation by a test function $v \in V_0$ and use integration by parts, the boundary integral vanishes because $v = 0$ on Γ. The need to choose test functions satisfying homogeneous Dirichlet boundary conditions can also be understood by considering the minimization problem that is equivalent to (15.40): find $u \in V_g$ such that $F(u) \leq F(w)$ for all $w \in V_g$, where $F(w) = \frac{1}{2}(\nabla w, \nabla w) - (f, w)$. The variational formulation (15.40) results from setting the derivative $\frac{d}{d\epsilon}F(u + \epsilon v)$ equal to zero, where $v \in V_0$ is a perturbation satisfying zero boundary conditions so that $u + \epsilon v \in V_g$.

We compute a finite element approximation on a triangulation $\mathcal{T}_h$, where we now also include the nodes on the boundary, denoting the internal nodes by $\mathcal{N}_h$ as above and the set of nodes on the boundary by $\mathcal{N}_b$. We compute an approximation U of the form

$$U = \sum_{N_j \in \mathcal{N}_b} \xi_j \varphi_j + \sum_{N_j \in \mathcal{N}_h} \xi_j \varphi_j, \tag{15.41}$$

where φ_j denotes the basis function corresponding to node N_j in an enumeration $\{N_j\}$ of all the nodes, and, because of the boundary conditions, $\xi_j = g(N_j)$ for $N_j \in \mathcal{N}_b$. Thus the boundary values of U are given by g on Γ and only the coefficients of U corresponding to the interior nodes remain to be found. To this end, we substitute (15.41) into (15.19) and compute inner products with all the basis functions corresponding to the interior nodes, which yields a square system of linear equations for the unknown coefficients of U:

$$\sum_{N_j \in \mathcal{N}_h} \xi_j (\nabla \varphi_j, \nabla \varphi_i) = (f, \varphi_i) - \sum_{N_j \in \mathcal{N}_b} g(N_j)(\nabla \varphi_j, \nabla \varphi_i), \quad N_i \in \mathcal{N}_h.$$

Note that the terms corresponding to the boundary values of U become data on the right-hand side of the system.

Problem 15.39. Show that V_g is not a vector space. Prove that the solution of the weak problem is unique.

Problem 15.40. Compute the discrete equations for the finite element approximation for $-\Delta u = 1$ on $\Omega = (0,1) \times (0,1)$ with boundary conditions $u = 0$ for $x_1 = 0$, $u = x_1$ for $x_2 = 0$, $u = 1$ for $x_1 = 1$ and $u = x_1$ for $x_2 = 1$ using the standard triangulation (Fig. 15.5).

15.4.2. Laplace's equation on a wedge-shaped domain

We consider Laplace's equation with Dirichlet boundary conditions in a wedge-shaped domain making an angle ω:

$$\begin{cases} -\Delta u = 0 & \text{in } \Omega = \{(r, \theta) : 0 \le r < 1, 0 < \theta < \omega\} \\ u(r, 0) = u(r, \omega) = 0, & 0 \le r < 1, \\ u(1, \theta) = \sin(\gamma \theta), & 0 \le \theta \le \omega, \end{cases} \quad (15.42)$$

where $\gamma = \pi/\omega$, see Fig. 15.16. The boundary conditions are chosen so that the exact solution u is given by the following simple explicit formula

$$u(r, \theta) = r^\gamma \sin(\gamma \theta). \quad (15.43)$$

Note that the solution satisfies homogenous Dirichlet boundary conditions on the straight sides joining the corner.

Problem 15.41. Verify the formula (15.43) by direct computation using the equation for the Laplacian in polar coordinates.

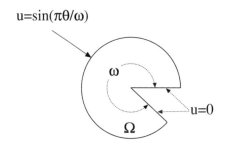

u=sin(πθ/ω)

u=0

Figure 15.16: A domain with an interior corner.

We noted in Section 15.0.5 that a solution of Laplace's equation in a domain (a harmonic function) is smooth inside the domain. We now show using the above example that a harmonic function may have a singularity at a corner of the boundary of the domain. Denoting the derivative with respect to r by D_r, we have from (15.43)

$$D_r u(r, \theta) = \gamma r^{\gamma-1} \sin(\gamma\theta), \quad D_r^2 u(r, \theta) = \gamma(\gamma - 1)r^{\gamma-2} \sin(\gamma\theta),$$

and so on, which shows that sufficiently high derivatives of u become singular at $r = 0$, with the number depending on γ or ω. For example if $\omega = 3\pi/2$, then $u(r, \theta) \approx r^{2/3}$ and $D_r u(r, \theta) \approx r^{-1/3}$ with a singularity at $r = 0$. The gradient ∇u corresponds to e.g. stresses in an elastic membrane or to an electric field. The analysis shows that these quantities become infinite at corners of angle $\omega > \pi$, which thus indicates extreme conditions at concave corners. If the boundary conditions change from Dirichlet to Neumann at the corner, then singularities may occur also at convex corners; see Problem 15.42.

More generally, a solution of Poisson's equation with smooth right hand side in a domain with corners, e.g. a polygonal domain, is a sum of terms of the form (15.43) plus a smooth function.

Problem 15.42. Solve the wedge problem with the Dirichlet condition replaced by a Neumann condition on one of the straight parts of the boundary.

15.4.3.   An example: an L-shaped membrane

We present an example that shows the performance of the adaptive algorithm on a problem with a corner singularity. We consider the Laplace equation in an L-shaped domain that has a non-convex corner at the origin satisfying homogeneous Dirichlet boundary conditions at the sides meeting at the origin and non-homogeneous conditions on the other sides, see Fig. 15.17. We choose the boundary conditions so that the exact solution is $u(r, \theta) = r^{2/3} \sin(2\theta/3)$ in polar coordinates (r, θ) centered at the origin, which has the typical singularity of a corner problem. We use the knowledge of the exact solution to evaluate the performance of the adaptive algorithm.

We compute using Femlab with energy norm control based on (15.34) to achieve an error tolerance of $TOL = .005$ using h refinement mesh modification. In Fig. 15.17, we show the initial mesh $\mathcal{T}_{h(0)}$ with 112 nodes and 182 elements. In Fig. 15.18, we show the level curves of

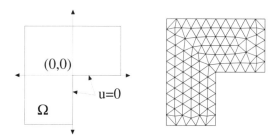

Figure 15.17: The L-shaped domain and the initial mesh.

the solution and the final mesh with 295 nodes and 538 elements that achieves the desired error bound. The interpolation constant was set to $C_i = 1/8$. The quotient between the estimated and true error on the final mesh was 1.5.

Since the exact solution is known in this example, we can also use the a priori error estimate to determine a mesh that gives the desired accuracy. We do this by combining the a priori error estimate (15.32) and the principle of equidistribution of error to determine $h(r)$ so that $C_i\|hD^2u\| = TOL$ while keeping h as large as possible (and keeping the number of elements at a minimum). Since $D^2u(r) \approx r^{-4/3}$, as long as $h \leq r$, that is up to the elements touching the corner, we determine that

$$\left(hr^{-4/3}\right)^2 h^2 \approx \frac{TOL^2}{M} \quad \text{or} \quad h^2 = TOL\, M^{-1/2} r^{4/3},$$

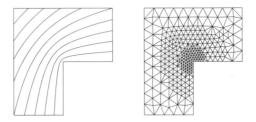

Figure 15.18: Level curves of the solution and final adapted mesh on
the L-shaped domain.

where M is the number of elements and h^2 measures the element area.
To compute M from this relation, we note that $M \approx \int_\Omega h^{-2}\,dx$, since
the number of elements per unit area is $O(h^{-2})$, which gives

$$M \approx M^{1/2}\mathrm{TOL}^{-1}\int_\Omega r^{-4/3}\,dx.$$

Since the integral is convergent (prove this), it follows that $M \propto \mathrm{TOL}^{-2}$,
which implies that $h(r) \propto r^{1/3}\,\mathrm{TOL}$. Note that the total number of
unknowns, up to a constant, is the same as that required for a smooth
solution without a singularity, namely TOL^{-2}. This depends on the
very local nature of the singularity in the present case. In general, of
course solutions with singulares may require a much larger number of
elements than smooth solutions do.

> **Problem 15.43.** Use Femlab to approximate the solution of the Poisson
> equation on the L-shaped domain using the stated boundary conditions.
> Start with a coarse triangulation and use a smallish error tolerance. Print
> out the final mesh and use a ruler to measure the value of h versus r roughly,
> and then plot the points on a log-log plot. Compute a line through the
> data and compare the slope of this to the relation $h \approx r^{1/3}\,\mathrm{TOL}$ based on
> the a priori result.

15.4.4.   Robin and Neumann boundary conditions

Next, we consider Poisson's equation with homogeneous Dirichlet conditions on part of the boundary and non-homogeneous Robin conditions

on the remainder:

$$\begin{cases} -\Delta u = f & \text{in } \Omega, \\ u = 0 & \text{on } \Gamma_1, \\ \partial_n u + \kappa u = g & \text{on } \Gamma_2, \end{cases} \tag{15.44}$$

where $\Gamma = \Gamma_1 \cup \Gamma_2$ is a partition of Γ into two parts and $\kappa \geq 0$. The natural trial space is

$$V = \left\{ v : v = 0 \text{ on } \Gamma_1 \text{ and } \int_\Omega \left(|\nabla v|^2 + v^2 \right) dx < \infty \right\},$$

where the trial functions satisfy the homogeneous Dirichlet condition but the Robin condition is left out. The test space is equal to the trial space, because of the homogeneous Dirichlet condition.

To find the variational formulation, we multiply the Poisson equation by a test function $v \in V$, integrate over Ω, and use Green's formula to move derivatives from u to v:

$$(f, v) = -\int_\Omega \Delta u \, v \, dx = \int_\Omega \nabla u \cdot \nabla v \, dx - \int_\Gamma \partial_n u v \, ds$$
$$= \int_\Omega \nabla u \cdot \nabla v \, dx + \int_{\Gamma_2} \kappa u v \, ds - \int_{\Gamma_2} g v \, ds,$$

where we use the boundary conditions to rewrite the boundary integral. We are thus led to the following variational formulation: find $u \in V$ such that

$$(\nabla u, \nabla v) + \int_{\Gamma_2} \kappa u v \, ds = (f, v) + \int_{\Gamma_2} g v \, ds \quad \text{for all } v \in V. \tag{15.45}$$

It is clear that a solution of (15.44) satisfies (15.45). Conversely, we show that a solution of (15.45) that has two continuous derivatives also satisfies the differential equation (15.44). We integrate (15.45) by parts using Green's formula to put all the derivatives onto u to get

$$-\int_\Omega \Delta u \, v \, dx + \int_\Gamma \partial_n u v \, ds + \int_{\Gamma_2} \kappa u v \, ds = \int_\Omega f v \, dx + \int_{\Gamma_2} g v \, ds$$

$$\text{for all } v \in V$$

or

$$\int_{\Omega} (-\Delta u - f) v \, dx + \int_{\Gamma_2} (\partial_n u + \lambda u - g) v \, ds = 0 \quad \text{for all } v \in V.$$
$$(15.46)$$

By first varying v inside Ω as above while keeping $v = 0$ on the whole of the boundary Γ, it follows that u solves the differential equation $-\Delta u = f$ in Ω. Thus (15.46) reduces to

$$\int_{\Gamma_2} (\partial_n u + \kappa u - g) v \, ds = 0 \quad \text{for all } v \in V.$$

The same argument works here; if $\partial_n u + \kappa u - g$ is non-zero, say positive, at some point of Γ, then it is positive in some small neighborhood of the point in Γ and choosing v to be a positive "hill" centered at the point and zero outside the neighborhood, gives a contradiction. Thus by varying v on Γ_2, we see that the Robin boundary condition $\partial_n u + \lambda u = g$ on Γ_2 must be satisfied (provided $\partial_n u + \kappa u - g$ is continuous).

We recall that boundary conditions like the Dirichlet condition that are enforced explicitly in the choice of the space V are called *essential boundary conditions*. Boundary conditions like the Robin condition that are implicitly contained in the weak formulation are called *natural boundary conditions*. (To remember that we must assume essential conditions: there are two "ss" in assume and essential.)

To discretize the Poisson equation with Robin boundary conditions on part of the boundary (15.44), we triangulate Ω as usual, but we number both the internal nodes and the nodes on Γ_2, where the Robin boundary conditions are posed. We do not number the nodes on Γ_1 where the homogeneous Dirichlet conditions are imposed. Nodes located where Γ_1 and Γ_2 meet should then be considered Dirichlet nodes. We then write U as in (15.41) with $\mathcal{N}_b$ denoting the nodes on Γ_2. In this problem, however, the coefficients of U corresponding to nodes in $\mathcal{N}_b$ are unknown. We substitute (15.41) into the weak form (15.45) and compute the inner products with all the basis functions corresponding to nodes in $\mathcal{N}_h \cup \mathcal{N}_b$ to get a square system. The boundary value g enters into the discrete equations as data on the right-hand side of the linear system for U.

Note that the stiffness matrix and load vector related to (15.45) contain contributions from both integrals over Ω and Γ_2 related to the basis functions corresponding to the nodes on the boundary Γ_2.

To illustrate, we compute the solution of Laplace's equation with a combination of Dirichlet, Neumann and Robin boundary conditions on the domain shown in Fig. 15.19 using Femlab. We show the boundary conditions in the illustration. The problem models e.g. stationary heat flow around a hot water pipe in the ground. We show the mesh that

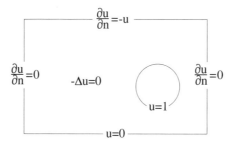

Figure 15.19: A problem with Robin boundary conditions.

Femlab used to compute the approximation so that the error in the L_2 norm is smaller than .0013 together with a contour plot of the approximation in Fig. 15.20. We notice that the level curves are parallel to a boundary with a homogeneous Dirichlet condition, and orthogonal to a boundary with a homogeneous Neumann condition.

Figure 15.20: The adaptive mesh and contour lines of the approximate solution of the problem shown in Fig. 15.19 computed with error tolerance .0013.

Problem 15.44. Compute the discrete system of equations for the finite element approximation of the problem $-\Delta u = 1$ in $\Omega = (0, 1) \times (0, 1)$ with $u = 0$ on the side with $x_2 = 0$ and $\partial_n u + u = 1$ on the other three sides of

Ω using the standard triangulation. Note the contribution to the stiffness matrix from the nodes on the boundary.

Problem 15.45. (a) Show that the variational formulation of the Neumann problem

$$\begin{cases} -\nabla \cdot (a\nabla u) + u = f & \text{in } \Omega, \\ a\partial_n u = g & \text{on } \Gamma, \end{cases} \tag{15.47}$$

where $a(x)$ is a positive coefficient, is to find $u \in V$ such that

$$\int_\Omega a\nabla u \cdot \nabla v \, dx + \int_\Omega uv \, dx = \int_\Omega fv \, dx + \int_\Gamma gv \, ds \quad \text{for all } v \in V, \tag{15.48}$$

where

$$V = \left\{ v : \int_\Omega a|\nabla v|^2 dx + \int_\Omega v^2 dx < \infty \right\}.$$

(b) Apply the finite element method to this problem and prove a priori and a posteriori error estimates. (c) Derive the discrete equations in the case of a uniform triangulation of a square and $a = 1$.

Problem 15.46. Apply the finite element method with piecewise linear approximation to the Poisson equation in three dimensions with a variety of boundary conditions. Compute the stiffness matrix and load vector in some simple case.

15.5. Error estimates in the L_2 norm

> Major scientific progress in different directions can only be gained through extended observation in a prolonged stay in a specific region, while observations during a balloon expedition cannot escape being of a superficial nature. (Nansen, in *Farthest North*, 1897)

So far in this chapter we have used the energy norm to measure the error. The main reason is that the energy norm arises naturally from the variational problem. However, it is often desirable to measure the error in different norms. In fact, specifying the quantities to be approximated and the norm in which to measure the error is a fundamentally important part of modeling, and directly affects the choice of approximation and error control algorithm.

As an example, we develop an error analysis in the L_2 norm. Actually, it is possible to derive an L_2 error estimate directly from the energy norm error estimates. In the two-point boundary value model problem (8.2) with $a = 1$, this follows by first expressing a function v defined on $[0, 1]$ and satisfying $v(0) = 0$ as the integral of its derivative:

$$v(y) = v(0) + \int_0^y v'(x)\, dx = \int_0^y v'(x)\, dx \quad \text{for } 0 \leq y \leq 1,$$

and then using Cauchy's inequality to estimate

$$|v(y)| \leq \int_0^1 |v'(x)|\, dx \leq \left(\int_0^1 1^2\, dx \right)^{1/2} \|v'\| = \|v'\|,$$

where $\| \cdot \|$ denotes the L_2 norm on $(0, 1)$. Squaring this inequality and integrating from 0 to 1 in x, we find

$$\|v\| \leq \|v'\|.$$

Applying this estimate with $v = U - u$ and recalling the a priori energy norm error estimate 8.1, we thus obtain the following L_2 error estimate for the two-point boundary value problem (8.2) with $a = 1$:

$$\|u - U\| \leq C_i \|h u''\|.$$

However, this estimate is not *optimal* because we expect the L_2 error of a good approximation of u in V_h, like for example the piecewise linear interpolant, to decrease quadratically in the mesh size h and not linearly as in the estimate. We now improve the estimate and show that the error of the finite element approximation indeed is optimal in order with respect to the L_2 norm. This is remarkable, because it requires the error in the derivative to be "in average" better than first order.

Problem 15.47. Prove that if e is zero on the boundary of the unit square Ω, then

$$\left(\int_\Omega |e|^2\, dx \right)^{1/2} \leq \left(\int_\Omega |\nabla e|^2\, dx \right)^{1/2}.$$

Hint: extend the proof of the corresponding result in one dimension. Use the result to obtain an error estimate in the L_2-norm for the finite element method for Poisson's equation with homogeneous Dirichlet boundary conditions.

15.5.1. Error analysis based on duality

An approach to error analysis in a general norm is to use the idea of
duality to compute the norm of a function by maximizing weighted average values, or inner products, of the function over a set of weights. For
example,

$$\|u\|_{L_2(\Omega)} = \max_{v \in L_2(\Omega), v \neq 0} \frac{\int_{\Omega} u\, v\, dx}{\|v\|_{L_2(\Omega)}},$$

which follows because Cauchy's inequality shows that the right-hand side
is bounded by the left-hand side, while choosing $v = u$ shows the equality. The fact that the norm of a function can be measured by computing
a sufficient number of average values is both fundamentally important
and widely applicable in a variety of situations. In fact, we already used
this technique in the analysis of the parabolic model problem discussed
in Chapter 9, though without much background. We now give a more
careful development.

We illustrate the idea behind a duality argument by first estimating
the error of a numerical solution of a linear $n \times n$ system of equations
$A\xi = b$. Recall that we discussed this previously in Chapter 7. We let
$\bar{\xi}$ denote a numerical solution obtained for instance using an iterative
method and estimate the Euclidean norm $|e|$ of the error $e = \xi - \bar{\xi}$. We
start by posing the *dual problem* $A^{\top}\eta = e$, where e is considered to be
the data. Of course, we don't know e but we will get around this. Using
the dual problem, we get the following *error representation* by using the
definition of the transpose,

$$|e|^2 = (e, A^{\top}\eta) = (Ae, \eta) = (A\xi - A\bar{\xi}, \eta) = (b - A\bar{\xi}, \eta) = (r, \eta)$$

where $r = b - A\bar{\xi}$ is the *residual error*. Suppose that it is possible to
estimate the solution η of the equation $A^{\top}\eta = e$ in terms of the data e
as

$$|\eta| \leq S|e|, \tag{15.49}$$

where S is a *stability factor*. It follows by Cauchy's inequality that

$$|e|^2 \leq |r||\eta| \leq S|r||e|,$$

or

$$|e| \leq S|r|.$$

This is an a posteriori error estimate for the error e in terms of the residual r and the stability factor S.

We can guarantee that (15.49) holds by defining the stability factor by

$$S = \max_{\theta \in \mathbb{R}^n, \theta \neq 0} \frac{|\zeta|}{|\theta|}$$

where ζ solves $A^T \zeta = \theta$.

The point of this example is to show how duality can be used to get an error representation in terms of the residual and the dual solution, from which the error can be estimated in terms of the residual and a stability factor. We use this approach repeatedly in this book, and also take advantage of the Galerkin orthogonality.

15.5.2. An a posteriori estimate for a two-point boundary value problem

We first prove an a posteriori error estimate in the $L_2(0,1)$ norm , denoted by $\|\cdot\|$, for the problem

$$\begin{cases} -(au')' + cu = f, & \text{in } (0,1), \\ u(0) = 0, \quad u(1) = 0, \end{cases} \tag{15.50}$$

where $a(x) > 0$ and $c(x) \geq 0$. We denote by U the cG(1) solution to the problem using the usual finite element space V_h of continuous piecewise linear functions. The *dual problem* takes just the same form as (15.50) because the given problem is symmetric:

$$\begin{cases} -(a\varphi')' + c\varphi = e, & \text{in } (0,1), \\ \varphi(0) = 0, \quad \varphi(1) = 0, \end{cases} \tag{15.51}$$

where $e = u - U$. We now use (15.50), (15.51), and the Galerkin orthogonality with the test function $v = \pi_h e \in V_h$, to obtain

$$\|e\|^2 = \int_0^1 e(-(a\varphi')' + c\varphi)\,dx = \int_0^1 (ae'\varphi' + ce\varphi)\,dx$$

$$= \int_0^1 (au'\varphi' + cu\varphi)\,dx - \int_0^1 (aU'\varphi' + cU\varphi)\,dx$$

$$= \int_0^1 f\varphi\,dx - \int_0^1 (aU'\varphi' + cU\varphi)\,dx$$

$$= \int_0^1 f(\varphi - \pi_h\varphi)\,dx - \sum_{j=1}^{M+1} \int_{I_j} (aU'(\varphi - \pi_h\varphi)' + cU(\varphi - \pi_h\varphi))\,dx.$$

Integrating by parts over each sub-interval I_j, we find that all the boundary terms disappear, and we end up with

$$\|e\|^2 \leq \|h^2 R(U)\| \|h^{-2}(\varphi - \pi_h\varphi)\|,$$

where $R(U) = f + (aU')' - cU$ on each sub-interval. Using an interpolation error estimate of the form $\|h^{-2}(\varphi - \pi_h\varphi)\| \leq C_i\|\varphi''\|$, and defining the strong stability factor by

$$S = \max_{\xi \in L_2(I)} \frac{\|\varphi''\|}{\|\xi\|} \tag{15.52}$$

where φ satisfies (15.51) with e replaced by ξ, we obtain the following a posteriori error estimate:

Theorem 15.4. *The finite element approximation U of (8.9) satisfies*

$$\|u - U\| \leq SC_i\|h^2 R(U)\|.$$

Note that the size of the stability factor S varies with the choice of the coefficients $a(x)$ and $c(x)$.

Problem 15.48.   Prove that if $a > 0$ and $c \geq 0$ are constant, then $S \leq a^{-1}$.

The implementation of an adaptive error control based on Theorem 15.51 is the same as for error control based on the energy norm. For an example, we choose $a = 0.01$, $c = 1$ and $f(x) = 1/x$ and compute using Femlab1d with the L_2 norm of the error bounded by $TOL = .01$.

We plot the finite element approximation, the residual, and the mesh size in Fig. 15.21. In this example, there are two sources of singularities in the solution. First, because the diffusion coefficient a is small, the solution may become steep near the boundaries, forming what are called *boundary layers*. Secondly, the source term f itself is large near $x = 0$ and undefined at 0. The singularity in the data f affects the residual, while the size of a affects both the residual and the stability factor S. The adaptive algorithm approximates the stability factor S by solving the dual problem (15.51) with e replaced by an approximation. In this example, $S \approx 37$.

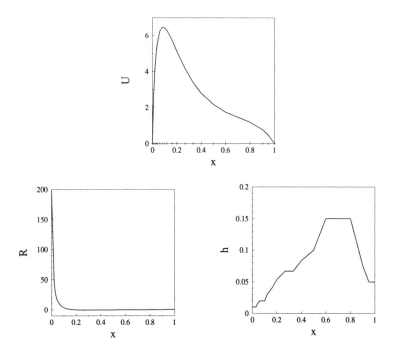

Figure 15.21: Finite element approximation, residual error, and mesh-size computed with adaptive error control based on the L_2 norm.

15.5.3. A priori error estimate for a two-point boundary value problem

We also prove an a priori error estimate in the L_2 norm assuming for simplicity that the mesh size h is constant, and $c = 0$. Note the presence of the weighted norm $\|\cdot\|_a$.

Theorem 15.5. *The finite element approximation U of (8.9) satisfies*

$$\|u - U\| \le C_i S_a \|h(u - U)'\|_a \le C_i^2 S_a \|h^2 u''\|_a,$$

where $S_a = \max_{\xi \ne 0} \|\varphi''\|_a / \|\xi\|$ with φ satisfying (15.51) with e replaced by ξ.

Proof. Assuming that φ satisfies (15.51) with $c = 0$, and using the Galerkin orthogonality (8.10) and an L_2 estimate for the interpolation error, we obtain

$$\|e\|^2 = \int_0^1 ae'\varphi'\, dx = \int_0^1 ae'(\varphi - \pi_h\varphi)'\, dx$$
$$\le \|he'\|_a \|h^{-1}(\varphi - \pi_h\varphi)'\|_a \le C_i \|he'\|_a \|\varphi''\|_a,$$

where $C_i = C_i(a)$. The proof is completed by using the definition of S_a and noting that multiplying the energy norm error estimate by h gives

$$\|he'\|_a \le C_i \|h^2 u''\|_a. \tag{15.53}$$

∎

This estimate generalizes to the case of variable h assuming that the mesh size h does not change too rapidly from one element to the next.

Problem 15.49. Prove that if $a > 0$ then $S_a \le 1/\sqrt{a}$. Note that S and S_a involve somewhat different norms, which is compensated by the presence of the factor a in $R(U)$.

15.5.4. A priori and a posteriori error estimates for the Poisson equation

We now carry through the same program for the Poisson equation in two dimensions. We here assume that the mesh function $h(x)$ is differentiable and there is a constant $\tau_1 > 0$ such that $\tau_1 h_K \le h(x) \le h_K$ for $x \in K$

for each K in $\mathcal{T}_h$. This may be realized by smoothing of the original piecewise constant mesh function.

The proofs are based on a basic strong stability (or elliptic regularity) estimate for the solution of the Poisson equation (15.18) giving an estimate of the strong stability factor S. In Chapter 21, we give the proof in the case of a convex domain with smooth boundary. In this case $S = 1$, and the stability estimate states that all second derivatives of a function u vanishing on the boundary of Ω can be bounded by the particular combination of second derivatives given by Δu.

Theorem 15.6. *If Ω is convex with polygonal boundary, or if Ω is a general domain with smooth boundary, then there is a constant S independent of f, such that the solution u of (15.18) satisfies*

$$\|D^2 u\| \leq S\|\Delta u\| = S\|f\|. \tag{15.54}$$

If Ω is convex, then $S = 1$.

The a priori error estimate is

Theorem 15.7. *Let Ω be convex with polygonal boundary or a general domain with smooth boundary. Then there exists a constant C_i only depending on τ and τ_1, such that the finite element approximation U of the Poisson problem (15.18) satisfies*

$$\|u - U\| \leq SC_i\|h\nabla(u - U)\|, \tag{15.55}$$

where S is defined in Theorem 15.6. Furthermore, if $|\nabla h(x)|_\infty \leq \mu$ for $x \in \Omega$ for some sufficiently small positive constant μ, then

$$\|h\nabla(u - U)\| \leq C_i\|h^2 D^2 u\|, \tag{15.56}$$

where C_i also depends on μ. In particular, if Ω is convex then

$$\|u - U\| \leq C_i\|h^2 D^2 u\|. \tag{15.57}$$

Proof. Letting φ solve the dual problem $-\Delta\varphi = e$ in Ω together with $\varphi = 0$ on Γ, we obtain by integration by parts, using the Galerkin orthogonality and the interpolation estimate Theorem 14.2

$$\|e\|^2 = (u - U, u - U) = (\nabla(u - U), \nabla\varphi)$$
$$= (\nabla(u - U), \nabla(\varphi - \pi_h\varphi)) \leq C_i\|h\nabla(u - U)\|\|D^2\varphi\|,$$

from which the first estimate follows using the strong stability result. The second estimate (15.56) follows directly from the energy norm error estimate if h is constant and we discuss the general result in the advanced companion volume. The final result (15.57) is obtained using the regularity estimate (15.54). ∎

The a posteriori error estimate is

Theorem 15.8. *There are constants C_i and S such that, if U is the finite element approximation of (15.18), then with the residual R defined as in Theorem 15.3,*

$$\|u - U\| \le SC_i\|h^2 R(U)\|. \qquad (15.58)$$

If Ω is convex, then $S = 1$.

Proof. With φ defined as in the previous proof, we have

$$\|e\|^2 = (\nabla(u - U), \nabla\varphi) = (f, \varphi) - (\nabla U, \nabla\varphi)$$
$$= (f, \varphi - \pi_h\varphi) - (\nabla U, \nabla(\varphi - \pi_h\varphi)).$$

The desired result follows by an argument similar to that used in the a posteriori energy norm estimate by estimating $\|h^{-2}(\varphi - \pi_h\varphi)\|$ in terms of $C_i\|D^2\varphi\|$ and using the strong stability estimate to close the loop. ∎

It is like an attempt, over and over again, to reveal the heart of things. (K. Jarret)

A poem should be equal to:
Not true ...
A poem should not mean
But be. (Archibald MacLeish)

16

The Heat Equation

The simpler a hypothesis is, the better it is. (Leibniz)

In this chapter, we consider the numerical solution of the *heat equation*, which is the prototype of a linear parabolic partial differential equation. Recall that we originally derived the heat equation in Chapter 13 to model heat flow in a conducting object. More generally, the same equation may be used to model *diffusion* type processes. From a quite different point of view, we begin this chapter by deriving the heat equation as a consequence of Maxwell's equations under some simplifying assumptions. After that, we recall some of the properties of solutions of the heat equation, focussing on the characteristic parabolic "smoothing" and stability properties. We then proceed to introduce a finite element method for the heat equation, derive a posteriori and a priori error estimates and discuss adaptive error control. The analysis follows the basic steps used in the analysis of the parabolic model problem in Chapter 9 and of Poisson's equation in Chapter 15.

16.1. Maxwell's equations

Thus then, we are led to the conception of a complicated mechanism capable of a vast variety of motion... Such a mechanism must be subject to the general laws of Dynamics, and we ought to be able to work out all the consequences of its motion, provided we know the form of the relation between the motions of the parts... We now proceed to investigate whether the properties of that which constitutes the electromagnetic field, deduced

from electromagnetic phenomena alone, are sufficient to explain
the propagation of light through the same substance. (Maxwell)

We met in the previous chapter a special case of Maxwell's equations in
the form of Poisson's equation for an electric potential in electrostatics.
Here, we consider another special case that leads to a parabolic problem
for a magnetic potential, which in the simplest terms reduces to the heat
equation. Another important special case gives rise to the wave equation
studied in Chapter 17.

It is remarkable that the complex phenomena of interaction between
electric and magnetic fields can be described by the relatively small set
of Maxwell's equations:

$$\begin{cases} \dfrac{\partial B}{\partial t} + \nabla \times E = 0, \\[2mm] -\dfrac{\partial D}{\partial t} + \nabla \times H = J, \\[2mm] \nabla \cdot B = 0, \quad \nabla \cdot D = \rho, \\[2mm] B = \mu H, \quad D = \epsilon E, \quad J = \sigma E, \end{cases} \tag{16.1}$$

where E is the electric field, H is the magnetic field, D is the electric
displacement, B is the magnetic flux, J is the electric current, ρ is the
charge, μ is the magnetic permeability, ϵ is the dielectric constant, and σ
is the electric conductivity. The first equation is referred to as *Faraday's
law*, the second is *Ampère's law*, $\nabla \cdot D = \rho$ is *Coulomb's law*, $\nabla \cdot B = 0$
expresses the absence of "magnetic charge", and $J = \sigma E$ is *Ohm's law*.
Maxwell included the term $\partial D / \partial t$ for purely mathematical reasons and
then using calculus predicted the existence of electromagnetic waves
before these had been observed experimentally. We assume to start
with that $\partial D / \partial t$ can be neglected; cf. Problem 16.1 and Problem 16.40.

Because $\nabla \cdot B = 0$, B can be written as $B = \nabla \times A$, where A is a
magnetic vector potential. Inserting this into Faraday's law gives

$$\nabla \times \left(\frac{\partial A}{\partial t} + E \right) = 0,$$

from which it follows that

$$\frac{\partial A}{\partial t} + E = \nabla V,$$

for some scalar potential V. Multiplying by σ and using the laws of Ohm and Ampère, we obtain a vector equation for the magnetic potential A:

$$\sigma \frac{\partial A}{\partial t} + \nabla \times \left(\mu^{-1} \nabla \times A \right) = \sigma \nabla V.$$

To obtain a scalar equation in two variables, we assume that $B = (B_1, B_2, 0)$ is independent of x_3. It follows that A has the form $A = (0, 0, u)$ for some scalar function u that depends only on x_1 and x_2, so that $B_1 = \partial u / \partial x_2$ and $B_2 = -\partial u / \partial x_1$, and we get a scalar equation for the scalar magnetic potential u of the form

$$\sigma \frac{\partial u}{\partial t} - \nabla \cdot \left(\mu^{-1} \nabla u \right) = f, \tag{16.2}$$

for some function $f(x_1, x_2)$. This is a parabolic equation with variable coefficients σ and μ. Choosing $\sigma = \mu = 1$ leads to the heat equation:

$$\begin{cases} \frac{\partial}{\partial t} u(x, t) - \Delta u(x, t) = f(x, t) & \text{for } x \in \Omega,\, 0 < t \leq T, \\ u(x, t) = 0 & \text{for } x \in \Gamma,\, 0 < t \leq T, \\ u(x, 0) = u_0(x) & \text{for } x \in \Omega, \end{cases} \tag{16.3}$$

where $\Omega \subset \mathbb{R}^2$ with boundary Γ, and we posed homogeneous Dirichlet boundary conditions.

Problem 16.1. What equation is obtained if $\partial D / \partial t$ is not neglected, but the x_3 independence is kept?

Problem 16.2. Show that the magnetic field H around a unit current along the x_3-axis is given by $\frac{1}{2\pi|x|}(-x_2, x_1, 0)$, where $|x| = (x_1^2 + x_2^2)^{\frac{1}{2}}$.

16.2. The basic structure of solutions of the heat equation

The structure of solutions of the heat equation is closely related to the properties of solutions of the initial value problems discussed in Chapter 10 and the boundary value problems discussed in Chapters 8 and 15.

16.2.1. Separation of variables and Fourier series

For some domains the method of separation of variables can be employed to find analytic solutions of the heat equation in terms of series expansions into eigenfunctions. We illustrate this approach for the one-dimensional, homogeneous heat equation

$$\begin{cases} \dot{u}(x,t) - u''(x,t) = 0 & \text{for } 0 < x < \pi, \, t > 0, \\ u(0,t) = u(\pi,t) = 0 & \text{for } t > 0, \\ u(x,0) = u_0(x) & \text{for } 0 < x < \pi. \end{cases} \qquad (16.4)$$

We start by seeking solutions of the differential equation and the boundary conditions in (16.4) of the form $u(x,t) = \varphi(x)\,\psi(t)$ with $\varphi(0) = \varphi(\pi) = 0$. Substituting this into (16.4) and separating the functions depending on x and t, gives

$$\frac{\dot{\psi}(t)}{\psi(t)} = \frac{\varphi''(x)}{\varphi(x)}.$$

Since x and t are independent variables, each fraction must be equal to the same constant $-\lambda \in \mathbb{R}$ and we are led to the eigenvalue problem

$$\begin{cases} -\varphi''(x) = \lambda\varphi(x) & \text{for } 0 < x < \pi, \\ \varphi(0) = \varphi(\pi) = 0, \end{cases} \qquad (16.5)$$

and the initial value problem

$$\begin{cases} \dot{\psi}(t) = -\lambda\psi(t) & \text{for } t > 0, \\ \psi(0) = 1, \end{cases} \qquad (16.6)$$

where $\psi(0)$ is normalized to 1. Thus, seeking solutions in the form of a product of functions of one variable decouples the partial differential equation into two ordinary differential equations. It is important to this technique that the differential equation is linear, homogeneous, and has constant coefficients.

The problem (16.5) is an eigenvalue problem with eigenfunctions $\varphi_j(x) = \sin(jx)$ and corresponding eigenvalues $\lambda_j = j^2$, $j = 1, 2, \ldots$ For each eigenvalue, we can solve (16.6) to get the corresponding solution $\psi(t) = \exp(-j^2 t)$. We obtain a set of solutions $\{\exp(-j^2 t)\sin(jx)\}$ of (16.4) with corresponding initial data $\{\sin(jx)\}$ for $j = 1, 2, \ldots$, which are called the *eigenmodes*. Each eigenmode decays exponentially as time passes and the rate of decay increases with the frequency j. We illustrate this in Fig. 16.1. Any finite linear combination of eigenmodes

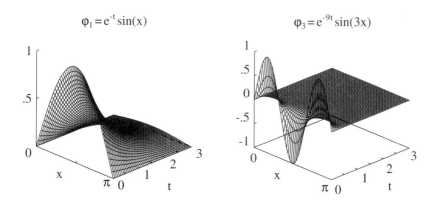

$$\varphi_1 = e^{-t}\sin(x) \qquad\qquad \varphi_3 = e^{-9t}\sin(3x)$$

Figure 16.1: The solutions of the heat equation corresponding to frequencies $j = 1$ and $j = 3$.

$$\sum_{j=1}^{J} a_j \exp(-j^2 t)\sin(jx),$$

with coefficients $a_j \in \mathbb{R}$, is a solution of the homogeneous heat equation corresponding to the initial data

$$u_0(x) = \sum_{j=1}^{J} a_j \sin(jx). \tag{16.7}$$

More generally, if the initial data u_0 has a convergent Fourier series,

$$u_0(x) = \sum_{j=1}^{\infty} u_{0,j}\sin(jx),$$

with Fourier coefficients given by $u_{0,j} = 2\pi^{-1}\int_0^\pi u_0(x)\sin(jx)\,dx$, then the function defined by

$$u(x,t) = \sum_{j=1}^{\infty} u_{0,j}\exp(-j^2 t)\sin(jx), \tag{16.8}$$

solves $\dot{u} - u'' = 0$. This is seen by differentiating the series term by term, which is possible because the coefficients $u_{0,j}\exp(-j^2 t)$ decrease very quickly with j as long as $t > 0$. Moreover $u(0) = u(\pi) = 0$, so to show that u is a solution of (16.4), we only have to check that $u(x,t)$ equals the initial data u_0 at $t = 0$. If we only require that $u_0 \in L_2(0, \pi)$, then it is possible to show that

$$\lim_{t \to 0} \|u(\cdot, t) - u_0\| = 0. \tag{16.9}$$

If u_0 has additional smoothness and also satisfies the boundary conditions $u_0(0) = u_0(\pi) = 0$ (which is not required if we only assume that $u_0 \in L_2(0, \pi)$), then the initial data is assumed in the stronger pointwise sense, i.e.

$$\lim_{t \to 0} u(x,t) = u_0(x) \quad \text{for } 0 < x < \pi. \tag{16.10}$$

Recalling that the rate at which a function's Fourier coefficients tends to zero reflect the smoothness of the function, we see from the solution formula (16.8) that a solution $u(x,t)$ of the homogeneous heat equation becomes smoother with increasing time. This is known as *parabolic smoothing*. We illustrate the smoothing in Fig. 16.2, where we plot the solution starting with the discontinuous function

$$u_0(x) = \begin{cases} x, & 0 \leq x \leq \pi/2, \\ x - \pi, & \pi/2 < x \leq \pi, \end{cases}$$

at various times (the solution formula is given in Problem 16.3). This corresponds well with intuition about a diffusive process in which sharp features are smoothed out for positive time. Nonsmooth functions have slowly decreasing Fourier coefficients, so that the Fourier coefficients of the high modes with j large are relatively large compared to those of smooth functions. As soon as $t > 0$, these high modes are damped rapidly because of the presence of the factor $\exp(-j^2 t)$, and the solution becomes smoother as t increases.

Problem 16.3. Verify the following formulas for the solutions of the heat equation corresponding to the indicated initial data:

1. $u_0(x) = x(\pi - x)$,

$$u(x,t) = \sum_{j=1}^{\infty} \frac{8}{(2j-1)^3} e^{-(2j-1)^2 t} \sin((2j-1)x).$$

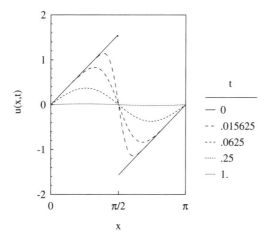

Figure 16.2: The evolution of discontinuous initial data for the heat equation.

2. $u_0(x) = \begin{cases} x, & 0 \le x \le \pi/2 \\ \pi - x, & \pi/2 < x \le \pi \end{cases}$,

$$u(x,t) = \sum_{j=1}^{\infty} \frac{4(-1)^{j+1}}{\pi(2j-1)^2} e^{-(2j-1)^2 t} \sin((2j-1)x).$$

3. $u_0(x) = \begin{cases} x, & 0 \le x \le \pi/2 \\ x - \pi, & \pi/2 < x \le \pi \end{cases}$,

$$u(x,t) = \sum_{j=1}^{\infty} \frac{(-1)^{j+1}}{j} e^{-4j^2 t} \sin(2jx).$$

Problem 16.4. Find a formula for the solution of (16.4) with the Dirichlet boundary conditions replaced by the Neumann conditions $u'(0) = 0$ and $u'(\pi) = 0$. Hint: the series expansion is in terms of cosine functions. Do the same with the boundary conditions $u(0) = 0$ and $u'(\pi) = 0$.

Problem 16.5. (a) Prove (16.10) assuming that $\sum_{j=1}^{\infty} |u_{0,j}| < \infty$. (b) Prove (16.9) assuming that $u_0 \in L_2(0, \pi)$, that is $\sum_{j=1}^{\infty} |u_{0,j}|^2 < \infty$.

Problem 16.6. (Strauss ([18])) Waves in a resistant

medium are described by the problem

$$\begin{cases} \ddot{u}(x,t) + c\dot{u}(x,t) - u''(x,t) = 0, & 0 < x < \pi, t > 0, \\ u(0,t) = u(\pi,t) = 0, & t > 0, \\ u(x,0) = u_0(x), & 0 < x < \pi, \end{cases}$$

where $c > 0$ is a constant. Write down a series expansion for the solution using separation of variables. Can you say something about the behavior of the solution as time passes?

Problem 16.7. Give the Fourier series formula for the solution of the homogeneous heat equation (16.3) posed on the unit square $\Omega = (0,1) \times (0,1)$. Hint: first use separation of variables to get an ordinary differential equation in t and an eigenvalue problem for the Laplacian in (x_1, x_2). Then, use separation of variables to decompose the eigenvalue problem for the Laplacian into independent eigenvalue problems for x_1 and x_2. Hint: see Chapter 15.

Problem 16.8. Consider the *backward heat eaquation*

$$\begin{cases} \dot{u}(x,t) + u''(x,t) = 0 & \text{for } 0 < x < \pi, t > 0, \\ u(0,t) = u(\pi,t) = 0 & \text{for } t > 0, \\ u(x,0) = u_0(x) & \text{for } 0 < x < \pi. \end{cases} \tag{16.11}$$

Write down a solution formula in the case u_0 is a finite Fourier series of the form (16.7). Investigate how the different components of u_0 get amplified with time. Why is the equation called the backward heat equation? Can you find a connection to image reconstruction?

16.2.2. The fundamental solution of the heat equation

The solution of the homogeneous heat equation

$$\begin{cases} \dot{u} - \Delta u = 0 & \text{in } \mathbb{R}^2 \times (0,\infty), \\ u(\cdot,0) = u_0 & \text{in } \mathbb{R}^2, \end{cases} \tag{16.12}$$

with u_0 equal to the delta function at the origin δ_0, is called the *fundamental solution* of the heat equation and is given by

$$u(x,t) = E(x,t) = \frac{1}{4\pi t} \exp\left(-\frac{|x|^2}{4t}\right). \tag{16.13}$$

Direct computation shows that $E(x,t)$ solves $\dot{E} - \Delta E = 0$ for $x \in \mathbb{R}^2$ and $t > 0$. Further $E(\cdot,t)$ approaches the delta function δ_0 as $t \to 0^+$ since

$E(x, t) \geq 0$, $\int_{\mathbb{R}^2} E(x, t)\, dx = 1$ for $t > 0$, and $E(x, t)$ rapidly decays as $|x|/\sqrt{t}$ increases, so that the support of $E(x, t)$ becomes more and more concentrated around $x = 0$ as $t \to 0^+$. In terms of a model of heat, $E(x, t)$ corresponds to choosing the initial conditions to be a "hot spot" at the origin. In Fig. 16.3 we plot $E(x, t)$ at three different times.

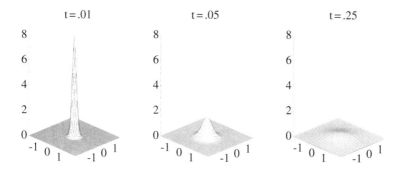

Figure 16.3: The fundamental solution $E(x, t)$ at three times.

Problem 16.9. Show that E defined by (16.13) solves $\dot{E} - \Delta E = 0$ for $t > 0$, and verify that $\int_{\mathbb{R}} E(x, t)\, dx = 1$.

Problem 16.10. Determine the fundamental solution of the heat equation in $\mathbb{R}^d$, d=1,3.

Problem 16.11. Give the formula for the fundamental solution $E_\epsilon(x, t)$ for the heat equation $\dot{u} - \epsilon \Delta u = 0$ in two space dimensions, where ϵ is a positive constant. Determine, as a function of ϵ and t, the diameter of the set of points x outside which $E_\epsilon(x, t)$ is essentially zero.

The solution of (16.12) can be expressed in terms of the fundamental solution and the initial data as follows:

$$u(x, t) = \frac{1}{4\pi t} \int_{\mathbb{R}^2} u_0(y) \exp\left(-\frac{|x - y|^2}{4t}\right) dy. \qquad (16.14)$$

Problem 16.12. Motivate this formula.

From the solution formula we see that the value $u(x, t)$ at a point $x \in \mathbb{R}^2$ and $t > 0$ is a weighted mean value of all the values $u_0(y)$ for $y \in \Omega$. The influence of the value $u_0(y)$ on $u(x,t)$ decreases with increasing distance $|x-y|$ and decreasing time t. In principle, information appears to travel with an *infinite speed of propagation* because even for very small time t there is an influence on $u(x,t)$ from $u_0(y)$ for $|x-y|$ arbitrarily large. However, the nature of the fundamental solution causes the influence to be extremely small if t is small and $|x-y|$ is large. In particular, the solution formula shows that if $u_0 \geq 0$ is concentrated around $x = 0$, say $u_0(x) \equiv 0$ for $|x| \geq d$ for some small $d > 0$, then $u(x,t)$ "spreads out" over a disk of radius proportional to $\sqrt{t}$ for $t > 0$ and rapidly decays to zero outside this disk.

Problem 16.13. (a) Write a code that inputs an x and t and then uses the composite trapezoidal rule to approximate the integrals in (16.14) when $u_0(x)$ is 1 for $|x| \leq 1$ and 0 otherwise and use the code to generate plots of the solution at several different times. (b) *(Harder.)* Verify the claim about the rate of spread of the solution.

16.3. Stability

Throughout the book, we emphasize that the stability properties of parabolic problems are an important characteristic. To tie into the previous stability results for parabolic-type problems, we prove a strong stability estimate for an abstract parabolic problem of the form: find $u(t) \in H$ such that

$$\begin{cases} \dot{u}(t) + Au(t) = 0 & \text{for } t > 0, \\ u(0) = u_0, \end{cases} \qquad (16.15)$$

where H is a vector space with inner product $(\cdot, \cdot)$ and norm $\|\cdot\|$, A is a positive semi-definite symmetric linear operator defined on a subspace of H, i.e. A is a linear transformation satisfying $(Aw, v) = (w, Av)$ and $(Av, v) \geq 0$ for all v and w in the domain of definition of A, and u_0 is the initial data. In the parabolic model problem of Chapter 10, $H = \mathbb{R}^d$ and A is a positive semi-definite symmetric $d \times d$ matrix. In the case of the heat equation (16.3), $A = -\Delta$ is defined on the infinite-dimensional space of functions v in $L^2(\Omega)$ which are square integrable and satisfy homogeneous Dirichlet boundary conditions.

Lemma 16.1. *The solution u of (16.15) satisfies for $T > 0$,*

$$\|u(T)\|^2 + 2 \int_0^T (Au(t), u(t))\, dt = \|u_0\|^2, \tag{16.16}$$

$$\int_0^T t\|Au(t)\|^2\, dt \le \frac{1}{4}\|u_0\|^2, \tag{16.17}$$

$$\|Au(T)\| \le \frac{1}{\sqrt{2}\,T}\|u_0\|. \tag{16.18}$$

Proof. The proof uses the same ideas used to show (10.26). Taking the inner product of (16.15) with $u(t)$, we obtain

$$\frac{1}{2}\frac{d}{dt}\|u(t)\|^2 + (Au(t), u(t)) = 0,$$

from which (16.16) follows.

Next, taking the inner product of the first equation of (16.15) with $tAu(t)$ and using the fact that

$$(\dot{u}(t), tAu(t)) = \frac{1}{2}\frac{d}{dt}\big(t(Au(t), u(t))\big) - \frac{1}{2}(Au(t), u(t)),$$

since A is symmetric, we find after integration that

$$\frac{1}{2}T(Au(T), u(T)) + \int_0^T t\|Au(t)\|^2\, dt = \frac{1}{2}\int_0^T (Au(t), u(t))\, dt,$$

from which (16.17) follows using (16.16) and the fact that $(Av, v) \ge 0$.

Finally, taking the inner product in (16.15) with $t^2 A^2 u(t)$, we obtain

$$\frac{1}{2}\frac{d}{dt}\big(t^2\|Au(t)\|^2\big) + t^2(A^2 u(t), Au(t)) = t\|Au(t)\|^2,$$

from which (16.18) follows after integration and using (16.17).

∎

Problem 16.14. Assuming that there is an $a > 0$ such that A is strictly positive-definite, so that $(Av, v) \ge a\|v\|^2$ for all v, show that the solution of $\dot{u} + Au = f$, $u(0) = u_0$, satisfies

$$\|u(T)\|^2 + a\int_0^T \|u(t)\|^2\, dt \le \|u_0\|^2 + \frac{1}{a}\int_0^T \|f\|^2\, dt.$$

Hint: use that $|(v, w)| \le (4\epsilon)^{-1}\|v\|^2 + \epsilon\|w\|^2$ for any $\epsilon > 0$.

In the case of a solution of the heat equation (16.3), these estimates read

$$\|u(T)\|^2 + 2 \int_0^T (\nabla u(t), \nabla u(t))\, dt \le \|u_0\|^2, \qquad (16.19)$$

$$\int_0^T t\|\Delta u(t)\|^2\, dt \le \frac{1}{4}\|u_0\|^2, \qquad (16.20)$$

$$\|\Delta u(T)\| \le \frac{1}{\sqrt{2\,T}}\|u_0\|. \qquad (16.21)$$

Problem 16.15. (a) Consider u and $\tilde{u}$ solving (16.3) with initial data $u_0(x)$ and $\tilde{u}_0(x) = u_0(x) + \epsilon(x)$ respectively. Show that the difference $\tilde{u} - u$ solves (16.3) with initial data $\epsilon(x)$. (b) Give estimates for the difference between u and $\tilde{u}$. (c) Prove that the solution of (16.3) is unique.

Recall that we call these *strong stability* estimates because they provide bounds on derivatives of the solution as well as the solution itself. Such estimates are related to parabolic smoothing. For example, (16.21) implies that the L_2 norm of the derivative $\dot{u}(T) = \Delta u(T)$ decreases (increases) like $1/T$ as T increases (decreases), which means that the solution becomes smoother as time passes.

Problem 16.16. Compute (exactly or approximately) the quantities on the left-hand sides of (16.16), (16.17), and (16.18) for the solutions of (16.4) computed in Problem 16.3. Compare to the bounds on the right-hand sides.

Problem 16.17. Prove the stability estimates of Lemma 16.1 applied to the one-dimensional heat equation (13.24) using the Fourier series formula for the solution.

16.4. A finite element method for the heat equation

The time discretization of the heat equation (16.3) is based on a partition $0 = t_0 < t_1 < \cdots < t_N = T$ of the time interval $I = [0, T]$ into sub-intervals $I_n = (t_{n-1}, t_n)$ of length $k_n = t_n - t_{n-1}$. We divide each *space-time slab* $S_n = \Omega \times I_n$ into space-time prisms $K \times I_n$, where $\mathcal{T}_n = \{K\}$ is a triangulation of Ω with mesh function h_n; see Fig. 16.4. Note that the space mesh may change from one time interval to the next. We construct a finite element method using approximations consisting

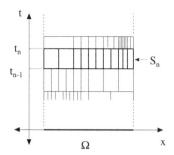

Figure 16.4: Space-time discretization for the cG(1)dG(r) method.

of continuous piecewise linear functions in space and discontinuous polynomials of degree r in time, which we call the cG(1)dG(r) method. We define the trial space $W_k^{(r)}$ to be the set of functions $v(x,t)$ defined on $\Omega \times I$ such that the restriction $v|_{S_n}$ of v to each space-time slab S_n is continuous and piecewise linear in x and a polynomial of degree r in t, that is, $v|_{S_n}$ belongs to the space

$$W_{kn}^{(r)} = \left\{ v(x,t) : v(x,t) = \sum_{j=0}^{r} t^j \psi_j(x), \ \psi_j \in V_n, \ (x,t) \in S_n \right\},$$

where $V_n = V_{h_n}$ is the space of continuous piecewise linear functions vanishing on Γ associated to $\mathcal{T}_n$. The "global" trial space $W_k^{(r)}$ is the space of functions v defined on $\Omega \times I$, such that $v|_{S_n} \in W_{kn}^{(r)}$ for $n = 1, 2, ..., N$. The functions in $W_k^{(r)}$ in general are discontinuous across the discrete time levels t_n and we use the usual notation $[w_n] = w_n^+ - w_n^-$ and $w_n^{+(-)} = \lim_{s \to 0+(-)} w(t_n + s)$.

Problem 16.18. Describe a set of basis functions for (a) $W_{kn}^{(0)}$ and (b) $W_{kn}^{(1)}$.

The cG(1)dG(r) method is based on a variational formulation of (16.3) as usual and reads: find $U \in W_k^{(r)}$ such that for $n = 1, 2, \ldots, N$,

$$\int_{I_n} \left((\dot{U}, v) + (\nabla U, \nabla v) \right) dt + ([U_{n-1}], v_{n-1}^+) = \int_{I_n} (f, v) \, dt$$

$$\text{for all } v \in W_{kn}^{(r)}, \quad (16.22)$$

where $U_0^- = u_0$ and $(\cdot, \cdot)$ is the $L_2(\Omega)$ inner product.

Using the discrete Laplacian Δ_n, se (15.27), we may write (16.22)in the case $r = 0$ as follows: find $U_n \in V_n$:

$$(I - k_n \Delta_n) U_n = P_n U_{n-1} + \int_{I_n} P_n f \, dt, \qquad (16.23)$$

where we set $U_n = U_n^- = U|_{I_n} \in V_n$, and P_n is the $L_2(\Omega)$-projection onto V_n. Note that the "initial data" $U_{n-1} \in V_{n-1}$ from the previous time interval I_{n-1} is projected into the space V_n. If $V_{n-1} \subset V_n$, then $P_n U_{n-1} = U_{n-1}$. In the case $r = 1$, writing $U(t) = \Phi_n + (t - t_{n-1})\Psi_n$ on I_n with Φ_n, $\Psi_n \in V_n$, then (16.22) becomes

$$\begin{cases} (I - k_n \Delta_n)\Phi_n + \left(I - \dfrac{k_n}{2}\Delta_n\right)\Psi_n = P_n U_{n-1} + \displaystyle\int_{I_n} P_n f \, dt, \\ \left(\dfrac{1}{2}I - \dfrac{k_n}{3}\Delta_n\right)\Psi_n - \dfrac{k_n}{2}\Delta_n \Phi_n = \displaystyle\int_{I_n} \dfrac{t - t_{n-1}}{k_n} P_n f \, dt, \end{cases} \qquad (16.24)$$

which gives a system of equations for Φ_n and Ψ_n.

Problem 16.19. Verify (16.23) and (16.24).

Problem 16.20. Writing $U(t) = \Phi_n(t_n - t)/k_n + \Psi_n(t - t_{n-1})/k_n$ on I_n with Φ_n, $\Psi_n \in V_n$, formulate equations for the cG(1)dG(1) approximation using the discrete Laplacian.

16.4.1. Constructing the discrete equations

To construct the matrix equation that determines U_n in the case $r = 0$ according to (16.23), we introduce some notation. We let $\{\varphi_{n,j}\}$ denote the nodal basis of V_n associated to the M_n interior nodes of $\mathcal{T}_n$ numbered in some fashion, so U_n can be written

$$U_n = \sum_{j=1}^{M_n} \xi_{n,j} \varphi_{n,j},$$

where the coefficients $\xi_{n,j}$ are the nodal values of U_n. We abuse notation to let $\xi_n = (\xi_{n,j})$ denote the vector of coefficients. We define the $M_n \times M_n$ *mass matrix* B_n, *stiffness matrix* A_n, and again abusing notation, the $M_n \times 1$ data vector b_n with coefficients

$$(B_n)_{ij} = (\varphi_{n,j}, \varphi_{n,i}), \quad (A_n)_{ij} = (\nabla \varphi_{n,j}, \nabla \varphi_{n,i}), \quad (b_n)_i = (f, \varphi_{n,i}),$$

for $1 \le i, j \le M_n$. Finally, we define the $M_n \times M_{n-1}$ matrix $B_{n-1,n}$ with coefficients

$$(B_{n-1,n})_{ij} = (\varphi_{n,j}, \varphi_{n-1,i}) \quad 1 \le i \le M_n, 1 \le j \le M_{n-1}. \tag{16.25}$$

The discrete equation for the cG(1)dG(0) approximation on I_n is

$$(B_n + k_n A_n)\xi_n = B_{n-1,n}\xi_{n-1} + b_n. \tag{16.26}$$

The coefficient matrix $B_n + k_n A_n$ of this system is sparse, symmetric, and positive-definite and the system can be solved using a direct or an iterative method.

Problem 16.21. Prove that $B_{n-1,n}\xi_{n-1} = B_n\hat{\xi}_{n-1}$ where $\hat{\xi}_{n-1}$ are the coefficients of $P_n U_{n-1}$ with respect to $\{\varphi_{n,j}\}$.

Problem 16.22. Specify the matrix equations for the cG(1)dG(1) method. Hint: consider (16.24).

Problem 16.23. Assume that $\Omega = (0, 1) \times (0, 1]$ and the standard uniform triangulation is used on each time step. Compute the coefficient matrix in (16.26).

Problem 16.24. (a) Formulate the cG(1)dG(r) with $r = 0, 1$, for the heat equation in one dimension with homogeneous Dirichlet boundary conditions. (b) Write out the matrix equations for the coefficients of U_n in the case of a uniform partition and $r = 0$. (c) Assume that $\mathcal{T}_n$ is obtained by dividing each element of $\mathcal{T}_{n-1}$ into two intervals. Compute $B_{n-1,n}$ explicitly. (d) Repeat (c) assuming that $\mathcal{T}_{n-1}$ has an even number of elements and that $\mathcal{T}_n$ is obtained by joining together every other neighboring pair of elements.

Problem 16.25. Repeat Problem 16.24 for the modified heat equation $\dot{u} - \Delta u + u = f$ with homogeneous Neumann boundary conditions.

16.4.2. The use of quadrature

In general it may be difficult to compute the integrals in (16.26) exactly, and therefore quadrature is often used to compute the integrals approximately. If K denotes an element of $\mathcal{T}_n$ with nodes $N_{K,1}$, $N_{K,2}$, and $N_{K,3}$

and area $|K|$, then we use the lumped mass quadrature for a function $g \in V_n$,

$$Q_K(g) = \frac{1}{3}|K| \sum_{j=1}^{3} g(N_{K,j}) \approx \int_K g(x)\, dx.$$

For the integration in time, we use the midpoint rule,

$$g\left(\frac{t_n + t_{n-1}}{2}\right) k_n \approx \int_{t_{n-1}}^{t_n} g(t)\, dt.$$

We define the approximations $\tilde{B}_n$, $\tilde{B}_{n-1,n}$, and $\tilde{b}_n$ by

$$(\tilde{B}_n)_{ij} = \sum_{K \in \mathcal{T}_n} Q_K(\varphi_{n,i}\varphi_{n,j}), \quad (\tilde{B}_{n-1,n})_{ij} = \sum_{K \in \mathcal{T}_n} Q_K(\varphi_{n,i}\varphi_{n-1,j}),$$

$$\text{and } (\tilde{b}_n)_i = \sum_{K \in \mathcal{T}_n} Q_K\big(f(\,\cdot\,, (t_n + t_{n-1})/2)\varphi_{n,i}(\cdot)\big) k_n,$$

for indices in the appropriate ranges. Note that the terms in the sums over $K \in \mathcal{T}_n$ for $\tilde{B}_n$ and $\tilde{B}_{n-1,n}$ are mostly zero, corresponding to the near orthogonality of the nodal basis functions. We find that $\tilde{\xi}_n$, the vector of nodal values of the cG(1)dG(0) approximation computed using quadrature, satisfies

$$(\tilde{B}_n + k_n A_n)\tilde{\xi}_n = \tilde{B}_{n-1,n}\tilde{U}_{n-1} + \tilde{b}_n. \tag{16.28}$$

If we use the rectangle rule with the right-hand end point of I_n instead, the resulting scheme is called the backward Euler-continuous Galerkin approximation.

Problem 16.26. Repeat Problem 16.23 using $\tilde{B}_n$, $\tilde{B}_{n-1,n}$, and $\tilde{b}_n$ instead of B_n, $B_{n-1,n}$, b_n respectively.

Problem 16.27. Repeat Problem 16.24 using $\tilde{B}_n$, $\tilde{B}_{n-1,n}$, and $\tilde{b}_n$ instead of B_n, $B_{n-1,n}$, b_n respectively.

Problem 16.28. Formulate the cG(1)dG(1) finite element method for the heat equation using the lumped mass quadrature rule in space and the two point Gauss quadrature rule for the time integration over I_n.

Problem 16.29. (a) Formulate the cG(1)dG(0) finite element method for the non-constant coefficient heat equation

$$\dot{u}(x,t) - (a(x,t)u'(x,t))' = f(x,t), \quad (x,t) \in (0,1) \times (0,\infty),$$

together with homogeneous Dirichlet boundary conditions and initial data u_0, using lumped mass quadrature rule in space and the midpoint rule in time to evaluate B_n, $B_{n-1,n}$, and any integrals involving a and f. (b) Assuming that $a(x, t) \geq a_0 > 0$ for all x and t, prove the modified mass and stiffness matrices are positive definite and symmetric. (c) Write down the matrix equations explicitly. (d) Assuming that the same space mesh is used for every time step, compute explicit formulas for $\tilde{B}_n$, $\tilde{A}_n$, and $\tilde{b}_n$.

16.5. Error estimates and adaptive error control

In this section, we state a posteriori and a priori error estimates for the cG(1)dG(0) method (16.22) and discuss an adaptive algorithm based on the a posteriori estimate. We also illustrate the performance of the algorithm in an example. The proofs of the error estimates are presented in the next section. For simplicity, we assume that Ω is convex so that the strong stability estimate (15.54) of Lemma 15.6 with stability constant $S = 1$ holds. We also assume that $u_0 \in V_1$; otherwise an additional term accounting for an initial approximation of u_0 appears in the estimates. We define $\tau = \min_n \tau_n$, where τ_n is the minimal angle of $\mathcal{T}_n$.

16.5.1. The error estimates

We begin by stating the a posteriori error estimate including residual errors associated to space discretization, time discretization, and mesh changes between space-time slabs. Here $\| \cdot \|$ denotes the $L_2(\Omega)$-norm and $\|v\|_J = \max_{t \in J} \|v(t)\|$.

Theorem 16.2. *There is a constant C_i only depending on τ such that for $N \geq 1$,*

$$\|u(t_N) - U_N\| \leq L_N C_i \max_{1 \leq n \leq N} \left(\|h_n^2 R_2(U)\|_{I_n} + \|h_n^2 f\|_{I_n} \right.$$

$$\left. + \|[U_{n-1}]\| + \|k_n f\|_{I_n} + \left\| \frac{h_n^2}{k_n} [U_{n-1}] \right\|^* \right),$$

where $u(t_N) = u(\cdot, t_N)$,

$$L_N = 2 + \max_{1 \leq n \leq N} \max \left\{ \left(\log\left(\frac{t_n}{k_n}\right) \right)^{1/2}, \log\left(\frac{t_n}{k_n}\right) \right\},$$

$$R_2(U) = \frac{1}{2} \max_{S \subset \partial K} h_K^{-1} |[\partial_S U]| \quad on \ K \in \mathcal{T}_n,$$

and the starred term is present only if $V_{n-1} \not\subseteq V_n$.

The two first terms on the right of (16.29) measure the residual error of the space discretization with f the contribution from the element interiors (there $\Delta U = 0$), and $R_2(U)$ the contribution from the jumps in the normal derivative $[\partial_S U]$ on elements edges S, cf. Chapter 15. The next two terms measure the residual error of the time discretization and finally the last term reflects the effect of changing from one mesh to the next. The case $V_{n-1} \not\subseteq V_n$ occurs e.g. when $\mathcal{T}_n$ is obtained from $\mathcal{T}_{n-1}$ by removing some nodes, introducing the L_2-projection $P_n U_{n-1} \in V_n$ of $U_{n-1} \in V_{n-1}$. The starred term is of the same order as the time residual term $\|[U_{n-1}]\|$ if h_n^2/k_n is kept bounded by a moderate constant, which usually may be arranged.

Problem 16.30. Draw examples in one space dimension that show a mesh coarsening in which $V_{n-1} \not\subseteq V_n$ and a mesh refinement in which $V_{n-1} \subseteq V_n$.

In the proof of the a priori error estimate, we use the following bounds on the change of mesh size on consecutive slabs. We assume there are positive constants γ_i, with γ_2 sufficiently small, such that for $n = 1, ..., N$,

$$\gamma_1 k_n \le k_{n+1} \le \gamma_1^{-1} k_n, \tag{16.30}$$

$$\gamma_1 h_n(x) \le h_{n+1}(x) \le \gamma_1^{-1} h_n(x) \quad \text{for } x \in \Omega, \tag{16.31}$$

$$\bar{h}_n^2 \le \gamma_2 k_n, \tag{16.32}$$

where $\bar{h}_n = \max_{x \in \bar{\Omega}} h_n(x)$, and (16.32) only enters if $V_{n-1} \not\subseteq V_n$. The a priori error estimate reads as follows:

Theorem 16.3. *If Ω is convex and γ_2 sufficiently small, there is a constant C_i depending only on τ and $\gamma_i, i = 1, 2$, such that for $N \ge 1$,*

$$\|u(t_N) - U_N\| \le C_i L_N \max_{1 \le n \le N} \left(k_n \|\dot{u}\|_{I_n} + \|h_n^2 D^2 u\|_{I_n} \right). \tag{16.33}$$

16.5.2. Adaptive error control

The a posteriori error bound can be used to estimate the error of a particular computation and also as the basis of an adaptive algorithm.

Suppose we seek an approximation $U(t)$ satisfying

$$\max_{0 \leq t \leq T} \|u(t) - U(t)\| \leq \text{TOL},$$

for a given error tolerance TOL, while using the least amount of computational work. We try to achieve this goal by computing a sequence of triangulations $\{\mathcal{T}_n\}$ and time steps $\{k_n\}$ so that for $n = 1, ..., N$, with $t_N = T$,

$$C_i L_N \max_{1 \leq n \leq N} \left(\|h_n^2 R_2(U_n)\| + \|[U_{n-1}]\| \right.$$
$$\left. + \|(k_n + h_n^2)f\|_{I_n} + \|h_n^2 k_n^{-1}[U_{n-1}]\|^* \right) = \text{TOL}, \quad (16.34)$$

while the total number of degrees of freedom is minimal. This is a nonlinear constrained minimization problem that we try to solve approximately using an iterative process based on the L_2 equidistribution strategy for elliptic problems described in Chapter 15 and the time step control described in Chapter 9. From the current time level t_{n-1}, we compute U_n using a predicted time step k_n and predicted mesh size h_n and then we check whether (16.34) holds or not. If not, we compute a new time step k_n and mesh size h_n using (16.34) seeking to balance the error contributions from space and time. It is relatively rare for the error control to require more than a few iterations.

We illustrate the adaptive error control using Femlab. We choose $\Omega = (-1, 1) \times (-1, 1)$ and approximate the solution of (16.3) with forcing

$$f(x, t) = \begin{cases} 10^3, & (x_1 + .5 - t)^2 + (x_2 + .5 - t)^2 < .1, \\ 0, & \text{otherwise}, \end{cases}$$

which in the context of a model of heat flow, amounts to swiping a hot blowtorch diagonally across a square plate. We compute the approximation using TOL=.05 and plot the results at the second, sixth, and tenth time steps in Fig. 16.5-Fig. 16.7. The time steps used are $k_1 \approx .017$, $k_2 \approx .62$, and $k_n \approx .1$ for $n \geq 3$. In Fig. 16.5, we can see the refined region centered around the heated region. At later times, we can see further refinement in the direction that the hot region moves and mesh coarsening in regions which have been passed by. Notice the shape of the refined region and the solution at later times indicating residual heat.

Problem 16.31. Implement an error estimation block in a code for the heat equation using the cG(1)dG(0) method. Construct several test problems with known solutions and compare the error bound to the true error.

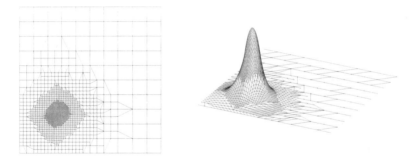

Figure 16.5: The approximation and mesh at $t \approx .64$.

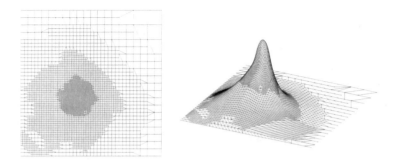

Figure 16.6: The approximation and mesh at $t \approx 1.04$.

16.6. Proofs of the error estimates

The proofs are based on a combination of the techniques used to prove the error estimates for the parabolic model problem in Chapter 9 and the Poisson problem of Chapter 15.

16.6.1. The a posteriori analysis

Let P_n be the L_2 projection into V_n, and π_k the L_2 projection into the piecewise constants on the time partition $\{t_n\}$, that is, $\pi_k v$ on I_n is the average of v on I_n. We use the following error estimate for P_n which is analogous to the interpolation error estimates discussed in Chapter 5.

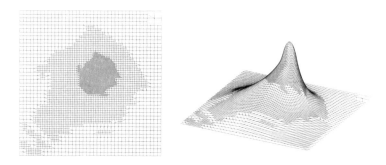

Figure 16.7: The approximation and mesh at $t \approx 1.44$.

Lemma 16.4. *There is a constant C_i only depending on τ such that if $\varphi = 0$ on Γ, then for all $w \in V_n$,*

$$|(\nabla w, \nabla(\varphi - P_n \varphi))| \le C_i \|h_n^2 R_2(w)\| \|D^2 \varphi\|.$$

In particular, if Ω is convex, then for all $w \in V_n$,

$$|(\nabla w, \nabla(\varphi - P_n \varphi))| \le C_i \|h^2 R_2(w)\| \|\Delta \varphi\|. \qquad (16.35)$$

The proof of this lemma is a little technical, so we put it off until the advanced book. Note that the second estimate follows from the first using (15.54).

We introduce the continuous dual problem

$$\begin{cases} -\dot{\varphi} - \Delta \varphi = 0 & \text{in } \Omega \times (0, t_N), \\ \varphi = 0 & \text{on } \Gamma \times (0, t_N), \\ \varphi_N(\cdot, t_N) = e_N & \text{in } \Omega, \end{cases} \qquad (16.36)$$

where $e_N = u(t_N) - U_N$. By the definition,

$$\|e_N\|^2 = (e_N, \varphi_N) + \sum_{n=1}^{N} \int_{I_n} (e, -\dot{\varphi} - \Delta \varphi) \, dt,$$

with $e = u - U$ and $\varphi_N = \varphi(\cdot, t_N)$. After integrating by parts in t over each interval I_n and using Green's formula in space, we get

$$\|e_N\|^2 = \sum_{n=1}^{N} \int_{I_n} (\dot{e}, \varphi) \, dt + \sum_{n=1}^{N} \int_{I_n} (\nabla e, \nabla \varphi) \, dt + \sum_{n=1}^{N} ([e_{n-1}], \varphi_{n-1}).$$

Using the facts that $\dot{u} - \Delta u = f$, $[u_n] = 0$, $\dot{U} \equiv 0$ on each I_n, and $U_0^- = u_0$ together with (16.22) with $v = \pi_k P_h \varphi \in W_k^{(0)}$, we obtain the *error representation*:

$$\|e_N\|^2 = \sum_{n=1}^{N} \int_{I_n} \left(\nabla U, \nabla(\pi_k P_h \varphi - \varphi)\right) dt$$
$$+ \sum_{n=1}^{N} \left([U_{n-1}], (\pi_k P_h \varphi)_{n-1}^+ - \varphi_{n-1}\right)$$
$$+ \int_0^T (f, \varphi - \pi_k P_h \varphi)\, dt = T_1 + T_2 + T_3.$$

This formula is analogous to the error representation for the model problem studied in Chapter 9. We now estimate the terms T_1, T_2 and T_3 by repeatedly using the splitting $\pi_k P_h \varphi - \varphi = (\pi_k - I)P_h\varphi + (P_h - I)\varphi$, where I is the identity, which is a way to split the time and space approximations. First, noting that

$$\int_{I_n} \left(\nabla U, \nabla(\pi_k P_h \varphi - P_h \varphi)\right) dt = \int_{I_n} (-\Delta_h U, \pi_k \varphi - \varphi)\, dt = 0, \quad 1 \le n \le N,$$

because U is constant on I_n, the term T_1 reduces to

$$T_1 = \sum_{n=1}^{N} \int_{I_n} (\nabla U, \nabla(P_h - I)\varphi)\, dt = \sum_{n=1}^{N} \left(\nabla U_n, \nabla(P_n - I)\int_{I_n} \varphi\, dt\right).$$

Recalling (16.35), we find that

$$|T_1| \le C_i \sum_{n=1}^{N} \|h_n^2 R_2(U_n)\| \left\|\Delta \int_{I_n} \varphi\, dt\right\|$$
$$\le C_i \max_{1 \le n \le N} \|h_n^2 R_2(U_n)\| \left(\int_0^{t_{N-1}} \|\Delta\varphi\|\, dt + 2\|\varphi\|_{I_N}\right),$$

where on the interval I_N, we used the fact that

$$\Delta \int_{I_N} \varphi\, dt = \int_{I_N} \Delta\varphi\, dt = \int_{I_N} \dot\varphi\, dt = \varphi(t_N) - \varphi(t_{N-1}).$$

To estimate T_2, we again use (16.35) to get

$$|([U_{n-1}], (P_n - I)\varphi_{n-1})| \le C_i\|h_n^2[U_{n-1}]\|^*\|\Delta\varphi_{n-1}\|,$$

where the star is introduced since the left-hand side is zero if $V_{n-1} \subset V_n$. Using the interpolation estimate $\|\varphi_{n-1} - (\pi_k\varphi)^+_{n-1}\| \le \min\{\int_{I_n} \|\dot\varphi\|\, dt, \|\varphi\|_{I_n}\}$ combined with the stability estimate $\|P_n v\| \le \|v\|$, we further have

$$\left|([U_{n-1}], ((\pi_k - I)P_h\varphi)^+_{n-1})\right| \le \|[U_{n-1}]\| \min\left\{\int_{I_n} \|\dot\varphi\|\, dt, \|\varphi\|_{I_n}\right\},$$

and we conclude that

$$|T_2| \le C_i \max_{1 \le n \le N} \|h^2_n[U_{n-1}]/k_n\|^* \sum_{n=1}^{N} k_n\|\Delta\varphi_{n-1}\|$$

$$+ \max_{1 \le n \le N} \|[U_{n-1}]\| \left(\int_0^{t_{N-1}} \|\dot\varphi\|\, dt + \|\varphi\|_{I_N}\right).$$

Finally to estimate T_3, we have arguing as in the previous estimates

$$\left|\sum_{n=1}^{N} \int_{I_n} (f, P_h\varphi - \pi_k P_h\varphi)\, dt\right|$$

$$\le \max_{1 \le n \le N} \|k_n f\|_{I_n} \left(\int_0^{t_{N-1}} \|\dot\varphi\|\, dt + \|\varphi\|_{I_N}\right),$$

$$\left|\sum_{n=1}^{N-1} \int_{I_n} (f, (I - P_h)\varphi)\, dt\right|$$

$$\le C_i \max_{1 \le n \le N-1} \|h^2_n f\|_{I_n} \left(\int_0^{t_{N-1}} \|\Delta\varphi\|\, dt\right),$$

and

$$\left|\int_{I_N} (f, (I - P_h)\varphi)\, dt\right| \le \|k_N f\|_{I_N}\|\varphi\|_{I_N}.$$

To complete the proof, we bound the different factors involving φ in the estimates above in terms of $\|e_N\|$ using the strong stability estimates (16.19)-(16.21) applied to the dual problem (16.36) with time reversed. We obtain with $w = \dot\varphi = \Delta\varphi$,

$$\int_0^{t_{N-1}} \|w\|\, dt \le \left(\int_0^{t_{N-1}} (t_N - t)^{-1}\, dt\right)^{1/2} \left(\int_0^{t_N} (t_N - t)\|w\|^2\, dt\right)^{1/2}$$

$$\le \left(\log(\frac{t_N}{k_N})\right)^{1/2} \|e_N\|,$$

$$\sum_{n=1}^{N-1} k_n \|w_{n-1}\| \le \sum_{n=1}^{N-1} \frac{k_n}{t_N - t_{n-1}} \|e_N\|$$

$$\le \int_0^{t_{N-1}} (t_N - t)^{-1}\, dt \|e_N\|,$$

and

$$k_N \|\Delta \varphi_{N-1}\| \le \|e_N\|.$$

Together, the above estimates prove the a posteriori error estimate.

Problem 16.32. (a) Write out the details of the proof in the case of the heat equation in one dimension with $\Omega = (0,1)$ and $r = 0$. (b) *(Hard.)* Do the same for $r = 1$.

Problem 16.33. *(Ambitious.)* Formulate and prove an a posteriori error estimate for the cG(1)dG(0) method that uses the lumped mass and midpoint quadrature rules as described above. Less ambitious is to do the same for the method that uses quadrature only to evaluate integrals involving f.

16.6.2. The a priori analysis

The a priori analysis follows the same line as the a posteriori analysis, after we introduce a discrete dual problem. The proof of the stability estimate on the solution of the discrete dual problem simplifies if $V_n \subset V_{n-1}$, and in particular, only assumption (16.30) is needed. We present this case below, and leave the general case to a later time.

The discrete strong stability estimate reads.

Lemma 16.5. *Assume that $V_{n-1} \subset V_n$ and that (16.30) holds. Then there is a constant C depending on γ_1 such that the solution U of (16.22) with $f \equiv 0$ satisfies for $N = 1, 2, ...,$*

$$\|U_N\|^2 + 2\sum_{n=1}^{N} \|\nabla U_n\|^2 k_n + \sum_{n=0}^{N-1} \|[U_n]\|^2 = \|U_0\|^2, \tag{16.37}$$

$$\sum_{n=1}^{N} t_n \|\Delta_n U_n\|^2 k_n \le C\|U_0\|^2, \tag{16.38}$$

and

$$\sum_{n=1}^{N} \|[U_{n-1}]\| \le C\left(2 + \left(\log\left(\frac{t_N}{k_1}\right)\right)^{1/2}\right)\|U_0\|. \tag{16.39}$$

Proof. We recall the equation satisfied by U:

$$(I - k_n\Delta_n)U_n = U_{n-1}, \tag{16.40}$$

where we used that $V_{n-1} \subset V_n$. Multiplying by U_n gives

$$\|U_n\|^2 + k_n\|\nabla U_n\|^2 = (U_{n-1}, U_n)$$

or

$$\frac{1}{2}\|U_n\|^2 + \|U_n - U_{n-1}\|^2 + k_n\|\nabla U_n\|^2 = \frac{1}{2}\|U_{n-1}\|^2,$$

which upon summation proves (16.37).

Next, multiplying (16.40) by $-t_n\Delta_n U_n$ gives

$$t_n\|\nabla U_n\|^2 + t_n\|\Delta_n U_n\|^2 k_n = t_n(\nabla U_{n-1}, \nabla U_n),$$

that is

$$\frac{1}{2}t_n\|\nabla U_n\|^2 + t_n\|\nabla(U_n - U_{n-1})\|^2 + t_n\|\Delta_n U_n\|^2 k_n$$
$$= \frac{1}{2}t_{n-1}\|\nabla U_{n-1}\|^2 + \frac{1}{2}\|\nabla U_{n-1}\|^2 k_n.$$

Summing over $n = 2, ..., N$ using that $k_n \leq \gamma_1 k_{n-1}$ and (16.37) proves (16.38) with the summation starting at $n = 2$. Finally, we note that

$$\sum_{n=2}^{N}\|[U_{n-1}]\| = \sum_{n=2}^{N}\|\Delta_n U_n\|k_n \leq \Big(\sum_{n=2}^{N}t_n\|\Delta_n U_n\|^2 k_n\Big)^{1/2}\Big(\sum_{n=2}^{N}\frac{k_n}{t_n}\Big)^{1/2}$$
$$\leq C\Big(\log(\frac{t_N}{k_1})\Big)^{1/2}\|U_0\|.$$

The term corresponding to $n = 1$ in (16.38) and (16.39) is estimated using the equation (16.40) with $n = 1$ and the fact that $\|U_1\| \leq \|U_0\|$. This concludes the proof. ∎

We can now complete the proof of the a priori error estimate. We first estimate $\|U_n - \tilde{U}_n\|$ where $\tilde{U}_n \in W_{kn}^{(0)}$ is the *average elliptic projection* of u defined for $n = 1, \dots, N$, by

$$\int_{I_n} (\nabla(u - \tilde{U}_n), \nabla v)\, dt = 0 \quad \text{for all } v \in W_{kn}^{(0)},$$

Using the estimate $\|\tilde{u}_n - \tilde{U}_n\| \le C_i\|h_n^2 D^2 u\|_{I_n}$ (see Chapter 15), where $\tilde{u}_n = \pi_k u|_{I_n}$ is the average of u on I_n, together with the obvious estimate $\|u(t_n) - \tilde{u}_n\| \le k_n\|\dot{u}\|_{I_n}$, we obtain the desired estimate for $\|u_n - U_n\|$.

We let $\Phi \in W_k^{(0)}$ be the solution of the discrete dual problem

$$-(\Phi_{n+1} - \Phi_n) - k_n\Delta_n\Phi_n = 0 \quad \text{for } n = N, ..., 1,$$

where $\Phi_{N+1} = U_N - \tilde{U}_N$. Multiplying by $\tilde{e}_n = U_n - \tilde{U}_n$ and summing over n gives the *error representation*

$$\|\tilde{e}_N\|^2 = (\tilde{e}_N, \Phi_{N+1}) - \sum_{n=1}^{N}(\tilde{e}_n, \Phi_{n+1} - \Phi_n) + \sum_{n=1}^{N}(\nabla\tilde{e}_n, \nabla\Phi_n)k_n$$

$$= \sum_{n=1}^{N}(\tilde{e}_n - \tilde{e}_{n-1}, \Phi_n) + \sum_{n=1}^{N}\int_{I_n}(\nabla\tilde{e}_n, \nabla\Phi_n)\,dt,$$

where we used a summation by parts formula and the assumption that $\tilde{e}_0 = 0$.

Problem 16.34. Show that the last formula holds.

Using the fact that for all $v \in W_{kn}^{(0)}$

$$(u_n - U_n - (u_{n-1} - U_{n-1}), v) + \int_{I_n}(\nabla(u - U), \nabla v)\,dt = 0,$$

the error representation takes the form

$$\|\tilde{e}_N\|^2 = \sum_{n=1}^{N}(\rho_n - \rho_{n-1}, \Phi_n) + \sum_{n=1}^{N}\int_{I_n}(\nabla\rho_n, \nabla\Phi_n)\,dt$$

$$= \sum_{n=1}^{N}(\rho_n - \rho_{n-1}, \Phi_n)$$

where $\rho = u - \tilde{U}$ and in the last step we used the definition of $\tilde{U}$. Summing by parts again, we get

$$\|\tilde{e}_N\|^2 = -\sum_{n=2}^{N}(\rho_{n-1}, \Phi_n - \Phi_{n-1}) + (\rho_N, \Phi_N),$$

using the assumption that $\rho_0 = 0$. Applying Lemma 16.5 to Φ (after reversing time) proves the desired result. Note that the assumption $V_n \subset V_{n-1}$ of the a priori error estimate corresponds to the assumption $V_{n-1} \subset V_n$ in the stability lemma, because time is reversed.

Problem 16.35. Consider the cG(1)dG(1) method for the homogeneous heat equation, i.e. (16.22) with $f \equiv 0$, under the assumption that $k_n \leq Ck_{n-1}$ for some constant C. (a) Show that $\|U_n^-\| \leq \|U_0^-\|$ for all $1 \leq n \leq N$. (b) Show that $\|U\|_{I_n} \leq 5\|U_0^-\|$ for all $1 \leq n \leq N$.

Problem 16.36. *(Hard.)* Referring to Problem 16.33, prove the corresponding a priori error estimate.

Think for yourself, 'cause I won't be with you. (George Harrison)

I see no essential difference between a materialism, which includes a soul as a complicated type of material particle, and a spiritualism that includes particles as a primitive type of soul. (Wiener)

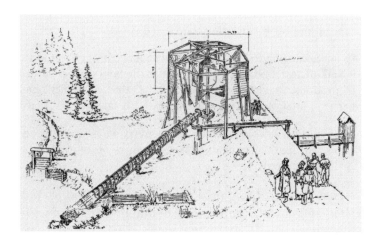

Proposal for a wind-driven pump by Leibniz

17

The Wave Equation

Go ahead and faith will come to you. (d'Alembert)

Souls act according to the laws of final causes, through apparitions, ends and means. Bodies act according to the laws of efficient causes or of motions. And these two kingdoms, that of efficient causes and that of final causes, are in harmony with each other. (Leibniz)

Those beautiful laws of physics are a marvellous proof of an intelligent and free being against the system of absolute and brute necessity. (Leibniz)

The wave equation is a basic prototype of a hyperbolic partial differential equation, and models propagation of different types of waves such as elastic waves in an elastic string, membrane, or solid, sound waves in a gas or fluid, or electromagnetic waves. The simplest model of wave propagation is an equation for transport in one direction, which we derive in the next section. After that, we derive the wave equation by examining the familiar model of the motion of a discrete system of masses and springs in the limit as the number of masses increases. We then recall some of the properties of solutions of the wave equation; contrasting their behavior to that of solutions of the heat equation, which is the other basic example of a time dependent partial differential equation. We continue with a discussion of the wave equation in higher dimensions, emphasizing the important fact that the behavior of solutions of the wave equation depends on the dimension. Finally, we discuss the approximate solution of the wave equation using a Galerkin finite element method.

17.1. Transport in one dimension

The simplest model for wave propagation is in fact the simplest of all partial differential equations. We model the convective transport of a pollutant suspended in water that is flowing at constant speed c through a pipe of uniform cross section assuming that there is no diffusion of the pollutant. We illustrate this in Fig. 17.1. Letting $u(x,t)$ denote the

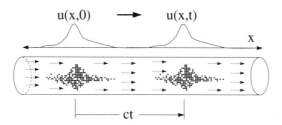

Figure 17.1: The transport of a pollutant suspended in a fluid flowing in a pipe.

concentration of the pollutant at the point x in the pipe at time t, the conservation of mass can be formulated in terms of integrals as

$$\int_0^{\bar{x}} u(x,t)\,dx = \int_{c(\bar{t}-t)}^{\bar{x}+c(\bar{t}-t)} u(x,\bar{t})\,dx \quad \text{for } \bar{x} > 0,\, \bar{t} \geq t.$$

This equation states that the amount of pollutant in the portion of the fluid occupying $[0,\bar{x}]$ at time t and $[c(\bar{t}-t), \bar{x}+c(\bar{t}-t)]$ at time $\bar{t}$ is the same. To obtain a differential equation expressing the conservation of mass, we first differentiate with respect to $\bar{x}$ to get $u(\bar{x},t) = u(\bar{x}+c(\bar{t}-t),\bar{t})$ and then differentiate with respect to $\bar{t}$ (or t) to get $0 = cu'(x,t) + \dot{u}(x,t)$, after letting $\bar{t} \to t$ and $\bar{x} \to x$.

Assuming that the pipe is infinitely long in order to avoid having to deal with what happens at the ends, we obtain the initial value problem: Find $u(x,t)$ such that

$$\begin{cases} \dot{u}(x,t) + cu'(x,t) = 0 & \text{for } x \in \mathbb{R},\, t > 0, \\ u(x,0) = u_0(x) & \text{for } x \in \mathbb{R}, \end{cases} \tag{17.1}$$

where c is a constant. The solution is $u(x,t) = u_0(x - ct)$, which simply says that the solution at time t is the initial data u_0 translated a distance

ct. The line $x - ct = \xi$ is called a *characteristic* line and c is called the *speed*. Since the value of the solution is constant, namely $u_0(\xi)$, at all points along the characteristic, we say that information travels along characteristics.

Problem 17.1. (a) Verify this formula. (b) Plot the solution corresponding to $u_0(x) = \sin(x)$ at times $t = 0, \pi/4, \pi/3, \pi/2$, and $23\pi/2$.

The transport problem (17.1) is the basic model of wave propagation. Below, we will see that the wave equation, which describes the propagation of vibrations in an elastic string, can be written as a system of transport equations. We will also meet the scalar transport model in the context of convection-diffusion problems in Chapter 19, where we consider the additional effect of diffusion.

We point out an interesting fact: the solution formula $u(x, t) = u_0(x - ct)$ is defined even if u_0 is discontinuous, though in this case, u obviously doesn't satisfy the differential equation at every point. Such initial data corresponds to a sharp signal, for example turning a light switch on and off. We can use the variational formulation of (17.1) to make sense of the solution formula when the data is nonsmooth, and we pick this up again later.

Problem 17.2. Plot the solution corresponding to $u_0(x) = \sin(x)$ for $0 \le x \le \pi$ and 0 otherwise at times $t = 0, \pi/4, \pi/3, \pi/2$, and $23\pi/2$.

One important difference between parabolic equations like the heat equation and hyperbolic equations like the transport and wave equations lies in the treatment of boundaries. It is natural to consider the transport equation with a boundary condition posed on the *inflow* boundary. If $c > 0$, then the inflow boundary is on the left. Choosing the boundary to be at $x = 0$ arbitrarily, we obtain

$$\begin{cases} \dot{u}(x,t) + cu'(x,t) = 0 & \text{for } x > 0, t > 0, \\ u(0,t) = g(t) & \text{for } t > 0, \\ u(x,0) = u_0(x) & \text{for } x > 0, \end{cases} \qquad (17.2)$$

where c is constant and $g(t)$ gives the inflow of material. By direct computation, we can verify that the solution satisfies

$$u(x,t) = \begin{cases} g(t - x/c), & x - ct \le 0, \\ u_0(x - ct), & x - ct > 0 \end{cases}$$

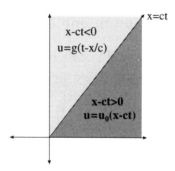

Figure 17.2: Solving the transport equation with a boundary condition on the inflow boundary when $c > 0$.

and we illustrate this in Fig. 17.2.

Problem 17.3. (a) Plot the solution of (17.2) for $u_0(x) = \sin(x)$ for $0 < x < \pi$ and 0 otherwise and $g(t) = t$ at $t = 0, \pi/6, \pi/4, \pi/3$, and $\pi/2$. (b) What does such boundary conditions mean interpreted in terms of the transport of a pollutant down a pipe?

Problem 17.4. Show that the solution of (17.2) for g given for $t \geq 0$ and u_0 given for $x > 0$ agrees with the solution of (17.1) corresponding to initial data $\bar{u}_0$ defined so that $\bar{u}_0(x) = u_0(x)$ for $x > 0$ and $\bar{u}_0(x) = g(-x/c)$ for $x \leq 0$ in the region $x \geq 0, t \geq 0$.

Problem 17.5. Find a formula for the solution of the homogeneous wave equation posed with a boundary condition on the left at a point x_0.

Note that once again the solution formula holds even though it may imply that u is discontinuous across the line $x = ct$. We can resolve this difficulty using the variational formulation as well.

The value of the solution at any *outflow boundary*, which is located on the right when $c > 0$, is determined from the initial data and therefore we cannot impose arbitrary values for the solution on an outflow boundary. In general, a hyperbolic problem posed on a finite domain may have inflow, outflow, or both kinds of boundaries and this is an important consideration in the design of numerical methods. This is a sharp contrast to the situation with the heat equation.

17.2.   The wave equation in one dimension

We begin by describing a physical system consisting of N weights each of mass m joined by $N + 1$ springs with equal length and spring constant. We choose coordinates so that the system occupies the interval $(0, 1)$ and assume that the springs at the ends are fixed and the masses are constrained to move horizontally along the x axis without friction. The rest position of the n'th weight is nh with $h = 1/(N + 1)$. We let $u_n(t)$ denote the displacement of the n'th weight from the rest position with $u_n > 0$ representing a displacement to the right. We illustrate this in Fig. 17.3. Below, we want to compare the motion of systems with

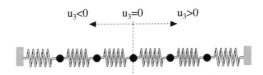

Figure 17.3: The coordinate system for a system of masses and springs.

different numbers of weights but totalling the same mass. Hence, we assume that $m = h$, so that as N increases, the total mass of the system tends to one.

Hamilton's principle states that the Lagrangian of the system, which is equal to the difference between the kinetic and potential energies integrated over an arbitrary time interval (t_1, t_2),

$$\int_{t_1}^{t_2} \left(\sum_{n=1}^{N} \frac{m}{2} (\dot{u}_n)^2 - \sum_{n=1}^{N+1} \frac{1}{2} h^{-1} (u_n - u_{n-1})^2 \right) dt,$$

where we set $u_0 = 0$ and $u_{N+1} = 0$, is stationary at the trajectory followed by the system. We assume that the spring constant is $1/h$, since it should scale with the length of the springs.

To obtain the differential equation for $u = (u_n(t))$, we add an arbitrary small perturbation to u_n in the direction of $v = (v_n)$, with $v_0 = v_{N+1} = 0$, to get $u_n + \epsilon v_n$ for $\epsilon \in \mathbb{R}$. Differentiating with respect to ϵ and setting the derivative equal to zero for $\epsilon = 0$, which corresponds to the Lagrangian being stationary at the solution u, and then varying v gives the following system

$$\ddot{u}_n - h^{-2}(u_{n-1} - 2u_n + u_{n+1}) = 0, \quad t > 0, \ n = 1, ..., N,$$
$$(17.3)$$

where $u_0 = 0$ and $u_{N+1} = 0$. The differential equation (17.3) is supplemented by initial conditions specifying the initial position and velocity of each weight.

We present an example with $N = 5$ in which the n'th weight is displaced a distance $.5h\sin(nh)$ to the right of the rest position and the initial speed is zero. We solve the system (17.3) using the Cards code keeping the error below .06. We show the position of the weights for a few times in Fig. 17.4.

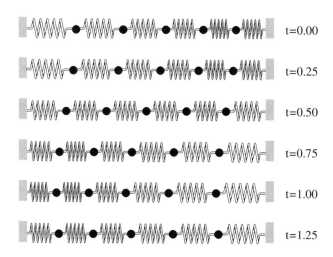

t=0.00

t=0.25

t=0.50

t=0.75

t=1.00

t=1.25

Figure 17.4: The evolution of the discrete system of masses and springs.

Problem 17.6. (a) Derive (17.3). (b) Change the system of equations (17.3) into a first order system by introducing new unknowns $v_n = \dot{u}_n$. (c) Solve the system keeping the error below .05 for $N = 5, 10, 15, ..., 55$ and compare the solutions. (d) Compute the solution for $N = 5$ where the masses start at the rest position with initial velocities $\{\sin(nh)\}$ and plot the results for $t = 0, .25, .5, .75, 1.0$ and 1.25.

Letting the number of weights N tend to infinity (with a corresponding decrease in the mass of each weight since $m = h$) in the discrete equation (17.3), we formally obtain the wave equation in one dimension:

$$\begin{cases} \ddot{u}(x,t) - u''(x,t) = 0 & \text{for } 0 < x < 1 \text{ and } t > 0, \\ u(0,t) = u(1,t) = 0 & \text{for } t > 0, \\ u(x,0) = u_0(x), \, \dot{u}(x,0) = \dot{u}_0(x) & \text{for } 0 < x < 1, \end{cases} \qquad (17.4)$$

where now with abuse of notation u_0 and $\dot{u}_0$ are given initial data. This is the initial value problem describing the longitudinal vibrations in an elastic string. It turns out that the same equation describes also the transversal vibration of an elastic string, like a string on a guitar.

17.2.1.   Sound waves in a tube

The wave equation (17.4) is also used to model the propagation of sound waves. We consider a long thin tube, represented by $\mathbb{R}$, filled with gas of density ρ, pressure p, and velocity u. The behavior of the gas is described by a set of nonlinear equations that result from the conservation of mass and Newton's law relating the rate of change of momentum to the pressure:

$$\begin{cases} \dot{\rho} + (u\rho)' = 0 & \text{in } \mathbb{R} \times (0, \infty), \\ \dot{m} + (um)' + p' = 0 & \text{in } \mathbb{R} \times (0, \infty), \end{cases} \qquad (17.5)$$

where $m = \rho u$ is the momentum. To derive a linear equation, we consider small fluctuations $\bar{\rho}$, $\bar{u}$ and $\bar{p}$ around a constant state of density ρ_0, pressure p_0 and zero velocity, so that $\rho = \rho_0 + \bar{\rho}$, $p = p_0 + \bar{p}$ and $u = 0 + \bar{u}$. We assume that $\bar{p} = c^2\bar{\rho}$, where c is a constant representing the speed of sound, substitute the new variables into (17.5), and drop quadratic terms in the resulting equation, since these are very small if the fluctuations are small, to obtain

$$\begin{cases} \dot{\bar{\rho}} + \rho_0 \bar{u}' = 0 & \text{in } \mathbb{R} \times (0, \infty), \\ \rho_0 \dot{\bar{u}} + c^2 \bar{\rho}' = 0 & \text{in } \mathbb{R} \times (0, \infty). \end{cases} \qquad (17.6)$$

Eliminating either $\bar{\rho}$ or $\bar{p}$ leads to the wave equations $\ddot{\bar{\rho}} - c^2\bar{\rho}'' = 0$ and $\ddot{\bar{p}} - c^2\bar{p}'' = 0$.

Problem 17.7. (a) Verify the derivation of (17.6). (b) Show that (17.6) implies that $\bar{\rho}$ and $\bar{p}$ satisfy the wave equation under the assumptions of the derivation.

17.2.2.   The structure of solutions: d'Alembert's formula

The general initial value problem for the wave equation,

$$\begin{cases} \ddot{u} - u'' = f & \text{in } \mathbb{R} \times (0, \infty), \\ u(x, 0) = u_0(x), \, \dot{u}(x, 0) = \dot{u}_0(x) & \text{for } x \in \mathbb{R}, \end{cases} \tag{17.7}$$

can be written as a system of transport equations by introducing the variable $w = \dot{u} - u'$ to get

$$\begin{cases} \dot{w} + w' = f & \text{in } \mathbb{R} \times (0, \infty), \\ \dot{u} - u' = w & \text{in } \mathbb{R} \times (0, \infty), \\ w(x, 0) = \dot{u}_0(x) - u_0'(x), \, u(x, 0) = u_0(x) & \text{for } x \in \mathbb{R}, \end{cases}$$

where the two transport equations in the new formulation correspond to transport of signals in opposite directions with speed one.

Problem 17.8.   Verify that the two problems have the same solution u.

It is therefore natural, following d'Alembert and Euler, to look for a solution $u(x, t)$ of (17.7) with $f \equiv 0$ of the form $u(x, t) = \varphi(x - t) + \psi(x + t)$, where $\varphi(x - t)$ corresponds to a wave propagating in the positive direction with speed one and $\psi(x + t)$ corresponds to a wave propagating with speed one in the negative direction. It is easy to see that a function of this form satisfies the wave equation $\ddot{u} - u'' = 0$.

Problem 17.9.   Verify this claim.

Determining the functions φ and ψ from the initial conditions, we find *d'Alembert's formula*:

$$u(x, t) = \frac{1}{2}(u_0(x - t) + u_0(x + t)) + \frac{1}{2}\int_{x-t}^{x+t} \dot{u}_0(y) \, dy. \tag{17.8}$$

Problem 17.10.   Prove (17.8).

Problem 17.11.   If the speed of the propagation of the waves is $c > 0$, then the corresponding wave equation takes the form $\ddot{u} - c^2 u'' = 0$. Derive d'Alembert's formula for this case.   Hint: seek a solution of the form $u(x, t) = \varphi(x - ct) + \psi(x + ct)$.

Using d'Alembert's formula, we can study the dependence of the solution on the initial data. For example, if $u_0(x)$ is an approximate "point" source supported in a small interval around $x = 0$ and $\dot{u}_0 \equiv 0$, then the solution $u(x, t)$ consists of two pulses propagating from $x = 0$ in the positive and negative directions with speed ± 1, see Fig. 17.5. This data corresponds to an elastic string being released at time zero with a displacement concentrated at 0 and with zero velocity. The d'Alembert formula shows that the solution $u(x, t)$ at a given time t is influenced only by the value of the initial data $u_0(x)$ at the points $x \pm t$, i.e. as for the transport equation, there is *sharp propagation* of the initial data u_0. The effect of an initial impulse in the derivative data $\dot{u}_0$ is different, as

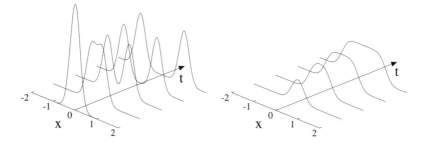

Figure 17.5: The evolution of solutions of the wave equation corresponding to an approximate "point" source in u_0 together with $\dot{u}_0 \equiv 0$ on the left and an approximate "point" source in $\dot{u}_0$ together with $u_0 \equiv 0$ on the right.

illustrated in Fig. 17.5. If $\dot{u}_0$ has support in a small interval centered at $x = 0$ and $u_0 \equiv 0$ then $u(x, t)$ is constant in most of the region $[x-t, x+t]$ and zero outside a slightly larger interval.

Problem 17.12. Define $g(x) = 10^8(x - .1)^4(x + .1)^4$ if $|x| < .1$ and 0 otherwise and show that g has continuous second derivatives. (a) Compute an explicit formula for the solution if $u_0(x) = g(x)$ and $\dot{u}_0 \equiv 0$ and plot the results for a few times. (b) Do the same if $u_0 \equiv 0$ and $\dot{u}_0(x) = g(x)$. (c) Referring to (b), given $t > 0$, determine the intervals on which u is constant.

The extension of the d'Alembert's formula to the nonhomogeneous problem (17.7) with $f \neq 0$ is

$$u(x,t) = \frac{1}{2}\big(u_0(x+t) + u_0(x-t)\big)$$

$$+ \frac{1}{2}\int_{x-t}^{x+t} \dot{u}_0(y)\,dy + \frac{1}{2}\iint_{\Delta(x,t)} f(y,s)\,dy\,ds, \quad (17.9)$$

where $\Delta = \Delta(x,t) = \{(y,s) : |x-y| \leq t-s, s \geq 0\}$ denotes the *triangle of dependence* indicating the portion of space-time where data can influence the value of the solution at the point (x,t), see Fig. 17.6. Turning the triangle of dependence upside-down gives the *triangle of*

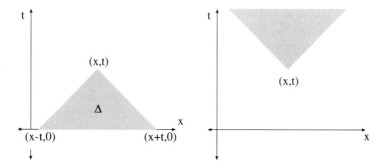

Figure 17.6: On the left, we show the triangle of dependence Δ of the point (x,t). On the right, we show the triangle of influence.

influence $\{(y,s) : |x-y| \leq s-t\}$ indicating the points (y,s) which can be influenced by the values of the data at (x,t).

Problem 17.13. Prove (17.9).

We can handle problems with boundaries by modifying d'Alembert's formula. For example, to find a formula for the homogeneous wave equation $\ddot{u} - u'' = 0$ for $x > 0$, $t > 0$ together with the boundary condition $u(0,t) = 0$ for $t > 0$ and initial conditions $u_0(x)$ and $\dot{u}_0(x)$ as above, we use d'Alembert's formula for the solution of the wave equation

$\ddot{w} - w'' = 0$ on $\mathbb{R} \times (0, \infty)$ together with odd initial data w_0 and $\dot{w}_0$, where w_0 is defined by

$$\bar{w}_0(x) = \begin{cases} -u_0(-x), & x < 0, \\ 0, & x = 0, \\ u_0(x), & x > 0, \end{cases}$$

and $\dot{w}_0$ is defined similarly. It is easy to verify that the solutions of the two problems agree in the region $x > 0$, $t > 0$. Using d'Alembert's formula and tracing the characteristic lines to their intersections with the x axis, see Fig. 17.7, we find that

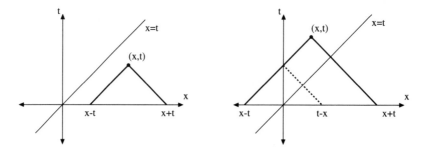

Figure 17.7: The two cases for applying d'Alembert's formula to the wave equation posed with a boundary at $x = 0$. We plot the characteristic lines for (x, t) with $x > t$ on the left and $x < t$ on the right. Note the reflection in the t axis of the point $x - t$ to $t - x$.

$u(x, t)$

$$= \begin{cases} \dfrac{1}{2}\big(u_0(x + t) + u_0(x - t)\big) + \dfrac{1}{2}\displaystyle\int_{x-t}^{x+t} \dot{u}_0(y)\, dy, & x > t \\[3mm] \dfrac{1}{2}\big(u_0(t + x) - u_0(t - x)\big) + \dfrac{1}{2}\displaystyle\int_{t-x}^{t+x} \dot{u}_0(y)\, dy, & x \le t. \end{cases} \quad (17.10)$$

Problem 17.14. (a) Verify (17.10). (b) Find a formula for the solution of the homogeneous wave equation posed with the Neumann boundary condition $u'(0, t) = 0$. Hint: extend the data to be even functions on $\mathbb{R}$.

Problem 17.15. Use d'Alembert's formula to construct the solution of the homogeneous wave equation posed on $(0, 1)$ with periodic boundary conditions.

Problem 17.16. Give a d'Alembert solution formula for the vibrating string problem (17.4). Hint: extend u_0 and $\dot{u}_0$ to be functions on $\mathbb{R}$.

The existence of the triangles of dependence and influence and the sharp propagation of the data are the result of the *finite speed of propagation* of solutions of the wave equation. This contrasts to the behavior of solutions of the heat equation, where the value of the solution at one point depends on the data at every point (although the exponential decay of the fundamental solution implies that the dependence is very small from point far away) and the diffusion of the data as time passes. One consequence is that it is more difficult to send recognizable signals by heating a conducting wire than sending sound waves down a pipe.

17.2.3. Separation of variables and Fourier's method

The technique of separation of variables and Fourier's method can be used to write the solution of the wave equation as a Fourier series. To simplify the notation, we pose (17.4) on $(0, \pi)$ instead of $(0, 1)$. In this case, the solution is

$$u(x, t) = \sum_{n=1}^{\infty} (a_n \sin(nt) + b_n \cos(nt)) \sin(nx),$$

$$(17.11)$$

where the coefficients a_n and b_n are determined from the Fourier series of the initial conditions:

$$u_0(x) = \sum_{n=1}^{\infty} b_n \sin(nx), \quad \dot{u}_0(x) = \sum_{n=1}^{\infty} n a_n \sin(nx).$$

Note that the time factor, $a_n \sin(nt) + b_n \cos(nt)$, in the Fourier series of the solution of the wave equation does not decrease exponentially as time increases like the corresponding factor in the Fourier series of a solution of the heat equation. Therefore, the solution of the wave equation generally does not become smoother as time passes.

Problem 17.17. Verify the solution formula (17.11) formally.

Problem 17.18. Compute the solution for (a) $u_0(x) = x(\pi - x)$, $\dot{u}_0(x) \equiv 0$, (b) $\dot{u}_0(x) = x(\pi - x)$, $u_0(x) \equiv 0$.

17.2.4. The conservation of energy

We saw that a solution of the heat equation tends to dissipate as time passes, with a corresponding decrease in the energy. In contrast, the total energy (the sum of kinetic and potential energies) of the solution u of the homogeneous wave equation (17.4) remains constant in time:

$$\|\dot{u}(\cdot, t)\|^2 + \|u'(\cdot, t)\|^2 = \|\dot{u}_0\|^2 + \|u_0'\|^2 \quad \text{for } t \geq 0,$$

where $\|\cdot\|$ denotes the $L_2(0, 1)$ norm as usual. To prove this, we multiply (17.12) by $2\dot{u}$, integrate over $(0, 1)$, and then integrate by parts to get

$$0 = \frac{\partial}{\partial t} \left(\int_0^1 \left(\dot{u}(x, t)^2 + u'(x, t)^2 \right) dx \right).$$

Problem 17.19. Provide the details of this derivation.

Problem 17.20. (a) Show that the only solution of (17.4) with $u_0 \equiv \dot{u}_0 \equiv 0$ is $u \equiv 0$. (b) Suppose that w solves (17.4) with initial data w_0 and $\dot{w}_0$. Estimate $u - w$, where u solves (17.4).

17.3. The wave equation in higher dimensions

Situations modelled by the wave equation in higher dimensions include the vibrations of a drum head and the propagation of sound waves in a volume of gas. Letting Ω denote a domain in $\mathbb{R}^d$, $d = 2$ or 3, with boundary Γ, the initial-boundary value problem for the wave equation is

$$\begin{cases} \ddot{u} - \Delta u = f & \text{in } \Omega \times (0, \infty), \\ u = 0 & \text{on } \Gamma \times (0, \infty), \\ u(x, 0) = u_0(x), \ \dot{u}(x, 0) = \dot{u}_0(x) & \text{for } x \in \Omega, \end{cases} \quad (17.12)$$

where f, u_0, and $\dot{u}_0$ are given functions. The wave equation is also posed on all of $\mathbb{R}^d$ in some models.

Before turning to the approximation of (17.12), we recall some of the properties of the solutions. We emphasize the important fact that the behavior of solutions of the wave equation depends on the dimension, and in particular, the behavior in two dimensions is significantly different that in three dimensions.

17.3.1. Symmetric waves

We begin by considering solutions of the homogeneous wave equation in $\mathbb{R}^d$ that are symmetric through the origin since this effectively reduces the problem to one dimension in space. In $\mathbb{R}^3$, these are called *spherically symmetric* waves. For simplicity, we assume that $\dot{u}_0 \equiv 0$. The wave equation (17.12) in spherical coordinates, assuming the solution depends only on r, i.e. the distance to the origin, reads

$$\ddot{u} - u_{rr} - \frac{2}{r} u_r = 0 \quad \text{for } r > 0,\, t > 0, \tag{17.13}$$

where $u_r = \partial u/\partial r$. Note the important factor two in the third term; by introducing the new unknown $v = ru$, this equation transforms into the one-dimensional wave equation,

$$\ddot{v} - v_{rr} = 0 \quad \text{for } r > 0,\, t > 0. \tag{17.14}$$

This equation is posed together with the boundary condition $v(0,t) = 0$ for $t > 0$ and initial conditions $v(r,0) = ru_0(r)$ and $\dot{v}(r,0) = 0$ for $r > 0$. Using (17.10) to write a formula for v and then changing back to u, we find that

$$u(r,t)$$
$$= \frac{1}{2} \begin{cases} \left(u_0(r+t) + u_0(r-t)\right) + \dfrac{t}{r}\left(u_0(r+t) - u_0(r-t)\right), & r \geq t, \\[2mm] \left(u_0(t+r) + u_0(t-r)\right) + \dfrac{t}{r}\left(u_0(t+r) - u_0(t-r)\right), & r < t, \end{cases}$$
$$\tag{17.15}$$

where we take $u(0,\cdot) = \lim_{r\to 0+} u(r,\cdot)$. From this, we conclude that the initial data propagates sharply outwards in the positive r direction as time passes. In particular, if u_0 has support in the ball $\{x : |x| \leq \rho\}$ for some $\rho > 0$, then at any point x with $x > \rho$, $u(x,t)$ is zero for $t < x - \rho$, then the solution is non-zero with values determined by u_0 for 2ρ time units, and finally after that the solution is once again zero.

Problem 17.21. Compute explicit formulas for the spherically symmetric solution corresponding to $u_0 \equiv 1$ for $|x| \leq 1$ and 0 otherwise and $\dot{u}_0 \equiv 0$. Hint: there are six regions in the (r,t) plane that have to be considered. Plot the solution as a function of the radius at several times.

We can also look for symmetric solutions of the wave equation in $\mathbb{R}^2$. Unfortunately in this case, the wave equation reduces to

$$\ddot{u} - u_{rr} - \frac{1}{r}u_r = 0 \quad \text{for } r > 0,\, t > 0, \tag{17.16}$$

and there is no simple change of variables that reduces this problem to the wave equation in one dimension.

Problem 17.22. Verify (17.13), (17.16), and (17.14).

Problem 17.23. (a) Verify (17.15). (b) Treat the problem where $\dot{u}_0$ is not assumed to be zero.

17.3.2. The finite speed of propagation

As suggested by the spherically symmetric case, there is a finite speed of propagation of information in solutions of the wave equation in higher dimensions. By this, we mean that the value of $u(x,t)$ depends only on the values of the data given in the *cone of dependence*

$$\Delta(x,t) := \{(y,s) \in \mathbb{R}^d \times \mathbb{R} : |y - x| \le t - s,\, s \ge 0\}.$$

The cone of dependence is the multi-dimensional counterpart to the triangle of dependence. Specifically, for any proper subdomain ω of Ω, we may define the enlarged region $\omega(t) = \{x \in \mathbb{R}^d : \operatorname{dist}(x, \omega) < t\}$ assuming for simplicity that $t \ge 0$ is not too large so that $\omega(t)$ is also contained in Ω, see Fig. 17.8. Then we prove the following estimate on the value of u in ω at time t in terms of the values of the data in $\omega(t)$:

Theorem 17.1. *For any proper subdomain ω of Ω and $t > 0$ such that $\omega(t) \subset \Omega$, the solution u of the homogeneous wave equation satisfies*

$$\|\dot{u}(\cdot,t)\|_{L_2(\omega)}^2 + \|\nabla u(\cdot,t)\|_{L_2(\omega)}^2 \le \|\dot{u}_0\|_{L_2(\omega(t))}^2 + \|\nabla u_0\|_{L_2(\omega(t))}^2.$$

Proof. We define the generalized cone of dependence $\Delta = \Delta(\omega,t) = \{\cup_{x \in \omega} \Delta(x,t)\}$, which is the union of all the cones of dependence $\Delta(x,t)$ with $x \in \omega$. Under the assumption on t, Δ is contained in the cylinder $\bar{\Omega} \times [0, t]$. We denote the exterior unit space-time normal to the boundary S of Δ by $n = (n_x, n_t)$, where n_x denotes the space components of n. To

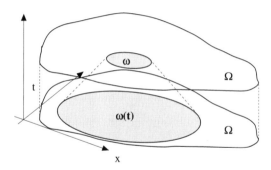

Figure 17.8: The generalized cone of dependence $\Delta(\omega, t)$ and enlarged region $\omega(t)$ associated to a subdomain ω of Ω.

obtain the desired estimate, we multiply (17.12) by $2\dot{u}$, integrate over Δ, and then integrate by parts to obtain

$$
\begin{aligned}
0 &= \int_\Delta (\ddot{u} - \Delta u) 2\dot{u} \, dx \, dt \\
&= \int_\Delta (2\ddot{u}\dot{u} + 2\nabla u \cdot \nabla \dot{u}) \, dx \, dt - \int_S n_x \cdot \nabla u 2\dot{u} \, ds \\
&= \int_\Delta \frac{d}{dt}((\dot{u})^2 + |\nabla u|^2) \, dx \, dt - \int_S n_x \cdot \nabla u 2\dot{u} \, ds \\
&= \int_S n_t((\dot{u})^2 + |\nabla u|^2) \, ds - \int_S n_x \cdot \nabla u 2\dot{u} \, ds.
\end{aligned}
$$

On the "sloping" sides of S, we have $n_t = |n_x| = 1/\sqrt{2}$ and thus by Cauchy's inequality, $n_t((\dot{u})^2 + |\nabla u|^2) - n_x \cdot \nabla u 2\dot{u} \geq 0$. We can therefore estimate the integral over the top part of S (with $n_t = 1$ and $n_x = 0$) corresponding to ω, in terms of the integral over the base of S corresponding to $\omega(t)$, and thus obtain the desired result.

∎

Problem 17.24. Write out the details of the last estimate.

Problem 17.25. Derive a version of Theorem 17.1 for the solution of (17.12) with $f \equiv 0$ without the restriction on t that keeps Δ inside the cylinder $\bar{\Omega} \times [0, t]$. Hint: define a generalized cone that includes part of the boundary of $\bar{\Omega} \times [0, t]$ when t is large.

Problem 17.26. Generalize the result of Lemma 17.1 to the case $f \neq 0$.

17.3.3.   The conservation of energy

Along with a finite speed of propagation, a solution u of (17.12) with $f = 0$ satisfies

$$\|\dot{u}(\cdot, t)\|^2 + \|\nabla u(\cdot, t)\|^2 = \|\dot{u}_0\|^2 + \|\nabla u_0\|^2, \quad t > 0,$$

where $\|\cdot\|$ denotes the $L_2(\Omega)$ norm.

Problem 17.27.   Prove this by modifying the proof in one dimension.

17.3.4.   Kirchhoff's formula and Huygens' principle

The generalization of d'Alembert's solution formula to the homogeneous wave equation (17.12) with $f = 0$ and $\Omega = \mathbb{R}^3$ is called *Kirchhoff's formula*, and was first derived by Poisson,

$$u(x,t) = \frac{1}{4\pi t} \int\limits_{S(x,t)} \dot{u}_0 \, ds + \frac{\partial}{\partial t} \left(\frac{1}{4\pi t} \int\limits_{S(x,t)} u_0 \, ds \right), \tag{17.17}$$

where $S(x,t) = \{y \in \mathbb{R}^3 : |y - x| = t\}$ is the sphere with radius t centered at x. This formula shows sharp propagation at speed one of both the initial data u_0 and $\dot{u}_0$, since the integrals involve only the surface $S(x,t)$ of the ball $B_3(x,t) = \{y \in \mathbb{R}^3 : |y - x| \leq t\}$, which is the set of points in $\mathbb{R}^3$ from which a signal of speed one may reach x within the time t. In other words, only the values of the data on the surface of the cone of dependence actually have an influence on the value at a point. The sharp wave propagation in three dimensions is referred to as *Huygens' principle*.

Problem 17.28.   Use (17.17) to write a formula for the solution of the wave equation with the data used in Problem 17.21.

A formula for the solution of the wave equation in two dimensions can be derived from (17.17) by considering the function to be a solution of the wave equation in three dimensions that happens to be independent of x_3. For $x \in \mathbb{R}^2$, we let $B_2(x,t) = \{y \in \mathbb{R}^2 : |y - x| \leq t\}$, which may be thought of as the projection of $B_3(x,t)$ onto the plane $\{x : x_3 = 0\}$. The solution is given by

$$u(x,t) = \frac{1}{2\pi} \int_{B_2(x,t)} \frac{\dot{u}_0(y)}{\left(t^2 - |y - x|^2\right)^{1/2}}\, dy$$

$$+ \frac{\partial}{\partial t}\left(\frac{1}{2\pi} \int_{B_2(x,t)} \frac{u_0(y)}{\left(t^2 - |y - x|^2\right)^{1/2}}\, dy\right).$$

Note that this formula involves integration over the entire ball $B_2(x,t)$ and not just the surface as in three dimensions. As a result, wave propagation in two dimensions is not as sharp as in three dimensions. If we strike a circular drumhead at the center, the vibrations propagate outwards in a circular pattern. The vibrations first hit a point a distance d from the center at time $t = d$ and that point continues to vibrate for all time afterwards. The amplitude of the vibrations decays roughly like $1/t$. We illustrate this in Fig. 17.10 where we show a finite element approximation to a related problem. See Strauss ([18]) for more details on wave propagation.

Problem 17.29. Write down a formula for the solution of the homogeneous wave equation in two dimensions corresponding to $u_0 = 1$ for $|x| \leq 1$ and 0 otherwise, and $\dot{u}_0 \equiv 0$.

17.4. A finite element method

To discretize (17.12), we first rewrite this scalar second order equation as a system of first order equations in time using the notation of Chapter 10 setting $u_1 = \dot{u}$ and $u_2 = u$: find the vector (u_1, u_2) such that

$$\begin{cases} \dot{u}_1 - \Delta u_2 = f & \text{in } \Omega \times (0, \infty), \\ -\Delta \dot{u}_2 + \Delta u_1 = 0 & \text{in } \Omega \times (0, \infty), \\ u_1 = u_2 = 0 & \text{on } \Gamma \times (0, \infty), \\ u_1(\cdot, 0) = \dot{u}_0,\ u_2(\cdot, 0) = u_0 & \text{in } \Omega. \end{cases} \tag{17.18}$$

We choose this formulation, and in particular write $\Delta u_1 = \Delta \dot{u}_2$ instead of $u_1 = \dot{u}_2$, because this brings (17.18) into a form that is analogous to the hyperbolic model problem of Chapter 10 with the positive coefficient a corresponding to $-\Delta$. Thus, we can use the same trick of cancellation that we used for the analysis of the hyperbolic model problem. In particular when $f \equiv 0$, if we multiply the first equation by u_1 and the second by u_2 and add, the terms $-(\Delta u_2, u_1)$ and $(\Delta u_1, u_2)$ cancel, leading to the conclusion that $\|u_1\|^2 + \|\nabla u_2\|^2$ is constant in time. In other words, we get energy conservation very easily.

The finite element functions we use to approximate the solution of (17.18) are piecewise linear polynomials in space and time that are continuous in space and "nearly" continuous in time. By nearly, we mean that the approximation is continuous unless the mesh changes from one time level to the next. We call this the cG(1) method. We discretize $\Omega \times (0, \infty)$ in the usual way, letting $0 = t_0 < \cdots < t_n < \cdots$ denote a partition of $(0, \infty)$ and to each time interval $I_n = (t_{n-1}, t_n]$ of length $k_n = t_n - t_{n-1}$, associate a triangulation $\mathcal{T}_n$ of Ω with mesh function h_n and a corresponding finite element space V_n of continuous piecewise linear vector functions in Ω that vanish on Γ. For $q = 0$ and 1, we define the space

$$W_{kn}^{(q)} = \left\{ (w_1, w_2) : w_j(x, t) = \sum_{r=0}^{q} t^r v_j^{(r)}(x), \ v_j^{(r)} \in V_n, \ j = 1, 2 \right\}$$

on the space-time slab $S_n = \Omega \times (t_{n-1}, t_n)$ and then the space $W_k^{(q)}$ of piecewise polynomial functions (v_1, v_2) such that $(v_1, v_2)|_{S_n} \in W_{kn}^{(q)}$ for $n = 1, 2, ..., N$. The functions in $W_k^{(q)}$ are forced to be continuous in space, but may be discontinuous in time.

The cG(1) method for (17.12) is based on the variational formulation of (17.18) as usual and reads: Find $U = (U_1, U_2) \in W_k^{(1)}$ such that for $n = 1, 2, \ldots,$

$$\begin{cases} \displaystyle\int_{t_{n-1}}^{t_n} \left((\dot{U}_1, w_1) + (\nabla U_2, \nabla w_1) \right) dt = \int_{t_{n-1}}^{t_n} (f, w_1) \, dt, \\ \displaystyle\int_{t_{n-1}}^{t_n} \left((\nabla \dot{U}_2, \nabla w_2) - (\nabla U_1, \nabla w_2) \right) dt = 0, \\ U_{1,n-1}^+ = P_n U_{1,n-1}^-, \quad U_{2,n-1}^+ = \pi_n U_{2,n-1}^-, \end{cases} \tag{17.19}$$

for all $w = (w_1, w_2) \in W_{kn}^{(0)}$, where $U_{1,0}^- = \dot{u}_0$, $U_{2,0}^- = u_0$, and

$$U_j(x, t)|_{S_n} = U_{j,n}^-(x)\frac{t - t_{n-1}}{k_n} + U_{j,n-1}^+(x)\frac{t - t_n}{-k_n}, \quad j = 1, 2.$$

Further, π_n is the *elliptic projection* into V_n defined by $(\nabla \pi_n w, \nabla v)$ $= (\nabla w, \nabla v)$ for all $v \in V_n$. Note that $\pi_n w \in V_h$ is the Galerkin approximation of the solution w of Poisson's equation on Ω with homogeneous Dirichlet boundary conditions.

Note that if the mesh is unchanged across t_{n-1}, i.e. $\mathcal{T}_n = \mathcal{T}_{n-1}$, then both P_n and π_n reduce to the identity and the approximation U is continuous across t_{n-1}. If $V_{n-1} \subset V_n$, which occurs for example when the mesh is refined using the customary strategies, then the coefficients of $U^+_{j,n-1}$, $j = 1, 2$, can be found by straightforward interpolation of $U^-_{j,n-1}$, i.e., for j=1,2,

$$U^+_{j,n-1}(N_{n,i}) = U^-_{j,n-1}(N_{n,i}),$$

where $\{N_{n,i}\}$ is the set of nodes in $\mathcal{T}_n$.

Problem 17.30. Prove this last claim.

In this case, the components U_j are continuous across t_{n-1} in the sense that

$$\lim_{t \to t^-_{n-1}} U_j(x, t) = \lim_{t \to t^+_{n-1}} U_j(x, t) \quad \text{for all } x \in \Omega.$$

However, when the mesh is changed so $V_{n-1} \not\subset V_n$, which typically happens when the mesh is coarsened, then U_j will in general be discontinuous across t_{n-1}. We illustrate this in Fig. 17.9.

Problem 17.31. Compute $U^+_{n-1} = \pi_n U^-_{n-1}$ for the example on the right in Fig. 17.9 assuming that $U^-_{n-1}(.25) = 1/2$, $U^-_{n-1}(.5) = 1/3$, and $U^-_{n-1}(.75) = 1$.

We use B_n and A_n to denote the $M_n \times M_n$ mass and stiffness matrices associated to the nodal basis $\{\varphi_{i,n}\}$ for V_n with dimension M_n, and further $A_{n-1,n}$ to denote the $M_n \times M_{n-1}$ matrix with coefficients

$$\left(A_{n-1,n}\right)_{i,j} = (\nabla \varphi_{i,n}, \nabla \varphi_{j,n-1}), \quad 1 \le i \le M_n, 1 \le j \le M_{n-1},$$

and let $B_{n-1,n}$ be defined by (16.25). Finally $\xi^-_{j,n}$ and $\xi^+_{j,n-1}$ denote the vectors of coefficients with respect to $\{\varphi_{i,n}\}$ of $U^-_{j,n}$ and $U^+_{j,n-1}$ for $j = 1, 2$. With this notation, (17.19) is equivalent to the set of matrix equations

$$\begin{cases} B_n\left(\xi^-_{1,n} - \xi^+_{1,n-1}\right) + k_n A_n\left(\xi^-_{2,n} + \xi^+_{2,n-1}\right)/2 = F_n, \\ A_n\left(\xi^-_{2,n} - \xi^+_{2,n-1}\right) - k_n A_n\left(\xi^-_{1,n} + \xi^+_{1,n-1}\right)/2 = 0, \\ B_n \xi^+_{1,n-1} = B_{n-1,n}\xi^-_{1,n-1}, \quad A_n \xi^+_{2,n-1} = A_{n-1,n}\xi^-_{2,n-1}, \end{cases} \quad (17.20)$$

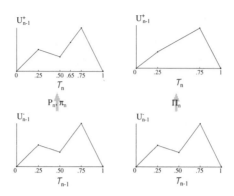

Figure 17.9: The effect of π_n in two cases of mesh changes. On the left, the mesh is refined so $V_{n-1} \subset V_n$ and π_n and P_n correspond to nodal interpolation. On the right, the mesh is coarsened and U_j is discontinuous across t_{n-1}.

where F_n is the data vector with coefficients

$$\left(F_n\right)_i = \int_{t_{n-1}}^{t_n} \left(f, \varphi_{i,n}\right) dt, \quad 1 \leq i \leq M_n.$$

Problem 17.32. Prove (17.20) is correct.

Problem 17.33. In the case the space mesh $\mathcal{T}_n$ does not change, show that (17.20) reduces to

$$\begin{cases} B(\xi_{1,n} - \xi_{1,n-1}) + k_n A(\xi_{2,n} + \xi_{2,n-1})/2 = F_n, \\ A(\xi_{2,n} - \xi_{2,n-1}) - k_n A(\xi_{1,n} + \xi_{1,n-1})/2 = 0, \end{cases}$$

where we have dropped the superscripts $+$ and $-$ on the coefficient vectors $\xi_{i,n}$ and the subscript n on A and B since U is continuous.

Problem 17.34. Formulate the cG(1) finite element method that uses the lumped mass quadrature rule in space and the midpoint rule in time to evaluate the integrals giving the approximation. Write out the discrete matrix equations for the approximation.

17.4.1. Energy conservation

One reason that we use the cG(1) method (17.19) is that the approximation conserves the total energy when $f \equiv 0$ provided $V_{n-1} \subset V_n$ for all n. To prove this, we choose $w_j = (U^+_{j,n-1} + U^-_{j,n})/2$ in (17.19) and add the two equations to get

$$\int_{t_{n-1}}^{t_n} (\dot{U}_1, U_1)\, dt + \int_{t_{n-1}}^{t_n} (\nabla \dot{U}_2, \nabla U_2)\, dt = 0,$$

because of the terms that cancel. This gives

$$\|U^-_{1,n}\|^2 + \|\nabla U^-_{2,n}\|^2 = \|U^-_{1,n-1}\|^2 + \|\nabla U^-_{2,n-1}\|^2.$$

$$(17.21)$$

In other words, the total energy of the cG(1) approximation is conserved from one time step to the next, just as holds for the solution of the continuous problem. When the mesh is changed so $V_{n-1} \not\subset V_n$, then the energy is only approximately conserved because each projection onto the new mesh changes the total energy.

Problem 17.35. Provide the details of the proof of (17.21).

Problem 17.36. Compute the change in energy in U in Problem 17.31.

17.5.   Error estimates and adaptive error control

In this section, we present a posteriori and a priori error analyses under some simplifying assumptions. The analysis of the cG(1) method for (17.18) is analogous to the analysis of the cG(1) method for the hyperbolic model problem in Chapter 10, but there are new technical difficulties in the case of the partial differential equation.

17.5.1.   An a posteriori error estimate

The adaptive error control is based on an a posteriori error estimate as usual. We prove the estimate under the assumptions that Ω is convex and the space mesh is kept constant, which simplifies the notation considerably. We use $\mathcal{T}_h$ to denote the fixed triangulation of mesh size $h(x)$ and we denote the corresponding finite element space by V_h. We use P_h to denote the L_2 projection into V_h and Δ_h to denote the discrete Laplacian on V_h. We use P_k to denote the L_2 projection into the set of piecewise constant functions on the partition $\{t_n\}$, and R_2, to denote the space residual associated to the discretization of the Laplacian as defined in Chapter 15. Finally, since U is continuous, we set $U_{j,n-1} = U_{j,n-1}^+ = U_{j,n-1}^-$. We shall prove the following a posteriori error estimate assuming that Ω is convex so that Theorem 15.6 applies.

Theorem 17.2. *There is a constant C_i such that for $N = 1, 2, ...,$*

$$\|u_2(t_N) - U_{2,N}\|$$

$$\leq C_i \bigg(\|h^2 R_2(U_{2,N})\| + \|h^2 R_2(U_{2,0})\|$$

$$+ \int_0^{t_N} \left(\|h(f - P_h f)\| + \|h^2 R_2(U_1)\| \right) dt$$

$$+ \int_0^{t_N} \left(\|k(f - P_k f)\| + \|k\Delta_h(U_2 - P_k U_2)\| \right.$$

$$\left. + \|k\nabla(U_1 - P_k U_1)\| \right) dt \bigg).$$

Note that the first four quantities on the right arise from the space discretization and the last three quantities arise from the time discretization. The integrals in time implies that errors accumulate at most linearly with time, as expected from the analysis of the model hyperbolic problem.

Proof. The proof is based on using the continuous dual problem to get an error representation formula. The dual problem is: For $N \geq 1$, find $\varphi = (\varphi_1, \varphi_2)$ such that

$$\begin{cases} -\dot{\varphi}_1 + \Delta\varphi_2 = 0 & \text{in } \Omega \times (0, t_N), \\ \Delta\dot{\varphi}_2 - \Delta\varphi_1 = 0 & \text{in } \Omega \times (0, t_N), \\ \varphi_1 = \varphi_2 = 0 & \text{on } \Gamma \times (0, t_N), \\ -\Delta\varphi_2(\cdot, t_N) = e_{2,N} & \text{in } \Omega, \\ \varphi_1(\cdot, t_N) = 0 & \text{in } \Omega, \end{cases} \quad (17.22)$$

where $e_2 = u_2 - U_2$. We multiply the first equation in (17.22) by $e_1 = u_1 - U_1$, the second by e_2, add the two together, integrate over $\Omega \times (0, t_N)$, integrate by parts, and finally use the Galerkin orthogonality of the approximation, to obtain

$$\|e_{2,N}\|^2 = \sum_{n=1}^{N} \Bigg(\left(f - \dot{U}_1, \varphi_1 - P_k P_h \varphi_1\right)_n$$

$$- \left(\nabla U_2, \nabla(\varphi_1 - P_k P_h \varphi_1)\right)_n - \left(\nabla(\dot{U}_2 - U_1), \nabla(\varphi_2 - P_k P_h \varphi_2)\right)_n \Bigg),$$

where $(\cdot, \cdot)_n$ denotes the $L_2(S_n)$ inner product. The goal is to distinguish the effects of the space and time discretizations by using the splitting $v - P_k P_h v = (v - P_h v) + (P_h v - P_k P_h v)$ and the orthogonalities of the L_2 projections P_h and P_k to obtain

$$\|e_{2,N}\|^2$$

$$= \sum_{n=1}^{N} \Bigg(\left(f - P_k f, P_h \varphi_1 - P_k P_h \varphi_1\right)_n + \left(f - P_h f, \varphi_1 - P_h \varphi_1\right)_n$$

$$- \left(\nabla U_2, \nabla(\varphi_1 - P_h \varphi_1)\right)_n$$

$$- \left(\nabla(U_2 - P_k U_2), \nabla(P_h \varphi_1 - P_k P_h \varphi_1)\right)_n$$

$$- \left(\nabla(\dot{U}_2 - U_1), \nabla(\varphi_2 - P_h \varphi_2)\right)_n$$

$$- \left(\nabla(\dot{U}_2 - U_1), \nabla(P_h \varphi_2 - P_k P_h \varphi_2)\right)_n \Bigg).$$

Finally, using the fact that $\varphi_1 = \dot{\varphi}_2$ and integrating by parts in t, we obtain

$$\|e_{2,N}\|^2 =$$

$$\sum_{n=1}^{N} \Bigg(\left(f - P_k f, P_h \varphi_1 - P_k P_h \varphi_1\right)_n + \left(f - P_h f, \varphi_1 - P_h \varphi_1\right)_n$$

$$- \left(\nabla(U_2 - P_k U_2), \nabla(P_h \varphi_1 - P_k P_h \varphi_1)\right)_n$$

$$+ \left(\nabla U_1, \nabla(\varphi_2 - P_h \varphi_2)\right)_n$$

$$+ \left(\nabla(U_1 - P_k U_1), \nabla(P_h \varphi_2 - P_k P_h \varphi_2)\right)_n \Bigg)$$

$$- \left(\nabla U_{2,N}, \nabla(\varphi_{2,N} - P_h \varphi_{2,N})\right) + \left(\nabla U_{2,0}, \nabla(\varphi_{2,0} - P_h \varphi_{2,0})\right).$$

To complete the proof, we use (16.4) and a standard estimate for $v - P_k v$ together with the following stability result for the dual problem (17.22). ∎

Lemma 17.3. *If Ω is convex, then the solution φ of (17.22) satisfies*

$$\|\ddot{\varphi}_2\|_{[0,t_N]} + \|D\dot{\varphi}_2\|_{[0,t_N]} + \|D^2\varphi_2\|_{[0,t_N]} \leq C\|e_{2,N}\|.$$
$$(17.23)$$

Proof. Multiplying the first equation in (17.22) by $\Delta\varphi_1$ and the second by $\Delta\varphi_2$ and adding, after using Greens formula, we obtain

$$\frac{d}{dt}(\|\nabla\varphi_1\|^2 + \|\Delta\varphi_2\|^2) = 0.$$

It follows using the initial conditions that

$$\|\nabla\varphi_1\|^2 + \|\Delta\varphi_2\|^2 = \|e_{2,N}\|^2.$$

The desired conclusion results from using the elliptic regularity estimate (15.6) and the fact that $\ddot{\varphi}_2 = \Delta\varphi_2$. ∎

Problem 17.37. (a) Fill in the details of the above proof. (b) (*Hard!*) Extend the a posteriori error estimate to the case $\mathcal{T}_n$ varies with n.

17.5.2. Adaptive error control

Following the ideas in the previous chapters, we can formulate an algorithm for adaptive error control by using the a posteriori error bound in Theorem 17.2 to give an estimate of the error on a given space-time mesh. We illustrate the use of the a posteriori error bound in two examples.[1]

In the first example, we compute the effects of a sharp strike at the center of a large square drumhead. We can model the problem for small amplitude vibrations by posing the wave equation on a finite domain $\Omega = [0,1] \times [0,1]$ with homogeneous Neumann boundary conditions. We assume the drumhead is initially at rest, i.e. $u(\cdot, 0) = \dot{u}(\cdot, 0) = 0$ and we model the strike by a source f located at the center of the square defined by

$$f(x,t) = \begin{cases} \sin^2(\pi t/T), & \text{for } t \leq .1, \ |x - (.5, .5)| \leq .1, \\ 0, & \text{otherwise.} \end{cases}$$

We plot the finite element approximation at time $t \approx .7$ in Fig. 17.10. We compute with a fixed time step $k_n \equiv .01$ for a relatively short time

---

[1]These computations are provided courtesy of M. G. Larson and A. J. Niklasson. See *Adaptive finite element methods for elasto-dynamics*, preprint, Department of Mathematics, Chalmers University of Technology, S41296 Göteborg, Sweden, for further details.

 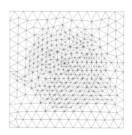

Figure 17.10: Plot of the finite element approximation and the corresponding space mesh for the model of the drumhead struck sharply at the center. In the plot of the displacement, a 15 level grey scale is used with black representing the largest and white the smallest displacement. The mesh on the right was adapted from an initial uniform mesh based on an a posteriori error bound.

so that the error due to time discretization is remains small. The space mesh is adapted according to the error control algorithm based on using an a posteriori error bound to equidistribute the error across the elements. The a posteriori analysis presented above can be changed to cover Neumann boundary conditions in a straightforward way.

The second example is a computation on a model of wave propagation in an inhomogeneous, linear elastic, viscously damped solid. We assume that the displacements are small and perpendicular to the (x_1, x_2) plane and that the solid is relatively long in the x_3 direction, which reduces the model to a scalar, two-dimensional wave equation for the shear waves propagating in the (x_1, x_2) plane.

In the specific example we present, the domain $\Omega = (0, 1) \times (0, 1)$ is composed of two isotropic materials joined along the line $x_2 = .59$. The material in the upper portion has a shear modulus that is five times larger than the material in the lower portion, so that the wave speed is five times greater in the upper portion of Ω. This gives the equation $\ddot{u} - a\Delta u = f$, where

$$a(x) = \begin{cases} 1, & x_2 \leq .59, \\ 5, & x_2 > .59. \end{cases}$$

We assume homogeneous Neumann (stress-free) boundary conditions and an approximate point source that is active for small time, so we define

$$f(x,t) = \begin{cases} \sin^2(\pi t/.07), & |x - (.4, .4)| \le .1 \text{ and } t \le .14, \\ 0, & \text{otherwise.} \end{cases}$$

This is the kind of forcing that might be found in nondestructive ultrasonic testing and seismology. The a posteriori error bound used to

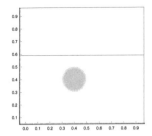 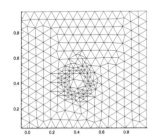

Figure 17.11: Density plot of the finite element approximation and the corresponding space mesh for the wave equation on an inhomogeneous material at time $t = .05$. The forcing has not reached maximum strength yet.

control the adaptivity is derived using techniques similar to those used to prove Theorem 17.2. We show contour plots of the approximation and the associated space meshes at times $t = .05, .15, .25$, and $.4$ in Fig. 17.11–Fig. 17.14. The material interface is marked with a horizontal line, and the difference in wave speed in the two materials is clear.

17.5.3.   An a priori error estimate

We state an a priori error estimate for the cG(1) method in the case $\mathcal{T}_n$ is constant and assuming that Ω is convex.

Theorem 17.4. *There is a constant C_i such that if U satisfies (17.19), then for $N = 1, 2, ...,$*

$$\|u_2(\cdot, t_N) - U_{2,N}\| \le C_i \int_0^{t_N} \left(\|k^2 \nabla \ddot{u}_2\| + \|k^2 \ddot{u}_1\| + \|h^2 D^2 \dot{u}_2\| \right) dt.$$

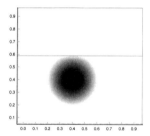
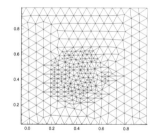

Figure 17.12: $t = .15$

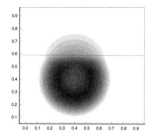
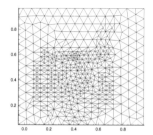

Figure 17.13: $t = .25$

We note that the a priori error estimate is of order $\mathcal{O}(h^2 + k^2)$ and like the a posteriori estimate, the integral in time corresponds to a linear rate of accumulation of errors.

Proof. To simplify the analysis, we only analyze the time discretization. This corresponds to setting V_n equal to the space of functions with square-integrable gradients on Ω and that vanish on Γ. We use $\hat{u}$ to denote the piecewise linear time interpolant of u. Since we already know how to estimate $\rho = u - \hat{u}$, we only have to estimate $e = \hat{u} - U$.

To this end, we use the discrete dual problem: Find $(\Phi_1, \Phi_2) \in W_k^{(1)}$ that satisfies for $n = N, n - 1, ..., 1$,

$$\begin{cases} -(v_1, \dot{\Phi}_1)_n - (\nabla v_1, \nabla \Phi_2)_n = 0, \\ -(\nabla v_2, \nabla \dot{\Phi}_2)_n + (\nabla v_2, \nabla \Phi_1)_n = 0, \end{cases} \tag{17.24}$$

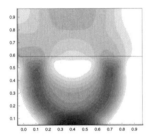 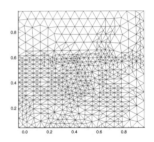

Figure 17.14: $t = .40$

for all $(v_1, v_2) \in W_{kn}^{(0)}$, where $\Phi_{1,N} \equiv 0$ and $-\Delta\Phi_{2,N} = e_{2,N}$ with $e_2 = \hat{u}_2 - U_2$. Because the test functions v_1 and v_2 are piecewise constant, we may replace Φ_1 and Φ_2 in the second terms in each equation by their mean values $\bar{\Phi}_{i,n} := (\Phi_{i,n-1} + \Phi_{i,n})/2$ on I_n. After this, we can replace the test functions v_1 and v_2 by arbitrary piecewise linear functions, because both $\dot{\Phi}_i$ and $\bar{\Phi}_i$ are piecewise constant. In particular, replacing v_1 by $e_1 = \hat{u}_1 - U_1$ and v_2 by e_2, adding the resulting equations and summing over n, and then integrating by parts in time, we obtain the error representation

$$\|e_{2,N}\|^2$$
$$= \sum_{n=1}^{N} \left(\left(\nabla \dot{e}_2, \nabla \Phi_2\right)_n + \left(\nabla e_2, \nabla \bar{\Phi}_1\right)_n + \left(\dot{e}_1, \Phi_1\right)_n - \left(\nabla e_1, \nabla \bar{\Phi}_2\right)_n \right)$$
$$= \sum_{n=1}^{N} \left(\left(\dot{e}_1, \bar{\Phi}_1\right)_n + \left(\nabla e_2, \nabla \bar{\Phi}_1\right)_n + \left(\nabla \dot{e}_2, \nabla \bar{\Phi}_2\right)_n - \left(\nabla e_1, \nabla \bar{\Phi}_2\right)_n \right)$$

$$= -\sum_{n=1}^{N} \left(\left(\dot{\rho}_1, \bar{\Phi}_1\right)_n + \left(\nabla \rho_2, \nabla \bar{\Phi}_1\right)_n \right.$$
$$\left. + \left(\nabla \dot{\rho}_2, \nabla \bar{\Phi}_2\right)_n - \left(\nabla \rho_1, \nabla \bar{\Phi}_0\right)_n \right).$$

We also replaced Φ_j by their mean values $\bar{\Phi}_j$ and then used Galerkin orthogonality to replace U by u. Since the terms involving $\dot{\rho}_i$, $i = 1, 2$

vanish, we arrive at the following error representation

$$\|e_{2,N}\|^2 = -\sum_{n=1}^{N}\left(\left(\nabla\rho_2, \nabla\bar{\Phi}_1\right)_n - \left(\rho_1, \Delta\bar{\Phi}_2\right)_n\right).$$

Choosing $v_1 = -\Delta\bar{\Phi}_1$ and $v_2 = -\Delta\bar{\Phi}_2$ in (17.24), we obtain the stability estimate:

$$\|\nabla\Phi_{1,n}\|^2 + \|\Delta\Phi_{2,n}\|^2 = \|e_{2,N}\|^2 \quad \text{for all } n \leq N.$$

Combining this with the error representation and then using standard estimates for the interpolation error ρ, we obtain the a priori error estimate.

∎

Problem 17.38. Supply the details of the proof.

Problem 17.39. Show that assuming the solution $u(x,t)$ of the wave equation $\ddot{u} - \Delta u = f$ has the form $u(x,t) = \exp(i\omega t)w(x)$, where $f(x,t) = \exp(i\omega t)g(x)$ and $\omega > 0$ is a given frequency, leads to the stationary Helmholtz's equation $-\Delta w - \omega^2 w = g$ for the amplitude $w(x)$. Show that a fundamental solution of Helmholtz's equation in $\mathbb{R}^3$ is given by $\frac{\exp(i\omega|x|)}{4\pi|x|}$. Solve Helmholtz's equation using Femlab on a bounded two-dimensional domain with suitable boundary conditions in a configuration of physical interest.

Problem 17.40. Derive the wave equation from Maxwell's equations under suitable assumptions.

> Let us pause in life's pleasures
> and count its many tears.
> For we all share sorrow with the poor.
> There's a song that will linger forever in our ears,
> "Hard times will come again no more".
> There's a song in the sigh of the weary,
> "Hard times, hard times come again no more.
> Many days you have lingered around our cabin door.
> Hard times come again no more." (S. Foster)

18

Stationary Convection-Diffusion Problems

> I have always found it difficult to read books that cannot be understood without too much meditation. For, when following one's own meditation, one follows a certain natural inclination and gains profit along with pleasure; but one is enormously cramped when having to follow the meditaton of others. (Leibniz)

In this chapter and the next, we consider a linear model for a problem that includes the effects of convection, diffusion, and absorption, which is an example of a *multi-physics* problem coupling several physical phenomena. We begin by deriving the model and discussing the basic properties of solutions. In this chapter, we continue by considering the discretization of the stationary case, starting with a discussion that explains why a straightforward application of Galerkin's method yields disappointing results for a convection dominated problem. We then present a modified Galerkin method that resolves the difficulties, that we call the *streamline diffusion finite element method* or *Sd method*. We continue with the time dependent case in the next chapter. The material of these two chapters lay the foundation for the application of the finite element method to incompressible and compressible fluid flow including reactive flow, multi-phase flow and free-boundary flow, developed in the advanced companion volume.

18.1. A basic model

We consider the transport of heat in a current flowing between two regions of a relatively large body of water, for example from a warm region to a cold region, taking into account the dissipation of the heat, the advection of the heat by the current, and the absorption of heat into the air. An example of such a physical situation is the North American Drift flowing from Newfoundland, where it continues the Gulf Stream, to the British Isles, where it splits into two branches. The North American Drift is responsible for the warm climate of Western Europe. Our interest is focused on the water temperature in the Drift at different locations at different times. The full problem takes place in three space dimensions, but we simplify the model to two dimensions assuming all functions are independent of the depth.

The model is a time-dependent scalar convection-diffusion-absorption problem posed on a *space-time* domain $Q = \Omega \times I$, where Ω is a polygonal domain in $\mathbb{R}^2$ with boundary Γ and $I = (0, T)$, of the form

$$\begin{cases} \dot{u} + \nabla \cdot (\beta u) + \alpha u - \nabla \cdot (\epsilon \nabla u) = f & \text{in } Q, \\ u = g_- & \text{on } (\Gamma \times I)_-, \\ u = g_+ \text{ or } \epsilon \partial_n u = g_+ & \text{on } (\Gamma \times I)_+, \\ u(\cdot, 0) = u_0 & \text{in } \Omega, \end{cases} \quad (18.1)$$

where u represents the temperature, $\beta = (\beta_1, \beta_2)$, α and $\epsilon > 0$ are functions of (x, t) representing the *convection velocity, absorption coefficient,* and *diffusion coefficient,* respectively. Further, $f(x, t)$, u_0, g, and u_0 are given data, and

$$(\Gamma \times I)_- = \{(x, t) \in \Gamma \times I : \beta(x, t) \cdot n(x) < 0\},$$
$$(\Gamma \times I)_+ = \{(x, t) \in \Gamma \times I : \beta(x, t) \cdot n(x) \geq 0\},$$

where $n(x)$ is the outward normal to Γ at point x, are the *inflow* and *outflow* parts of the space-time boundary $\Gamma \times I$, respectively. We illustrate this in Fig. 18.1.

Problem 18.1. Let $\Omega = (0, 1) \times (0, 1)$, $I = (0, 1)$, and $\beta = (\cos(\frac{\pi}{2} t + \frac{\pi}{4}), \sin(\frac{\pi}{2} t + \frac{\pi}{4}))$. Identify the inflow and outflow boundaries of Q.

The model is the result of expressing conservation of heat as

$$\frac{\partial}{\partial t}(\lambda u) + \nabla \cdot q + \alpha u = f,$$

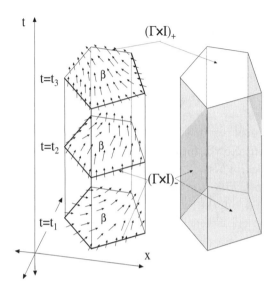

Figure 18.1: The space time domain Q indicating the inflow and outflow boundaries. The inflow boundary is shaded in the
figure on the right.

where q is the heat flow and λ the heat capacity, and assuming that the
constitutive law is the following generalization of Fourier's law (13.20)

$$q = \beta\lambda u - \epsilon\nabla u.$$

Setting the heat capacity $\lambda = 1$, gives (18.1). This model is a natural
extension of the model for heat flow considered in Chapter 13 with the
addition of terms corresponding to convection of heat with the current
β and absorption of heat at the rate α.

Using the identity

$$\nabla \cdot (\beta u) = \beta \cdot \nabla u + (\nabla \cdot \beta)u,$$

we may replace the convection term $\nabla \cdot (\beta u)$ by $\beta \cdot \nabla u$ by modifying the
term αu to $(\alpha + \nabla \cdot \beta)u$.

The model (18.1) models a variety of phenomena with the variable u
representing a quantity subject to convection, diffusion and absorption.
Another example is the evolution of a contaminant dropped into fluid
running in a pipe, see Fig. 18.2, where u represents the concentration of

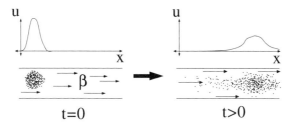

Figure 18.2: The convection and diffusion of a dye inside a water pipe. $u(x,t)$ represents the concentration of the dye at (x,t).

the contaminant in the fluid. A system of the form (18.1) may also serve as a simple model for fluid flow described by the Navier-Stokes equations, in which case u represents mass, momentum and energy. Thus, (18.1) is a fundamental model.

Problem 18.2. The motion of the rotor of an electrical motor gives rise to an additional contribution to the electric field E of the form $\beta \times B$ where β is the velocity and B the magnetic flux. Show that introducing this term into the derivation of (16.2) leads to the convection-diffusion equation

$$\sigma \frac{\partial u}{\partial t} + \sigma \beta \cdot \nabla u - \nabla \cdot \left(\frac{1}{\mu}\nabla u\right) = f.$$

18.2. The stationary convection-diffusion problem

We begin by considering the stationary convection-diffusion-absorption problem associated to (18.1),

$$\begin{cases} \beta \cdot \nabla u + \alpha u - \nabla \cdot (\epsilon \nabla u) = f & \text{in } \Omega, \\ u = g_- & \text{on } \Gamma_-, \\ u = g_+ \text{ or } \epsilon \partial_n u = g_+ & \text{on } \Gamma_+, \end{cases} \qquad (18.2)$$

with all functions independent of time, and α modified to include $\nabla \cdot \beta$ as indicated above. In this case, the definitions of the inflow and outflow boundaries Γ_- and Γ_+ are given by

$$\Gamma_- = \{x \in \Gamma : \beta(x) \cdot n(x) < 0\} \text{ and } \Gamma_+ = \{x \in \Gamma : \beta(x) \cdot n(x) \geq 0\},$$

see Fig. 18.3. We first discuss basic features of solutions of the problem

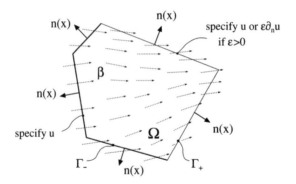

Figure 18.3: The notation for a stationary convection-diffusion problem.

(18.2) and then consider the computation of approximate solutions using finite element methods. Special care has to be taken in the design of the finite element method, because direct application of Galerkin's method to (18.2) when the convection is the dominant feature leads to numerical solutions with spurious oscillations, which is illustrated in Problem 18.6 below.

18.2.1. Convection versus diffusion

Generally, the relative size of ϵ and β govern the qualitative nature of (18.2). If $\epsilon/|\beta|$ is small, then (18.2) is *convection dominated* and has hyperbolic character. If $\epsilon/|\beta|$ is not small, then (18.2) is *diffusion dominated* and has elliptic character. Thus, the problem (18.2) changes character from hyperbolic to elliptic as $\epsilon/|\beta|$ increases. In the diffusion dominated case the material on elliptic problems in Chapter 15 is applicable since the convection terms are dominated by the diffusion terms.

We now focus on the convection-dominated hyperbolic case and then first consider the extreme case with $\epsilon = 0$.

18.2.2. The reduced problem

The *reduced problem* with $\epsilon = 0$ takes the form

$$
\begin{cases}
\beta \cdot \nabla u + \alpha u = f & \text{in } \Omega, \\
u = g_- & \text{on } \Gamma_-,
\end{cases}
\tag{18.3}
$$

where u is specified only on the inflow boundary Γ_-. The reduced problem couples convection and absorption.

The *streamlines* associated to the stationary convection velocity field $\beta(x)$ are curves $x(s)$, parametrized by $s \geq 0$, satisfying

$$\begin{cases} \dfrac{dx}{ds} = \beta(x(s)) & \text{for } s > 0, \\ x(0) = \bar{x}, \end{cases} \tag{18.4}$$

for the streamline starting at $\bar{x}$, see Fig. 18.4. This is the path followed by a particle starting at $\bar{x}$ that is convected with velocity $\beta(x)$. In this interpretation, s is time and dx/ds represents the particle velocity. A

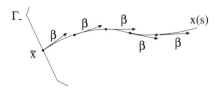

Figure 18.4: A streamline has tangent vector $\beta(x(s))$ at every point $x(s)$.

streamline is *closed* if the particle returns to the point of departure, i.e. $x(s) = \bar{x}$ for some $s > 0$. A problem with closed streamlines requires special care, so we assume for now that there aren't any. The reduced equation becomes an ordinary differential equation along a streamline since by the chain rule,

$$\frac{d}{ds} u(x(s)) + \alpha(x(s)) u(x(s)) = (\beta \cdot \nabla u + \alpha u)(x(s)) = f(x(s)), \quad s > 0,$$

where the inflow data $g_-(\bar{x})$ at $\bar{x} \in \Gamma_-$ gives the "initial data" $u(x(0))$. The solution of the reduced problem (18.3) therefore can be found by solving for each streamline $x(s)$ an ordinary differential equation of the form (9.1):

$$\dot{v}(s) + a(s)v(s) = \bar{f}(s), \quad s > 0, \ v(0) = g_-(x(0)),$$

where $v(s) = u(x(s))$, $a(s) = \alpha(x(s))$ and $\bar{f}(s) = f(x(s))$, corresponding to "solving along streamline starting at inflow". We note that the case of non-negative absorption with $\alpha(x) \geq 0$ corresponds to the parabolic case with $a(s) \geq 0$.

We conclude that in the reduced problem without diffusion information is propagated sharply along streamlines from the inflow boundary to the outflow boundary. We see in particular that if there is a discontinuity in the inflow data at some point $\bar{x}$ on the inflow boundary Γ_-, then the solution of (18.3) will in general be discontinuous across the entire streamline staring at $\bar{x}$. As an example, the solution of the problem

$$\begin{cases} \dfrac{\partial u}{\partial x_1} = 0 & \text{in } x \in \Omega, \\ u(0, x_2) = \begin{cases} 0, & 0 < x_2 < 1/2, \\ 1, & 1/2 \leq x_2 < 1, \end{cases} \end{cases}$$

corresponding to (18.2) with $\beta = (1, 0)$, $\alpha = 0$ and $\Omega = [0, 1] \times [0, 1]$, is given by

$$u(x_1, x_2) = \begin{cases} 0, & 0 < x_2 < 1/2, \ 0 < x_1 < 1, \\ 1, & 1/2 \leq x_2 < 1, \ 0 < x_1 < 1, \end{cases}$$

with a discontinuity across the streamline $x(s) = (s, 1/2)$.

Problem 18.3. Suppose $\beta = (1, 1 - x_1)$ and $\Omega = [0, 1] \times [0, 1]$. (a) Plot Ω and identify the inflow and outflow boundaries. (b) Compute the streamlines corresponding to each point on the inflow boundary (Hint: there are two cases). Plot enough of the streamlines so that you can describe the "flow" over Ω.

Problem 18.4. Solve the problem $x_1 \frac{\partial u}{\partial x_1} + x_2 \frac{\partial u}{\partial x_2} = 0$ on $\Omega = \{x : 1 < x_1, x_2 < 2\}$, with some choice of inflow data.

18.2.3. Layers of difficulty

The features of the reduced problem with $\epsilon = 0$ are present also in the hyperbolic case with $\epsilon/|\beta|$ small positive but now the presence of positive diffusion makes the solution continuous and "spreads out" a discontinuity over a *layer* in the solution, which is a narrow region where the solution changes significantly. For example, a discontinuity across a streamline becomes a *characteristic layer* of width $O(\sqrt{\epsilon})$, see Fig. 18.5. Further, if Dirichlet boundary conditions are specified on the outflow boundary Γ_+ in the case $\epsilon > 0$, then in general the solution u of (18.2)

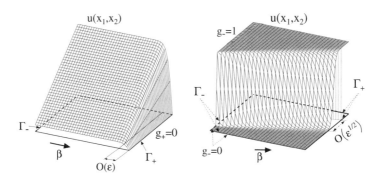

Figure 18.5: Illustrations of an outflow (on the left) and a characteristic layer caused by a discontinuity in g_- (on the right).

has an outflow *boundary layer* of width $O(\epsilon)$ close to Γ_+ where u changes rapidly to meet the boundary condition; see Fig. 18.5.

To give a concrete example of an outflow layer we consider the one-dimensional analog of (18.2), which takes the form

$$\begin{cases} -(\epsilon u')' + \beta u' + \alpha u = f & \text{for } 0 < x < 1, \\ u(0) = 0, \ u(1) = 0, \end{cases} \tag{18.5}$$

in the case of homogeneous Dirichlet boundary conditions. We present computational results in Fig. 18.6 for the case $\epsilon = 0.02$, $\beta = 1$, $\alpha = 0$ and $f = 1$ using L_2 norm error control on the tolerance level .02. The flow is from left to right with inflow at $x = 0$ and outflow at $x = 1$. Note the outflow layer in u in the boundary layer near $x = 1$ resulting from the convection in the positive x direction and how the mesh is refined in that region.

Problem 18.5. Show that the width of an outflow layer is approximately of order ϵ by explicitly solving the one-dimensional convection-diffusion problem $-\epsilon u'' + u' = 0$ for $0 < x < 1$ with $u(0) = 1$, $u(1) = 0$.

We now present a problem showing that Galerkin's method may go berserk under certain conditions. We urge the reader to do this problem before continuing.

Problem 18.6. Consider the continuous Galerkin cG(1) method for the one-dimensional problem $-\epsilon u'' + u' = 0$ in $(0, 1)$ with $u(0) = 1$, $u(1) =$

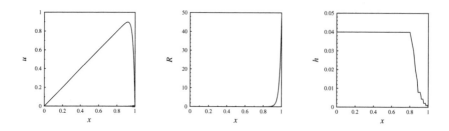

Figure 18.6: Solution, error, and meshsize for (18.5) with $\epsilon = .02$, $\beta = 1$, $\alpha = 0$, $f = 1$, and TOL$=.02$.

0. (a) Write down the discrete equations for the cG(1) approximation computed on a uniform mesh with M interior nodes. (b) With $\epsilon = 0.01$, compute the approximation for $M = 10$ and $M = 11$ and compare to the true solution. (c) Compute the approximation with $M \approx 100$ and compare with the exact solution. (d) Write out the discrete equations when $\epsilon = h/2$. Explain why this scheme is called the *upwind method* for the reduced problem. How is the convection term approximated by Galerkin's method? Compare with the upwind method. Compare the nature of propagation of effects (in particular the outflow boundary condition) in Galerkin's method with $\epsilon > 0$ much smaller that h and the upwind method.

18.3. The streamline diffusion method

Convection dominated problems present difficulties for computation that are not present in diffusion dominated problems, mainly because the stability properties of convection dominated problems cause the standard Galerkin finite element method to be non-optimal compared to interpolation. Recall that Galerkin methods are typically optimal for elliptic and parabolic problems, and in general for diffusion dominated problems. However, the standard Galerkin method for convection dominated problems may be far from optimal if the exact solution is nonsmooth, in which case the Galerkin approximations contain "spurious" oscillations not present in the true solution. This is illustrated in Problem 18.6. The oscillations occur whenever the finite element mesh is too coarse to resolve layers, which typically is the case in the early stages of an adaptive refinement process. The oscillations result from a lack of stability of the standard Galerkin finite element method for convection dominated problems, and may have disastrous influence on the performance of an adaptive method leading to refinements in large regions where no refinement is needed.

We conclude that it is important to improve the stability properties of the Galerkin finite element method. However, this has to be done cleverly, because additional stability is often obtained at the price of decreased accuracy. For example, increasing artificially the diffusion term (e.g. by simply setting $\epsilon = h$) will increase the stability of the Galerkin method, but may also decrease accuracy and prevent sharp resolution of layers. Thus, the objective is to improve stability without sacrificing accuracy.

We consider two ways of enhancing the stability of the standard Galerkin finite element method:

(a) introduction of weighted least squares terms

(b) introduction of artificial viscosity based on the residual.

We refer to the Galerkin finite element method with these modifications as the streamline diffusion, or Sd-method, and motivate this terminology below. The modification (a) adds stability through least squares control of the residual and the modification (b) adds stability by the introduction of an elliptic term with the size of the diffusion coefficient, or *viscosity*, depending on the residual with the effect that viscosity is added where the residual is large, i.e., typically where the solution is nonsmooth. Both modifications enhance stability without a strong effect on the accuracy because both modifications use the residual.

18.3.1. Abstract formulation

We begin by describing the Sd-method for an abstract linear problem of the form

$$Au = f, \tag{18.6}$$

for which the standard Galerkin finite element method reads: compute $U \in V_h$ such that

$$(AU, v) = (f, v) \quad \text{for all } v \in V_h, \tag{18.7}$$

where A is a linear operator on a vector space V with inner product $(\cdot, \cdot)$ and corresponding norm $\| \cdot \|$, and $V_h \subset V$ is a finite element space. Typically, A is a convection-diffusion differential operator, $(\cdot, \cdot)$ is the L_2 inner product over some domain Ω. We assume that A is positive semi-definite, i.e. $(Av, v) \geq 0$ for all $v \in V$.

The *least squares method* for (18.6) is to find $U \in V_h$ that minimizes the residual over V_h, that is

$$\|AU - f\|^2 = \min_{v \in V_h} \|Av - f\|^2.$$

This is a convex minimization problem and the solution U is characterized by

$$(AU, Av) = (f, Av) \quad \text{for all } v \in V_h. \tag{18.8}$$

We now formulate a Galerkin/least squares finite element method for (18.6) by taking a weighted combination of (18.7) and (18.8): compute $U \in V_h$ such that

$$(AU, v) + (\delta AU, Av) = (f, v) + (\delta f, Av) \quad \text{for all } v \in V_h, \tag{18.9}$$

where $\delta > 0$ is a parameter to be chosen. Rewriting the relation (18.9) as

$$(AU, v + \delta Av) = (f, v + \delta Av) \quad \text{for all } v \in V_h, \tag{18.10}$$

we can alternatively formulate the Galerkin/least squares method as a *Petrov-Galerkin method*, which is a Galerkin method with the space of test functions being different from the space of trial functions V_h. In our case, the test functions have the form $v + \delta Av$ with $v \in V_h$.

Adding the artificial viscosity modification (b) yields (with a typical choice of diffusion operator) the Sd-method in abstract form: find $U \in V_h$ such that

$$(AU, v + \delta Av) + (\hat{\epsilon} \nabla U, \nabla v) = (f, v + \delta Av) \quad \text{for all } v \in V_h, \tag{18.11}$$

where $\hat{\epsilon}$ is the artificial viscosity defined in terms of the residual $R(U) = AU - f$ through

$$\hat{\epsilon} = \gamma_1 h^2 |R(U)|, \tag{18.12}$$

with γ_1 a positive constant to be chosen, and $h(x)$ the local mesh size of V_h.

Choosing $v = U$ in (18.11) we see that the modifications improve the stability of the approximation as compared to (18.7).

Problem 18.7. Assume $(Av, v) \geq c\|v\|^2$ for some positive constant c. (a) Choose $v = U$ in (18.11) and derive a stability result for U. (b) Compare the result from (a) to the stability result obtained by choosing $v = U$ in (18.7). How does the stability result from (a) depend on δ and γ_1?

18.3.2. The streamline diffusion method for a convection-diffusion problem

We now formulate the streamline diffusion method for (18.2) with constant ϵ and homogeneous Dirichlet boundary conditions using the standard space of piecewise linear functions $V_h \subset V = H_0^1(\Omega)$ based on a triangulation $\mathcal{T}_h$ of Ω: compute $U \in V_h$ such that

$$(AU, v + \delta Av) + (\hat{\epsilon}\nabla U, \nabla v) = (f, v + \delta Av) \quad \text{for all } v \in V_h, \tag{18.13}$$

where $(\cdot, \cdot)$ is the $L_2(\Omega)$ inner product,

$$Aw = \beta \cdot \nabla w + \alpha w, \ \delta = \frac{1}{2}\frac{h}{|\beta|},$$

$$\hat{\epsilon}(U, h) = \max\{\epsilon, \gamma_1 h^2 | f - (\beta \cdot \nabla U + \alpha U)|, \gamma_2 h^{3/2}\},$$

where the γ_j are positive constants to be specified. We obtain (18.13) by multiplying the terms in (18.2) that appear in the reduced equation by the modified test function $v + \delta(\beta \cdot \nabla v + \alpha v)$, which corresponds to a least squares modification of the convection/absorption terms, while multiplying the diffusion term in (18.2) by v after replacing ϵ by $\hat{\epsilon}$. If ϵ is variable or higher order polynomials are used, then the diffusion term should be included in the least squares modification.

In general, $\hat{\epsilon}$ depends on U and the discrete problem (18.13) is nonlinear, even though the continuous problems (18.2) and (18.3) are linear. When iterative methods are used to solve the discrete equations, the additional complication in solving the discrete equations due to the nonlinearity introduced by $\hat{\epsilon}$ has little effect on the computational overhead. The artificial viscosity $\hat{\epsilon}$ is proportional to $|f - (\beta \cdot \nabla U + \alpha U)|$, which plays the role of the residual. For simplicity, the jump terms related to the diffusion term has been left out; see the statement of Theorem 18.5.

The size of the artificial viscosity $\hat{\epsilon}$ relative to the mesh size h (assuming $\epsilon \leq h$) gives a measure of the smoothness of the exact solution u. In regions where u is smooth, $\hat{\epsilon} \approx h^{3/2}$, while in outflow layers in general $\hat{\epsilon} = \gamma_1 h^2 |f - (\beta \cdot \nabla U + \alpha U)| \propto h$, because there $|f - (\beta \cdot \nabla U + \alpha U)| \propto h^{-1}$ on a general mesh. In characteristic layers, typically $|f - (\beta \cdot \nabla U + \alpha U)| \approx h^{-1/2}$ so that again $\hat{\epsilon} \propto h^{3/2}$. Thus, we distinguish two basic cases:

(a) u is "smooth" with $\hat{\epsilon} \propto h^{3/2}$, including characteristic layers,

(b) u is non-smooth with $\hat{\epsilon} \propto h$, typically resulting in outflow layers.

We assume for the sake of simplicity that $\hat{\epsilon} = \epsilon$, which can be guaranteed during a computation by adding this requirement to the stopping criterion in the adaptive algorithm. The case $\hat{\epsilon} > \epsilon$ typically occurs in initial stages of adaptive refinements when the mesh is coarse. We focus on the case with $h^2 \leq \epsilon \leq h$. If ϵ is larger than h then all layers are resolved by the mesh, and if ϵ is smaller than h^2 then the mesh is much too coarse.

In Fig. 18.7, we present the results of a computation using the adaptive streamline diffusion method on the convection-diffusion problem with $\Omega = (0,1) \times (0,1)$, $\beta = (2,1)$, $\alpha = 0$, $\epsilon = 0.01$, and discontinuous inflow data $u(0,y) \equiv 1$, $0 \leq y \leq 1$ and $u(x,0) \equiv 0$, $0 < x < 1$. Note the form and thickness of the layers and the corresponding shape of the adapted mesh.

Problem 18.8. Plot Ω for this computation and identify the streamlines and the inflow and outflow boundaries.

18.4. A framework for an error analysis

We describe the basic ingredients of the analysis of the streamline diffusion method. The goal is an a posteriori error estimate that can be used to guide the mesh adaptivity in order to control the error. After presenting the general points, we analyze a specific case in the following section.

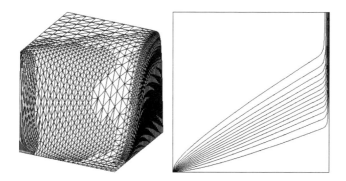

Figure 18.7: A surface plot with the mesh indicated and associated level curves of the approximate solution obtained using the streamline diffusion method for a convection-diffusion problem with both outflow and characteristic layers.

18.4.1. Basic stability estimates

We assume that

$$\alpha - \frac{1}{2}\nabla \cdot \beta \geq c > 0, \tag{18.14}$$

for some constant c. In the case of non-closed streamlines this condition may be satisfied by a change of variable; cf. Problem 18.10. The weak stability estimate for the solution u of (18.2) has the form:

$$\|\sqrt{\epsilon}\nabla u\| + \|u\| \leq C\|f\|, \tag{18.15}$$

with $C = (\sqrt{c} + 2)/(2c)$, and $\|\cdot\|$ denotes the $L_2(\Omega)$ norm. In what follows, the exact value of C changes, but it is always a constant that depends on the constant c in (18.14). The estimate (18.15) follows after multiplying the differential equation in (18.2) by u, integrating over Ω, and using the fact that $(\beta \cdot \nabla u, u) = -\frac{1}{2}(\nabla \cdot \beta u, u)$.

A corresponding stability estimate for the discrete problem (18.13) is obtained by choosing $v = U$, which gives

$$\|\sqrt{\bar{\epsilon}}\nabla U\| + \|\sqrt{\delta}(\beta \cdot \nabla U + \alpha U)\| + \|U\| \leq C\|f\|. \tag{18.16}$$

We note that the control of the $\|\sqrt{\delta}(\beta \cdot \nabla U + \alpha U)\|$ term results from the least squares modification of the streamline diffusion method, and that the artificial viscosity $\hat{\epsilon}$ is present in the gradient term. The $\|U\|$ term allows the $\|\sqrt{\delta}(\beta \cdot \nabla U + \alpha U)\|$ term to be replaced by $\|\sqrt{\delta}\beta \cdot \nabla U\|$, yielding a weighted control of the streamline derivative $\beta \cdot \nabla U$. This control corresponds to adding diffusion in the streamline direction with coefficient δ, and this is the motivation for the name "streamline diffusion method".

Below, we also use an analog of the stability estimate (18.16) with $\hat{\epsilon} = \epsilon = 0$ that has the following form: for piecewise continuous functions w with $w = 0$ on Γ_-,

$$\|\sqrt{\delta}Aw\|^2 + c\|w\|^2 \leq (Aw, w + \delta Aw), \tag{18.17}$$

where as above $A = \beta \cdot \nabla w + \alpha w$. This estimate follows from the following version of Green's formula after noting that the boundary term is guaranteed to be non-negative if $w = 0$ on Γ_-, because $\beta \cdot n \geq 0$ on Γ_+.

Lemma 18.1.

$$(\beta \cdot \nabla w, w) = -\frac{1}{2}(\nabla \cdot \beta\, w, w) + \frac{1}{2}\int_\Gamma w^2 \beta \cdot n\, ds. \tag{18.18}$$

We note that the stability estimate (18.17) requires w to be specified (to be zero) on the inflow boundary Γ_-. The estimate gives a motivation why it is natural to specify data on Γ_-, rather than on Γ_+, in the case $\epsilon = 0$.

Problem 18.9. Provide the details in the derivations of (18.15), (18.16), (18.17) and (18.18).

Problem 18.10. Show that the equation $u'(s) = f(s)$, where $s \in \mathbb{R}$, takes the form $v'(s) + v(s) = \exp(-s)f(s)$ using the change of dependent variable $v(s) = \exp(-s)u(s)$.

18.4.2. A strong stability estimate

In addition to the weak stability estimate (18.15), we also use the following estimate for a dual continuous problem which can be written in the form (18.2) with Neumann outflow boundary conditions:

$$\|\beta \cdot \nabla u + \alpha u\| + \|\epsilon D^2 u\| \leq C\|f\|, \tag{18.19}$$

where C is a moderately sized constant that does not depend in a significant way on ϵ if Ω is convex. We refer to this estimate as a *strong stability estimate* because second derivatives are bounded, in addition to the control of the term $\beta \cdot \nabla u + \alpha u$. The "price" of the second derivative control is a factor ϵ^{-1}, which is natural from the form of the equation. Since ϵ is small, the "price" is high, but nevertheless there is a net gain from using this estimate, because the presence of the second derivatives brings two powers of h to compensate the ϵ^{-1} factor.

We are not able to prove the analog of the strong stability estimate (18.19) for the discrete problem, which would be useful in deriving a priori error estimates. Instead, we use (18.16) as a substitute, yielding a weighted control of $\beta \cdot \nabla U + \alpha U$ with the weight $\sqrt{\delta}$.

We summarize the effects of the two modifications used to create the streamline diffusion method: the least squares modification gives added control of the derivative in the streamline direction with a weight $\sqrt{\delta}$, while the artificial viscosity modification gives control of the gradient ∇U with the weight $\sqrt{\hat{\epsilon}}$.

18.4.3. Basic forms of the error estimates

Assuming that $\hat{\epsilon} = \epsilon$, the a posteriori error estimate for the streamline diffusion method (18.13) has the form:

$$\|u - U\| \leq C_i S_c \left\| \frac{h^2}{\epsilon} R(U) \right\|, \tag{18.20}$$

where $S_c \approx 1$ and $R(U)$ is the residual of the finite element solution U defined in terms of the differential equation in a natural way. We note the presence of the factor h^2/ϵ that results from combining strong stability with Galerkin orthogonality. In many cases, we have $h^2/\epsilon \ll 1$. For example if $\epsilon \approx h^{3/2}$, which corresponds to a "smooth" exact solution such as a solution with a characteristic layer, then (18.20) reduces to

$$\|u - U\| \leq C \|h^{1/2} R(U)\|.$$

If $\epsilon \approx h$, which corresponds to a "non-smooth" exact solution such as a solution with an outflow layer, then (18.20) reduces to

$$\|u - U\| \leq C \|h R(U)\|.$$

To understand the gain in (18.20), compare it to the "trivial" a posteriori error estimate

$$\|u - U\| \leq C\|R(U)\| \tag{18.21}$$

that follows directly from the weak stability estimate and even holds for non-Galerkin methods. This estimate is almost useless for error control in the case the exact solution is non-smooth, because the right-hand side in general increases with decreasing mesh size until all layers are resolved.

The a priori error estimate takes the form

$$\|h^{1/2} R(U)\| + \|u - U\| \leq C_i S_{c,h} \|h^{3/2} D^2 u\|,$$

where $S_{c,h} \approx 1$. In the case of a smooth solution, the a posteriori and a priori error estimates match and both are non-optimal with a loss of $h^{1/2}$, while in the non-smooth case with $\hat{\epsilon} = h$, the a posteriori estimate appears in optimal form.

Problem 18.11. Prove (18.21). Assuming that $R(U) \approx h^{-1}$ in an outflow layer of width of order h, estimate $\|R(U)\|$ and discuss what would happen if an adaptive method tried to control $\|u - U\|$ by using (18.21). Do the same in the case of a characteristic layer assuming $R(U) \approx h^{-1/2}$ in a layer of width $h^{1/2}$.

18.5.   A posteriori error analysis in one dimension

We consider the one-dimensional convection-diffusion-absorption problem (18.5) with $\beta = 1$, $\alpha = 1$ and ϵ a small positive constant:

$$\begin{cases} -\epsilon u'' + u' + u = f & \text{in } (0,1), \\ u(0) = u(1) = 0. \end{cases} \tag{18.22}$$

This problem in general has an outflow layer of width $O(\epsilon)$ at $x = 1$ where the solution rapidly changes to adjust to the imposed outflow boundary value $u(1) = 0$.

For simplicity, we consider the streamline diffusion method for (18.22) without the artificial viscosity modification, which takes the form : Compute $U \in V_h$ such that

$$(U' + U, v + \delta(v' + v)) + (\epsilon U', v') = (f, v + \delta(v' + v)) \qquad \text{for all } v \in V_h, \tag{18.23}$$

where $\delta = h/2$ when $\epsilon < h$ and $\delta = 0$ otherwise, V_h is the usual space of continuous piecewise linear functions that vanish at $x = 0, 1$, and $(\cdot, \cdot)$ the $L_2(\Omega)$ inner product. We note that the streamline diffusion method is essentially obtained by multiplication by the modified test function $v + \delta(v' + v)$. The modification has a stabilizing effect, which is manifested by the presence of the positive term $(U' + U, \delta(U' + U))$, obtained by choosing $v = U$ in (18.23). If δ is increased, the stability is improved but at the cost of accuracy. If δ is decreased, then the reverse is true. Choosing $\delta \approx h/2$ gives the best compromise and results in a satisfactory combination of stability and accuracy.

We now prove an L_2 a posteriori error estimate for the streamline diffusion method (18.23). For simplicity, we consider a case with $h \leq \epsilon$ and $\delta = 0$.

Theorem 18.2. *There is a constant C independent of ϵ and h such that the solution U of (18.23) satisfies the following estimate for all $\epsilon \geq 0$*

$$\|u - U\| \leq C_i \left\| \frac{h^2}{\epsilon^*}(f - U_x) \right\| + |\epsilon u'(0)| + |\epsilon U'(0)|,$$

where $\epsilon^*(x) = h^{1/2}\epsilon$ on the interval of the subdivison underlying V_h with left-hand endpoint $x = 0$, $\epsilon^* = \epsilon$ elsewhere, and $\|\cdot\|$ denotes the $L_2(\Omega)$ norm

Proof. Let φ be the solution of the dual problem

$$\begin{cases} -\epsilon\varphi'' - \varphi' + \varphi = g & \text{for } 0 < x < 1, \\ \varphi'(0) = 0, \varphi(1) = 0, \end{cases} \tag{18.24}$$

with the direction of the convection from right to left, which is opposite to that of (18.22). We pose the dual problem with Dirichlet inflow boundary condition at the inflow at $x = 1$ and it is convenient to choose a Neumann outflow condition at the outflow at $x = 0$. Choosing $g = u - U$ in (18.24), multiplying by $u - U$ and integrating by parts, we get and using the Galerkin orthogonality,

$$\|u - U\|^2 = \int_0^1 (f - U' - U)(\varphi - \pi_h\varphi) \, dx$$
$$- \int_0^1 \epsilon U'(\varphi - \pi_h\varphi)' \, dx + \epsilon u'(0)\varphi(0),$$

where $\pi_h \varphi \in V_h$ interpolates φ at the interior mesh points. Using the following stability result this proves the desired result, up to the small modification of ϵ required because in general $\varphi(0) \neq 0$, while $\pi_h \varphi(0) = 0$. ∎

Lemma 18.3. *There is a constant C such that if φ solves (18.24), then*

$$|\varphi(0)| \leq \|g\| \quad \text{and} \quad \|\epsilon \varphi''\| \leq \|g\|. \tag{18.25}$$

Proof. Multiplication with φ and integration gives

$$\int_0^1 (\epsilon \varphi')^2 \, dx + \int_0^1 \varphi^2 \, dx + \frac{1}{2}\varphi(0)^2 \leq \frac{1}{2}\int_0^1 g^2 \, dx + \frac{1}{2}\int_0^1 \varphi^2 \, dx,$$

which proves the estimate for $|\varphi(0)|$. Next, multiplication with $-\epsilon \varphi''$ gives

$$\int_0^1 (\epsilon \varphi'')^2 \, dx + \int_0^1 \varphi' \epsilon \varphi'' \, dx + \int_0^1 \epsilon(\varphi')^2 \, dx = -\int_0^1 g \epsilon \varphi''.$$

Since

$$2\int_0^1 \varphi' \varphi'' \, dx = \varphi'(1)^2 \geq 0,$$

this proves the desired estimate for $\epsilon \varphi''$ by Cauchy's inequality. ∎

Problem 18.12. Determine the *Green's function* $g_z(x)$ for the boundary value problem

$$\begin{cases} -\epsilon u'' + bu' = f, & 0 < x < 1, \\ u(0) = u(1) = 0, \end{cases} \tag{18.26}$$

where b is constant. This is the function $g_z(x)$ defined for $0 < z < 1$ that satisfies

$$\begin{cases} -\epsilon g_z'' - bg_z' = \delta_z, & 0 < x < 1, \\ g_z(0) = g_z(1) = 0, \end{cases}$$

where δ_z denotes the delta function at z. Prove the representation formula

$$u(z) = \int_0^1 g_z(x) f(x) \, dx, \quad 0 < z < 1, \tag{18.27}$$

where $u(x)$ is the solution of (18.26). Consider first the case $\epsilon = 1$ and $b = 0$, and then the case $\epsilon > 0$ and $b = 1$, paying particular attention to the limit $\epsilon \to 0$.

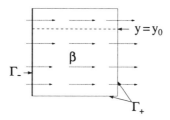

Figure 18.8: The model problem.

18.6. Error analysis in two dimensions

We prove error estimates for a model problem of the form (18.2) with $\beta = (1, 0)$, $\alpha = 1$, ϵ constant and $\Omega = (0, 1) \times (0, 1)$. For convenience, we denote the coordinates in $\mathbb{R}^2$ by (x, y), and we write

$$u_x = \partial u / \partial x = \beta \cdot \nabla u = (1, 0) \cdot \nabla u,$$

and formulate the model problem (see Fig. 18.8) as follows:

$$\begin{cases} u_x + u - \epsilon \Delta u = f & \text{in } \Omega, \\ u = 0 & \text{on } \Gamma. \end{cases} \tag{18.28}$$

This problem is a direct extension of the one-dimensional model problem (18.5) to two dimensions. Solutions of (18.28) can have an outflow layer of width $O(\epsilon)$ along Γ_+ and also characteristic layers of width $O(\sqrt{\epsilon})$ along characteristics $\{(x, y) : y = y_0\}$ that do not occur in the corresponding one-dimensional problem.

18.6.1. Strong stability analysis

We use the following strong stability estimate for the associated dual problem with homogeneous Neumann outflow boundary data.

Lemma 18.4. *The solution φ of the dual problem*

$$\begin{cases} -\varphi_x + \varphi - \epsilon \Delta \varphi = g, & \text{in } \Omega, \\ \varphi = 0, & \text{on } \Gamma_+, \\ \varphi_x = 0, & \text{on } \Gamma_-, \end{cases} \tag{18.29}$$

satisfies the stability estimates

$$\left(\|\varphi\|^2 + 2\|\epsilon^{1/2}\nabla\varphi\|^2 + \int_{\Gamma_-} \varphi^2 \, ds \right)^{1/2} \leq \|g\|,$$

$$(18.30)$$

$$\|\epsilon D^2\varphi\| \leq \|g\|.$$

$$(18.31)$$

Proof. Multiplying (18.29) by 2φ and integrating, we obtain (18.30) after using the fact that $-2(\varphi_x, \varphi) = \int_{\Gamma_-} \varphi^2 ds$. Next, multiplying (18.29) by $-\epsilon\Delta\varphi$ and integrating, we obtain

$$\|\epsilon\Delta\varphi\|^2 + (\epsilon\nabla\varphi, \nabla\varphi) + (\epsilon\varphi_{xx}, \varphi_x) + (\epsilon\varphi_{yy}, \varphi_x) = (f, -\epsilon\Delta\varphi).$$

Since $\varphi_x = 0$ on Γ_-, we have

$$2(\varphi_{xx}, \varphi_x) = \int_{\Gamma_+} \varphi_x^2 \, ds.$$

On the two sides of Ω with $y = 0$ and 1, $\varphi_x = 0$, while $\varphi_y = 0$ on Γ_+. This gives

$$2(\varphi_{yy}, \varphi_x) = -2(\varphi_y, \varphi_{xy}) = \int_{\Gamma_-} \varphi_y^2 \, ds.$$

We conclude that

$$\|\epsilon\Delta\varphi\|^2 \leq (f, -\epsilon\Delta\varphi) \leq \|f\|\|\epsilon\Delta u\|.$$

The desired estimate follows using the elliptic regularity result $\|D^2\varphi\| \leq \|\Delta\varphi\|$, see (15.54). ∎

18.6.2.   The a posteriori error estimate

We prove an a posteriori error estimate in the case $\delta = 0$ and $\hat{\epsilon} = \epsilon$ constant. The proof when $\delta \approx h$ is obtained by a simple modification.

Theorem 18.5. *There is a constant C such that*

$$\|u - U\| \leq C\left(\left\| \frac{h^2}{\epsilon^*} R(U) \right\| + \|\epsilon\partial_n u\|_{\Gamma_-} + \|\epsilon\partial_n U\|_{\Gamma_-} \right),$$

$$(18.32)$$

where $R(U) = R_1(U) + R_2(U)$ with

$$R_1(U) = |f - U_x - U|$$

and

$$R_2(U) = \frac{\epsilon}{2} \max_{S \subset \partial K} h_K^{-1} |[\partial_S U]| \quad on \ K \in \mathcal{T}_h, \tag{18.33}$$

where $[\partial_S v]$ denotes the jump across the side $S \subset \partial K$ in the normal derivative of the function v in V_h, and $\epsilon^* = \epsilon h^{1/2}$ on K if $K \cap \Gamma_- \neq 0$ and $\epsilon^* = \epsilon$ otherwise.

Proof. Letting φ denote the solution of the dual problem (18.29) with $g = e = u - U$, we obtain the following error representation by using Galerkin orthogonality and the equations defining u and U,

$$\begin{aligned}
\|e\|^2 &= (e, -\varphi_x + \varphi - \epsilon \Delta \varphi) \\
&= (e_x + e, \varphi) + (\epsilon \nabla e, \nabla \varphi) \\
&= (u_x + u, \varphi) + (\epsilon \nabla u, \nabla \varphi) - (U_x + U, \varphi) - (\epsilon \nabla U, \nabla \varphi) \\
&= (f, \varphi) + \int_{\Gamma_-} \epsilon \partial_n u \, \varphi \, ds - (U_x + U, \varphi) - (\epsilon \nabla U, \nabla \varphi) \\
&= (f - U_x - U, \varphi - \pi_h \varphi) - (\epsilon \nabla U, \nabla(\varphi - \pi_h \varphi)) + \int_{\Gamma_-} \epsilon \partial_n u \, \varphi \, ds,
\end{aligned}$$

from which the desired estimate follows by standard interpolation error estimates and Lemma 18.4.

Problem 18.13. Supply the details to finish the proof.

Problem 18.14. Prove a similar result when $\delta \approx h$.

We note that the $*$ modification of ϵ is required to deal with the incompatibility of boundary conditions for φ and functions in V_h on Γ_-.

∎

18.6.3. The a priori error estimate

We prove the a priori error estimate (18.6) in the simplified case that $\epsilon = \hat{\epsilon} = 0$. The Dirichlet boundary condition is specified only on the inflow boundary. Using Galerkin orthogonality in the analog of (18.17) for the error $e = u - U$ with $A = \frac{\partial}{\partial x} + I$, we get

$$\begin{aligned}
\|\sqrt{\delta} Ae\|^2 + c\|e\|^2 &\leq (Ae, e + \delta Ae) = (Ae, u - \pi_h u + \delta A(u - \pi_h u)) \\
&\leq \frac{1}{2}\|\sqrt{\delta} Ae\|^2 + \|\delta^{-1/2}(u - \pi_h u)\|^2 + \|\sqrt{\delta} A(u - \pi_h u)\|^2,
\end{aligned}$$

where as usual $\pi_h u$ denotes the nodal interpolant of u. Choosing $\delta = h$ and using standard interpolation results, yields

$$\frac{1}{2}\|\sqrt{h}Ae\|^2 + c\|e\|^2 \leq C_i^2\|h^{3/2}D^2 u\|^2.$$

We state the final result, which extends directly to the case with ϵ small, as a theorem.

Theorem 18.6. *If $\alpha - \frac{1}{2}\nabla \cdot \beta \geq c > 0$ and $\epsilon \leq h$, then*

$$\|u - U\| \leq CC_i\|h^{3/2}D^2 u\|.$$

18.6.4.   The propagation of information

It is possible to prove a "local" form of an a priori error estimate for the streamline diffusion method in which the L_2 error over a domain $\Omega_1 \subset \Omega$ that excludes layers is estimated in terms of the L_2 norm of $h^{3/2}D^2 u$ over a slightly larger domain Ω_2 that also excludes layers. The upshot is that the presence of e.g. an outflow layer where the error may be locally large if ϵ is small, does not degrade the accuracy away from the layer. This is because in the streamline diffusion method, effects are propagated more or less along streamlines from inflow to outflow in the direction of the "wind" β just as effects are propagated in the continuous problem. In particular, the streamline diffusion method does not have the spurious propagation in the opposite direction to the wind that occurs in the standard Galerkin method.

Problem 18.15.   (a) Consider the problem $-\epsilon u'' + u' + u = f$ for $0 < x < 1$, together with $u(0) = 1$, $u(1) = 0$. Let $\psi(x)$ be a positive weight function on I such that $0 \leq -\psi' \leq C\psi/\epsilon$, with C a suitable constant. Prove a stability estimate of the form $\|\sqrt{\psi}u\| \leq C\|\sqrt{\psi}f\|$. Use this estimate to draw a conclusion on the decay of information in the "upwind" direction. Hint: multiply by ψu. (b) (*Hard*) Extend to the streamline diffusion method. Hint: multiply by $\pi_h(\psi U)$ and estimate the effect of the perturbation $\psi U - \pi_h(\psi U)$.

18.7.   79 A.D.

The following figure, adapted with the permission of the National Geographic Society, shows the ash fall resulting from the eruption of Mount

Vesuvius in 79 A.D. This is an example of a full scale convection-diffusion problem with the convection velocity corresponding to a wind from north-west and an approximate delta function source. The level curves of the ash downfall (levels at 0.1, 1 and 2 meters are faintly shaded) give a measure of the concentration of ash in the atmosphere in various directions from the crater. Note the pronounced propagation of ash in the direction of the wind due to convection. The propagation against the wind due to diffusion is much smaller.

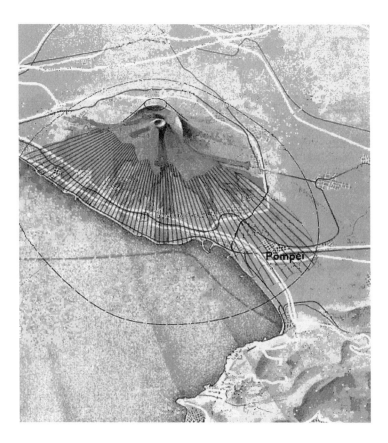

19

Time Dependent
Convection-Diffusion Problems

> The fact that in nature "all is woven into one whole", that space, matter, gravitation, the forces arising from the electromagnetic field, the animate and inanimate are all indissolubly connected, strongly supports the belief in the unity of nature and hence in the unity of scientific method. (Weyl)

We return to the time-dependent problem (18.1), considering mainly the *convection dominated* case with $\epsilon/|\beta|$ small since the case when $\epsilon/|\beta|$ is not small can be analyzed by extending the results for the heat equation presented in Chapter 16.

The cG(1)dG(r) method for the heat equation, using cG(1) in space and dG(r) in time on space-time slabs, can be applied to (18.1) with modifications like those used to create the streamline diffusion method for stationary convection diffusion problems (18.2). We discuss this approach briefly in Section 19.3. However, it turns out that using space-time meshes that discretize space and time independently is not optimal for convection dominated problems. It is better to use a mesh that is *oriented* along characteristics or space-time particle paths. We illustrate this in Fig. 19.1. We refer to a finite element method using oriented meshes as a *characteristic Galerkin* method, or a chG method.

In particular, we study the chG(0) method obtained applying the cG(1)dG(0) method on a mesh oriented along particle paths in space and time inside each slab. In its most elementary form, the chG(0) method reduces on each space-time slab to an L_2 projection from the

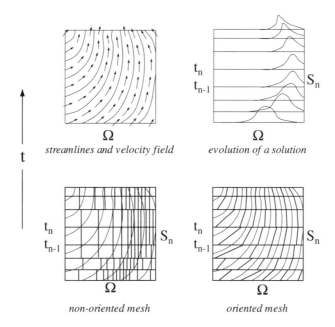

Figure 19.1: The upper figures show the space-time domain with the flow field in space-time, the space-time streamlines, and an illustration of the evolution of a solution. The two lower figures show non-oriented and oriented space-time meshes.

previous mesh onto the current mesh followed by an exact transport in the directions of the characteristics. The main computational work is spent on the L_2 projections. However for the purpose of analysis, it is more useful to view the chG method as a Galerkin method on a space-time mesh oriented in space-time along space-time particle paths. In addition, this opens the way to generalizations in which the space-time meshes are oriented in different ways.

We begin by describing the two fundamental ways to represent solutions of a convection problem, called respectively *Euler coordinates* and *Lagrange coordinates*.

19.1.   Euler and Lagrange coordinates

We describe the coordinates systems in the context of measuring the temperature in the North Atlantic stream. Dr. Euler and Dr. Lagrange each lead a team of assistants provided with boats and thermometers. Dr. Euler's assistants anchor their boats at specific locations and measure the temperature of the water continuously as it flows past their positions. Dr. Lagrange's assistants, on the other hand, drift with the current while measuring the temperature. An assistant to Dr. Euler measures the temperature of the water as it is affected by the current in contrast to an assistant to Dr. Lagrange who measures the temperature of the same "piece" of water, albeit at different positions. In order to correlate the measurements of the two groups, it is necessary to record the stationary positions of Dr. Euler's assistants and to keep track of the changing positions of Dr. Lagrange's assistants.

To simplify the mathematical description of the two sets of coordinates, we consider the model problem,

$$\begin{cases} \dot{u} + \beta \cdot \nabla u - \epsilon \Delta u = f & \text{in } Q = \mathbb{R}^2 \times (0, \infty), \\ u(x,t) \to 0 & \text{for } t > 0 \text{ as } |x| \to \infty, \\ u(\cdot, 0) = u_0 & \text{in } \mathbb{R}^2, \end{cases} \quad (19.1)$$

where we assume that β is smooth, f and u_0 have *bounded support*, which means that they are zero outside some bounded set, and $\epsilon \geq 0$ is constant. This means in particular that we avoid here discussing complications rising from boundaries in space.

19.1.1.   Space-time particle paths

The *space-time particle paths*, or *characteristics*, corresponding to the convection part $\dot{u} + \beta \cdot \nabla u$ of (19.1) are curves $(x,t) = (x(\bar{x}, \bar{t}), t(\bar{t}))$ in space and time parameterized by $\bar{t}$, where $x(\bar{x}, \bar{t})$ and $t(\bar{t})$ satisfy

$$\begin{cases} \dfrac{dx}{d\bar{t}} = \beta(x, \bar{t}) & \text{for } \bar{t} > 0, \\ \dfrac{dt}{d\bar{t}} = 1 & \text{for } \bar{t} > 0, \\ x(\bar{x}, 0) = \bar{x}, \quad t(0) = 0. \end{cases} \quad (19.2)$$

This is analogous to the stationary case with the operator $\beta \cdot \nabla$ replaced by $1 \cdot \partial/\partial t + \beta \cdot \nabla$, where the coefficient of the time-derivative is one and

t acts as an extra coordinate. Here, the time coordinate has a special role and in fact $t = \bar{t}$. The projection of the space-time particle path into space is given by the curve $x(\bar{x}, \bar{t})$ satisfying

$$\begin{cases} \dfrac{dx}{d\bar{t}} = \beta(x, \bar{t}) & \text{for } \bar{t} > 0, \\ x(\bar{x}, 0) = \bar{x} \end{cases} \qquad (19.3)$$

which is the time-dependent analog of the particle path in the stationary case and gives the path in space of a particle moving with speed $\beta(x, t)$.

Note that for time-dependent velocity fields, it is important to distinguish between particle paths and streamlines, unlike the case of stationary velocities when the two concepts are the same. Streamlines are related to a time-independent velocity, for instance we may "freeze" the velocity $\beta(x, t)$ for $t = \hat{t}$ and consider the *streamlines* of $\beta(x, \hat{t})$ that solve $dx/dt = \beta(x, \hat{t})$. The streamlines are therefore different from the particle paths, which satisfy $dx/dt = \beta(x, \bar{t})$, if $\beta(x, \bar{t})$ depends on $\bar{t}$.

It is also important to distinguish between a space-time particle path $(x(\bar{x}, \bar{t}), \bar{t})$ and its projection into space $x(\bar{x}, \bar{t})$. Space-time particle paths are essential for the construction of the oriented space-time mesh we describe below.

Problem 19.1. Compute and plot the space-time particle paths if (a) $\beta = (x_1, 1)$. (b) $\beta = (-x_2, x_1)$. (c) $\beta = (\sin(t), \cos(t))$.

19.1.2. Changing from Lagrange to Euler coordinates

The solution of (19.2) defines a map $(\bar{x}, \bar{t}) \to (x, t)$ by setting $(x, t) = (x(\bar{x}, \bar{t}), \bar{t})$ where $x(\bar{x}, \bar{t})$ is the position at time $\bar{t}$ of a particle starting at $\bar{x}$ at time zero. Because particle paths fill up space-time and cannot cross, the map is invertible. We illustrate this in Fig. 19.2. We refer to (x, t) as the Euler coordinates and $(\bar{x}, \bar{t})$ as the Lagrange coordinates. An observer using Euler coordinates is anchored at a fixed location x in space and observes the change of some quantity, such as the temperature at x, as time passes, where the change may be caused by the convection bringing new "particles" to the point x. On the other hand, the Lagrange coordinate system moves with the velocity field so that there is no convection relative to the moving coordinate system. The coordinate $\bar{x}$ then acts as a "label" or "marker" attached to "particles" moving along streamlines, where the $\bar{x}$ denotes the position of a particle

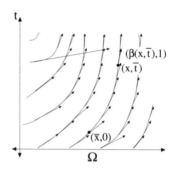

Figure 19.2: The vector field $(\beta, 1)$ and the corresponding streamlines define a map between the Euler and Lagrange coordinate systems.

at time zero, and $x = x(\bar{x}, \bar{t})$ is its position at time $\bar{t}$. In the context of the Dr. Euler and Dr. Lagrange's assistants, the mapping describes the positions of the Dr. Lagrange's crew as they are transported by the current.

Problem 19.2. Compute the coordinate map between Euler and Lagrange coordinates corresponding to β in Problem 19.1.

Given a function $u(x, t)$ in Euler coordinates, we define a corresponding function $\bar{u}(\bar{x}, \bar{t})$ in Lagrange coordinates by $\bar{u}(\bar{x}, \bar{t}) = u(x(\bar{x}, \bar{t}), \bar{t})$. By the chain rule,

$$\frac{\partial \bar{u}}{\partial \bar{t}} = \frac{\partial u}{\partial t} + \beta \cdot \nabla u,$$

since $\dfrac{dx}{d\bar{t}} = \beta(x, \bar{t})$ and $t = \bar{t}$. Thus, the convection equation

$$\frac{\partial u}{\partial t} + \beta \cdot \nabla u = f \tag{19.4}$$

in the Euler coordinates (x, t), which is (19.1) with $\epsilon = 0$, takes the simple form

$$\frac{\partial \bar{u}}{\partial \bar{t}} = \bar{f} \tag{19.5}$$

in the *global* Lagrange coordinates $(\bar{x}, \bar{t})$, where $\bar{f}(\bar{x}, \bar{t}) = f(x(\bar{x}, \bar{t}), \bar{t})$. We conclude that in global Lagrange coordinates, the convection term disappears and the original partial differential equation (19.4) reduces to a set of first order ordinary differential equations with respect to t indexed by the "marker" $\bar{x}$. In particular, if $f = 0$ then $\bar{u}(\bar{x}, t)$ is independent of time. This makes the job easy for a Lagrange assistant in the sense that if $f = 0$ then it is sufficient to measure the temperature at time equal to zero since the temperature following particles is constant. The Euler assistants on the other hand have to measure the temperature continuously at their fixed location since it may vary even though $f = 0$. Of course, the Lagrange assistants have to keep track of their positions as time passes.

Problem 19.3. Compute the solution of $\dot{u} + xu' = f$ for $x \in \mathbb{R}$ and $t > 0$ with

$$f(t) = \begin{cases} t(1-t), & 0 \leq t \leq 1, \\ 0, & 1 < t \end{cases} \quad \text{and} \quad u_0(x) = \begin{cases} 0, & |x| > 1, \\ 1, & |x| \leq 1. \end{cases}$$

by computing the characteristics and changing to Lagrange coordinates.

Problem 19.4. Compute the solution of $\dot{u} + (x_1, t) \cdot \nabla u = 0$ for $(x_1, x_2) \in \mathbb{R}^2$ and $t > 0$ with

$$u_0(x) = \begin{cases} 1, & (x_1, x_2) \in [0, 1] \times [0, 1], \\ 0, & \text{otherwise.} \end{cases}$$

by computing the characteristics and changing to Lagrange coordinates.

Because of the simplicity of (19.5), it is tempting to use Lagrange coordinates. But there is a hook: the Lagrange coordinates have to be computed by solving (19.2) and this is as difficult to solve as the original convection-diffusion problem formulated in Euler coordinates. However, using a kind of "local" Lagrange coordinates, we can avoid solving the equations (19.2) for the global characteristics, while keeping the advantage of the simple form (19.5) in the Lagrange description. The Lagrange coordinates associated to (19.1) underlie the construction of the space-time mesh on each slab S_n used in the chG(0) method in the sense that the space-time mesh in the chG(0) method is oriented approximately along the characteristics of the flow locally on S_n, as shown in Fig. 19.1.

19.2.   The characteristic Galerkin method

The characteristic chG(0) method is based on piecewise constant approximation along space-time characteristics and piecewise linear approximation in space. As usual we let $\{t_n\}$ be an increasing sequence of discrete time levels and associate to each time interval $I_n = (t_{n-1}, t_n)$ a finite element space V_n of piecewise linear continuous functions on a triangulation $\mathcal{T}_n$ of $\mathbb{R}^2$ with mesh function h_n. We use S_n to denote the space-time slab $\mathbb{R}^2 \times I_n$.

19.2.1.   Approximate particle paths

We let $\beta_n^h \in V_n$ denote the nodal interpolant of $\beta_n = \beta(\cdot, t_{n-1})$ and introduce the *approximate space-time particle path* $(x_n(\bar{x}, \bar{t}), \bar{t})$ in S_n, where

$$\begin{cases} \dfrac{dx_n}{d\bar{t}} = \beta_n^h(\bar{x}) & \text{in } I_n, \\ x_n(\bar{x}, t_{n-1}) = \bar{x}, \end{cases}$$

or

$$x_n(\bar{x}, \bar{t}) = \bar{x} + (\bar{t} - t_{n-1})\beta_n^h(\bar{x}) \quad \text{for } \bar{t} \in I_n. \tag{19.6}$$

The approximate particle path $(x_n(\bar{x}, \bar{t}), \bar{t})$ is a straight line segment with slope $\beta_n^h(\bar{x})$ starting at $\bar{x}$. We illustrate this in Fig. 19.3.

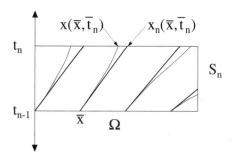

Figure 19.3: Exact and approximate particle paths in S_n.

Problem 19.5. Suppose that $\beta = (x_1, 1)$. Plot some of the particle paths and corresponding approximate particle paths for $0 \le t \le .1$ associated to mesh points on the standard uniform triangulation of the square.

19.2.2. The coordinate mapping

We introduce the coordinate map $F_n : S_n \to S_n$ defined by

$$(x, t) = F_n(\bar{x}, \bar{t}) = (x_n(\bar{x}, \bar{t}), \bar{t}) \quad \text{for } (\bar{x}, \bar{t}) \in S_n,$$

where the coordinates $(\bar{x}, \bar{t})$ acts like local Lagrange coordinates on S_n. We illustrate this in Fig. 19.4. We call β_n^h the *tilting velocity* for F_n.

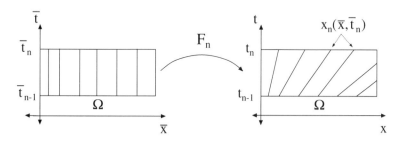

Figure 19.4: The map F_n between local Lagrange and Euler coordinates takes a non-oriented grid in $(\bar{x}, \bar{t})$ to an oriented grid in (x, t).

Note that these coordinates are similar but not the same as the global Lagrange coordinates unless β is constant.

Denoting the Jacobian with respect to $\bar{x}$ by $\overline{\nabla}$, we have from (19.6)

$$\overline{\nabla} x_n(\bar{x}, \bar{t}) = I + (\bar{t} - t_{n-1}) \overline{\nabla} \beta_n^h(\bar{x}),$$

where I denotes the identity, It follows from the inverse function theorem that the mapping $F_n : S_n \to S_n$ is invertible if

$$k_n \|\overline{\nabla} \beta_n^h\|_{L_\infty(R^2)} \leq c, \tag{19.7}$$

with c a sufficiently small positive constant. This condition guarantees that approximate particle paths don't cross in S_n.

Problem 19.6. Give an argument showing that F_n is invertible under the condition (19.7)

19.2.3.   The finite element spaces for chG(0)

We introduce the space-time finite element space

$$\overline{W}_{kn} = \{\bar{v} : \bar{v}(\bar{x}, \bar{t}) = \bar{w}(\bar{x}), \ (\bar{x}, \bar{t}) \in S_n \text{ for some } \bar{w} \in V_n\}.$$

To each function $\bar{v}(\bar{x}, \bar{t})$ defined on S_n, we associate a function $v(x, t)$ on S_n by

$$v(x, t) = \bar{v}(\bar{x}, \bar{t}) \quad \text{for } (x, t) = F_n(\bar{x}, \bar{t}).$$

The analog of $\overline{W}_{kn}$ in (x, t) coordinates is

$$W_{kn} = \left\{ v : v(x, t) = \bar{v}(\bar{x}, \bar{t}), \ (x, t) = F_n(\bar{x}, \bar{t}) \in S_n \text{ for some } \bar{v} \in \overline{W}_{kn} \right\}.$$
$$(19.8)$$

A function v belongs to W_{kn} if the limit v_{n-1}^+ is a continuous piecewise linear function on $\mathcal{T}_n$ and $v(x, t)$ is constant on the straight lines $x = \bar{x} + (t - t_{n-1})\beta_n^h(\bar{x})$ for t in I_n. The corresponding space-time mesh on S_n consists of the elements

$$\mathcal{T}_n^\beta = \{K : K = F_n(\bar{K} \times I_n) \text{ for some } \bar{K} \in \mathcal{T}_n\},$$

which are prisms in space-time "tilted" in the direction of β_n^h illustrated in Fig. 19.5. We use W_k to denote the space of functions v on Q such

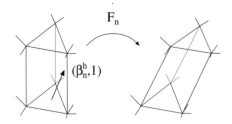

Figure 19.5: The normal and the tilted prism elements.

that $v|_{S_n} \in W_{kn}$ for $n = 1, 2, \ldots$.

Problem 19.7. Assume that $\beta = (x_1, 1)$ and that the standard triangulation is used to discretize the square. Draw a few of the "tilted prisms" for $S_1 = \Omega \times [0, k_1]$.

There are two space meshes associated to each time level t_{n-1}: the mesh $\mathcal{T}_n$ associated to S_n, that is the "bottom mesh" on the slab S_n, and $\mathcal{T}_n^- = \{F_{n-1}(\bar{K} \times \{t_{n-1}\}); \bar{K} \in \mathcal{T}_{n-1}\}$, that is the "top mesh" on the previous slab S_{n-1}, which results from letting the previous "bottom mesh" $\mathcal{T}_{n-1}$ be transported by the flow. The two meshes $\mathcal{T}_n$ and $\mathcal{T}_n^-$ may or may not coincide. In case they do not match, the L_2 projection is used to interpolate a function on $\mathcal{T}_n^-$ into V_n. Depending on the regularity of the velocity field β, it is possible to maintain matching meshes over a certain length of time simply by choosing $\mathcal{T}_n = \mathcal{T}_n^-$, until the mesh $\mathcal{T}_n^-$ is so distorted that this becomes infeasible. At the other extreme, we may use the same mesh $\mathcal{T}_n$ for all slabs S_n and perform the projection from $\mathcal{T}_n^-$ to $\mathcal{T}_n$ at every time step. We illustrate this in Fig. 19.6.

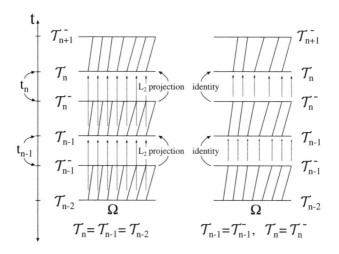

Figure 19.6: Two possibilities for constructing grids on succeeding slabs S_n.

19.2.4. The characteristic Galerkin method

The characteristic Galerkin method chG(0) reads: Compute $U \in W_k$ such that for $n = 1, 2, ...,$

$$\int_{I_n} (L(U), v)\, dt \;+\; ([U_{n-1}], v_{n-1}^+) + \int_{I_n} (\hat{\epsilon}\nabla U, \nabla v)\, dt$$

$$= \int_{I_n} (f, v)\, dt \quad \text{for all } v \in W_{kn}, \quad (19.9)$$

with

$$L(U) = \dot{U} + \beta \cdot \nabla U \quad \text{on } S_n,$$

$$\hat{\epsilon} = \max\{\epsilon, \gamma_1 h^2 R(U), \gamma_2 h^{3/2}\} \quad \text{on } S_n,$$

$$R(U) = |L(U) - f| + |[U_{n-1}]|/k_n \quad \text{on } S_n,$$

where γ_1 and γ_2 are non-negative constants to be specified and $[U_{n-1}]$ is extended to S_n as a constant along the characteristics $x_n(\bar{x}, \cdot)$. We have chosen the streamline diffusion parameter $\delta = 0$ because, as we shall see, the use of tilted elements effectively reduces the convection term, so that no streamline diffusion is needed unless β is non-smooth.

Rewriting (19.9) in local Lagrange coordinates on S_n displays the effect of the orientation. Extending β_n^h to S_n by setting $\beta_n^h(x, t) = \beta_n^h(\bar{x})$ if $(x, t) = F_n(\bar{x}, \bar{t})$, the chain rule implies

$$\frac{\partial v}{\partial t} + \beta \cdot \nabla v = \frac{\partial v}{\partial t} + \beta_n^h \cdot \nabla v + (\beta - \beta_n^h) \cdot \nabla v$$

$$= \frac{\partial \bar{v}}{\partial \bar{t}} + (\bar{\beta} - \bar{\beta}_n^h) \cdot J_n^{-1} \overline{\nabla} \bar{v}$$

$$= \frac{\partial \bar{v}}{\partial \bar{t}} + \bar{\alpha} \cdot \overline{\nabla} \bar{v},$$

where $J_n(\bar{x}, \bar{t}) = \dfrac{\partial x}{\partial \bar{x}}(\bar{x}, \bar{t})$ and $\bar{\alpha} = J_n^{-T}(\bar{\beta} - \bar{\beta}_n^h)$. Now, (19.7) implies that there is a constant C such that

$$|\bar{\alpha}| \le C|\bar{\beta} - \bar{\beta}_n^h| \quad \text{on } S_n,$$

so that $|\bar{\alpha}| \le C(k_n + h_n^2)$ if β is smooth. Reformulated in $(\bar{x}, \bar{t})$-coordinates, the characteristic Galerkin method takes the form: for $n = 1, 2, \ldots$, compute $\bar{U} = \bar{U}|_{S_n} \in \bar{W}_{kn}$ such that,

$$
\int_{I_n} \left(\frac{\partial \bar{U}}{\partial \bar{t}} + \bar{\alpha} \cdot \overline{\nabla} \bar{U}, \bar{v} |J_n| \right) dt + \left([\bar{U}_{n-1}], \bar{v}_{n-1}^+ |J_n| \right) + \int_{I_n} \left(\hat{\epsilon} \widehat{\nabla} \bar{U}, \widehat{\nabla} \bar{v} |J_n| \right) dt
$$
$$
= \int_{I_n} (\bar{f}, \bar{v} |J_n|) \, dt \quad \text{for all } \bar{v} \in \bar{W}_{kn}, \quad (19.10)
$$

where $\widehat{\nabla} = J_n^{-1} \overline{\nabla}$.

Comparing (19.9) and (19.10), we see that using the oriented space-time elements transforms the original problem with velocity β on each slab S_n to a problem with small velocity $\bar{\alpha}$ to which the cG(1)dG(0) method is applied on a tensor-product mesh in $(\bar{x}, \bar{t})$ coordinates with no tilting. Thus, the tilting essentially eliminates the convection term, which both improves the precision and facilitates the solution of the corresponding discrete system. The price that is payed is the L_2 projection at mesh changes.

Remark 19.2.1. The chG(0) method directly extends to the higher order chG(r) method with $r \geq 1$ by using an approximate velocity β_n^h on S_n defined by

$$
\bar{\beta}_n^h(\bar{x}, \bar{t}) = \sum_{j=0}^{r} \bar{t}^j \beta_{nj}^h(\bar{x})
$$

where $\beta_{nj}^h(\bar{x}) \in V_n$. The corresponding approximate characteristics are given by $x(\bar{x}, \bar{t}) = \bar{x} + \sum_{j=1}^{r+1} \frac{(\bar{t} - t_{n-1})^j}{j} \beta_{nj}^h(\bar{x})$.

Problem 19.8. Prove the last statement.

19.3. The streamline diffusion method on an Euler mesh

The cG(1)dG(r) method for the heat equation extends to (18.1) using the streamline diffusion and artificial viscosity modifications of Section 18.2. This corresponds to using a non-oriented space-time mesh. The coresponding cG(1)dG(r) streamline diffusion method is based on the space W_k^r of functions on Q which on each slab S_n belong to the space W_{kn}^r defined by

$$
W_{kn}^r = \{ v : v(x, t) = \sum_{j=0}^{r} t^j v_j(x), \ v_j \in V_n, \ (x, t) \in S_n \}.
$$

The method takes the form: compute $U \in W_k^r$ such that for $n = 1, 2, \ldots$, and for $v \in W_{kn}^r$,

$$\int_{I_n} \left(L(U), v + \delta L(v) \right) dt + \int_{I_n} \left(\hat{\epsilon} \nabla U, \nabla v \right) dt + \left([U_{n-1}], v_{n-1}^+ \right)$$

$$= \int_{I_n} \left(f, v + \delta L(v) \right) dt \quad (19.11)$$

where

$$L(w) = w_t + \beta \cdot \nabla w,$$

$$\delta = \frac{1}{2} (k_n^{-2} + h_n^{-2} |\beta|^2)^{-1/2},$$

$$\hat{\epsilon} = \max\{\epsilon, \gamma_1 h^2 R(U), \gamma_2 h^{3/2}\},$$

$$R(U) = |L(U) - f| + |[U_{n-1}]|/k_n \quad \text{on } S_n,$$

for positive constants γ_i. Note that the streamline diffusion modification $\delta L(v)$ only enters in the integrals over the slab S_n.

19.3.1. Two examples

We present two examples to illustrate the advantages gained in using the chG method, that is the Sd method on an oriented space-time mesh, as compared to the Sd method on a non-oriented space-time mesh. These examples bring up the point that comparing numerical results purely by comparing the errors in the L_2 norm may not give a complete picture. This is obvious after a moment of thought since a norm does not contain as much information about a function as a picture of the function. In the following examples, we compare results using the Sd method on non-oriented and oriented space-time meshes in computations with roughly the same accuracy in the L_2 norm. We will see, that the quality in the "picture norm" differs considerably.

The first example is a common quite difficult test problem with pure convection. The initial data consisting of a cylinder with a slit shown in Fig. 19.7, which is rotated counterclockwise by $\beta = (\sin(t), \cos(t))$ until time $t = \pi$, or a rotation of 180 degrees. We first plot in Fig. 19.8 the results obtained by using the cG(1)dG(1) method on a non-oriented space-time grid reaching one half rotation after 251 constant time steps. Next, in Fig. 19.9 we plot the solution obtained using the chG(0) method.

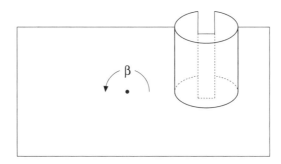

Figure 19.7: The initial data for the first example.

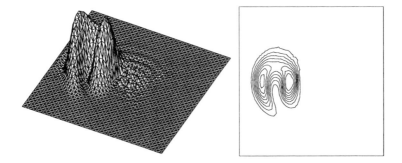

Figure 19.8: The approximation and associated level curves from the cG(1)dG(1) streamline diffusion method on a fixed space-time grid applied to the data shown in Fig. 19.7.

The mesh was tilted in space-time according to the rotating velocity field locally on each space-time slab and an L_2 projection back to a fixed uniform space mesh was performed at each time step, following the principle illustrated to the left in Fig. 19.6. The solution after a half revolution, is visibly much better than the previous computation, and also computationally much less expensive, because only 21 constant time steps were used and piecwise constants in time where used instead of piecewise linears.

The next problem, called the Smolarkiewicz example, is a very demanding test problem. The initial data is a cone of height 1 and base radius 15 centered in the rectangular region $\Omega = (25, 75) \times (12.5, 87.5)$.

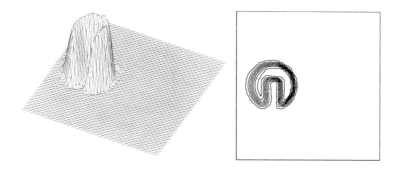

Figure 19.9: The approximation and associated level curves from the characteristic Galerkin chG(0) method applied to the data shown in Fig. 19.7.

The cone is convectively "folded" in the velocity field

$$\beta = \frac{8\pi}{25}\left(\sin\left(\frac{\pi x}{25}\right)\sin\left(\frac{\pi x}{25}\right),\,\left(\cos\left(\frac{\pi x}{25}\right)\cos\left(\frac{\pi x}{25}\right)\right)\right),$$

which is periodic in x and y with six "vortex cells" inside Ω. We illustrate this in Fig. 19.10. We compute the approximations using 1000 fixed time

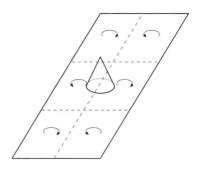

Figure 19.10: The initial data for the Smolarkiewicz problem. We also plot the convective vortex cells.

steps to reach the final time $t = 30$. In the first case, we use the chG(0) method without mesh change according to the principle on the right of Fig. 19.6 with $\mathcal{T}_n^- = \mathcal{T}_n$, so that no L_2 projections from changing the space mesh at a discrete time level were required. We plot the result in

Fig. 19.11. Note the extreme mesh distortion that develops as a result of avoiding projections into new, less distorted space meshes. In the second computation, shown in Fig. 19.12, the mesh was changed into a new uniform mesh every one hundred time steps. This limits the mesh distortion but introduces L_2 projection errors at the mesh changes that gradually destroy sharp features of the solution.

19.4. Error analysis

We analyze the chG(0) method applied to the model problem

$$\begin{cases} u_t + \beta \cdot \nabla u - \epsilon \Delta u = f & \text{in } \mathbb{R}^2 \times (0, \infty), \\ u(x,t) \to 0 & \text{for } t > 0 \text{ as } |x| \to \infty, \\ u(\cdot, 0) = u_0 & \text{on } \mathbb{R}^2, \end{cases} \qquad (19.12)$$

where $\beta = (\beta_1, \beta_2)$ and $\epsilon \geq 0$ are constant, and f and u_0 are given data with bounded support.

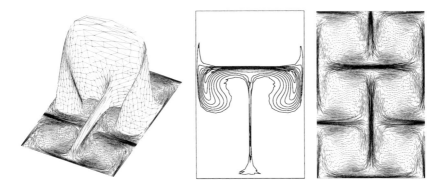

Figure 19.11: The approximation, level curves, and mesh resulting from the characteristic Galerkin chG(0) method applied to the Smolarkiewicz problem at $t = 30$. In this computation, the mesh passively follows the flow for all times and no L_2 projections are used.

The transformation between the Euler (x, t) and Lagrange coordinates $(\bar{x}, \bar{t})$ is simply $(x, t) = (\bar{x} + \bar{t}\beta, \bar{t})$ in this case. Reformulating

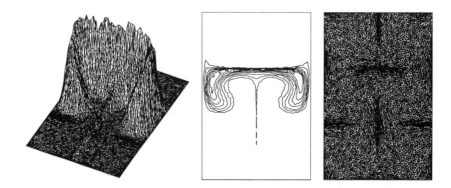

Figure 19.12: The approximation, level curves, and mesh resulting
from the characteristic Galerkin chG(0) method applied
to the Smolarkiewicz problem at $t = 30$. In this computation, an L_2 projection into a uniform mesh is used
every one hundred time steps to limit the mesh distortion.

(19.12) in Lagrange coordinates for $\bar{u}(\bar{x}, \bar{t}) = u(x, t)$, after noting that

$$\frac{\partial \bar{u}}{\partial \bar{t}} = \frac{\partial}{\partial \bar{t}} u(\bar{x} + \bar{t}\beta, \bar{t}) = \frac{\partial u}{\partial t} + \beta \cdot \nabla u,$$

we obtain

$$\begin{cases} \dfrac{\partial \bar{u}}{\partial \bar{t}} - \epsilon \Delta \bar{u} = \bar{f} & \text{in } \mathbb{R}^2 \times (0, \infty), \\ \bar{u}(\bar{x}, \bar{t}) \to 0 & \text{for } \bar{t} > 0 \text{ as } |\bar{x}| \to \infty, \\ \bar{u}(\bar{x}, 0) = u_0(\bar{x}) & \bar{x} \in \mathbb{R}^2. \end{cases} \tag{19.13}$$

We see that the Lagrange formulation (19.13) is the familiar heat equation with constant diffusion coefficient ϵ and the characteristic Galerkin
chG(0) method for (19.12) is simply the cG(1)dG(0) method for (19.13).

Before presenting the analysis, we write out the chG(0) method for
(19.12) explicitly. By construction, the functions v in W_{kn} are constant
in the direction β so that $v_t + \beta \cdot \nabla v = 0$ for $v \in W_{kn}$. Thus, the chG(0)
method for (19.12) reduces to: compute $U \in W_k$ such that

$$([U_{n-1}], v_{n-1}^+) + \int_{I_n} (\epsilon \nabla U, \nabla v) \, dt = \int_{I_n} (f, v) \, dt \quad \text{for all } v \in W_{kn}, \tag{19.14}$$

where

$$\hat{\epsilon} = \max\left\{\epsilon, \gamma_1 h^2(|[U_{n-1}]|/k_n + |f|), \gamma_2 h^{3/2}\right\} \quad \text{on } S_n,$$

with $h(x,t) = h_n(x - (t - t_{n-1})\beta)$, where h_n is the mesh function for V_n, and $[U_{n-1}]$ is similarly extended. If $f = 0$ (and ϵ is small), then (19.14) can be written: compute $U_{n-1}^+ \in V_n$ such that

$$\int_{\mathbb{R}^2} U_{n-1}^+ v \; dx + \int_{\mathbb{R}^2} \tilde{\epsilon} \nabla U_{n-1}^+ \cdot \nabla v \; dx = \int_{\mathbb{R}^2} U_{n-1}^- v \; dx \quad \text{for all } v \in V_n, \tag{19.15}$$

with $\tilde{\epsilon} = \gamma_1 h_n^2 |[U_{n-1}]|$ and $U_0^- = u_0$.

19.4.1. Projection and transport

This leads to an alternate formulation of the chG(0) method. Introducing the translation operator $\tau_n : \tau_n v(x) = v(x - k_n\beta)$ and the nonlinear projection $\tilde{P}_n$ into V_n defined by

$$(\tilde{P}_n w, v) + (\tilde{\epsilon} \nabla \tilde{P}_n w, \nabla v) = (w, v) \quad \text{for all } v \in V_n,$$

where $\tilde{\epsilon} = \gamma_1 h_n^2 |w - \tilde{P}_n w|$, we can write (19.15) using the notation $U_n = U_n^-$ as

$$U_n = \tau_n \tilde{P}_n U_{n-1}, \tag{19.16}$$

and $U_0 = u_0$.

Problem 19.9. Assuming that $\gamma_1 = 0$, show that (19.16) reduces to

$$U_n = \tau_n P_n U_{n-1}, \tag{19.17}$$

where P_n is the L_2-projection into V_n.

Thus, the chG(0) method in the simplest case may be viewed as an algorithm of the form "projection then exact transport". This view is useful for understanding some properties of the chG(0) method, but the chG(0) method is not derived from this concept because this complicates the extension to more complex situations with β variable and ϵ positive.

19.4.2.   A direct a priori error analysis in a simple case

We first derive an a priori error estimate for the chG(0) method in the simple case with $f = \epsilon = 0$, where the solution of (19.13) is simply given by $\bar{u}(\bar{x}, t) = u_0(\bar{x})$ and that of (19.12) by

$$u(x, t) = u_0(x - t\beta).$$

Using the formulation (19.17), we write the error as

$$u_n - U_n = \tau_n(u_{n-1} - P_n U_{n-1}) = \tau_n(u_{n-1} - P_n u_{n-1} + P_n(u_{n-1} - U_{n-1})).$$

Using the facts $\|\tau_n v\| = \|v\|$ and $\|P_n v\| \leq \|v\|$, we obtain by Pythagoras' theorem

$$\|u_n - U_n\|^2 = \|u_{n-1} - P_n u_{n-1}\|^2 + \|P_n(u_{n-1} - U_{n-1})\|^2$$
$$\leq \|u_{n-1} - P_n u_{n-1}\|^2 + \|u_{n-1} - U_{n-1}\|^2.$$

Iterating this inequality and using a standard error estimate for the L_2 projection, we obtain

$$\|u_N - U_N\| \leq \left(\sum_{n=1}^{N} \|u_{n-1} - P_n u_{n-1}\|^2\right)^{1/2}$$
$$\leq C_i \left(\sum_{n=1}^{N} \|h_n^2 D^2 u_{n-1}\|^2\right)^{1/2} \leq C_i \sqrt{N} h^2,$$

provided u is smooth and we set $h = \max_n h_n$. This estimate is slightly sub-optimal because of the factor $\sqrt{N}$. In the generic case with $k_n \approx h$, we conclude that

$$\|u_N - U_N\| \leq C_i h^{3/2}. \tag{19.18}$$

An optimal result can be derived if the viscosity is positive, as we show in the next section.

Problem 19.10.   Assuming that the time steps are constant $k_n = k = T/N$, prove that

$$\|u_N - U_N\| \leq C_i \sqrt{N/T} \, \|(I - P)u\|_{L_2(Q)}, \tag{19.19}$$

where $P = P_n$ on S_n. This result is also sub-optimal in comparison with the accuracy of the L_2 projection.

Remark 19.4.1. The orientation of the space-time mesh in the characteristic Galerkin method is chosen according to the flow velocity. In general, we could choose an arbitrary mesh translation velocity. We refer to this variant as the *oriented streamline diffusion method*. For example, if the solution is constant in time, this suggests an orientation with zero velocity, which in general is different from orientation along the flow velocity.

19.4.3. An error analysis based on error estimates for parabolic problems

The a priori and a posteriori results for the cG(1)dG(0) method for the heat equation apply to the chG(0) method for (19.12) written in the Lagrange form (19.10). We write out the a priori and a posteriori error estimates, which translate to corresponding optimal estimates for the chG(0) method immediately.

Theorem 19.1. *If $\mu k_n \epsilon \geq h_n^2$, μ sufficiently small, $\bar{\alpha} = 0$ and $\hat{\epsilon} = \epsilon$, then*

$$\|\bar{u}(\cdot, t_N) - \bar{U}_N\| \leq L_N C_i \max_{1 \leq n \leq N} \left(k_n \left\| \frac{\partial \bar{u}}{\partial \bar{t}} \right\|_{I_n} + \|h_n^2 D^2 \bar{u}\|_{I_n} \right), \tag{19.20}$$

and

$$\|\bar{u}(\cdot, t_N) - \bar{U}_N\| \leq L_N C_i \max_{1 \leq n \leq N} \left(\|k R_{0k}(\bar{U})\|_{I_n} + \left\| \frac{h_n^2}{\epsilon k_n} [\bar{U}_{n-1}] \right\|^\star \right.$$
$$\left. + \|h_n^2 R(\bar{U})\|_{I_n} \right), \tag{19.21}$$

where $L_N = (\max((log(t_N/k_N))^{1/2}, log(t_N/k_N)) + 2$, $R_{0k}(\bar{U}) = |f| + |[\bar{U}]|/k$, $R(\bar{U}) = \frac{1}{\epsilon}|f| + R_2(\bar{U})$ with R_2 defined by (18.33), and a star indicates that the corresponding term is present only if $V_{n-1} \not\subseteq V_n$.

The assumption that
$$k_n \epsilon \geq h_n^2$$

means that $\epsilon > 0$ is needed to get optimal estimates. In the case $k_n = h$ and $\epsilon \approx h^{\frac{3}{2}}$, the estimates reduce to (19.18) if $\partial \bar{u}/\partial \bar{t}$ is small. In the case of pure convection with $f = 0$, when $\frac{\partial \bar{u}}{\partial \bar{t}} = 0$, (19.20) reduces to

$$\|u_N - U_N\| \leq C_i \|(I - P)u\|_{[0, t_N]},$$

where $(I - P)u = (I - P_n)u$ on I_n. This shows that the chG(0) method in the convection dominated case is optimal compared to projection if the viscosity is not too small, cf. (19.19).

Problem 19.11. Prove Theorem 19.1.

Leibniz's spirit of inquiry is apparent even in his report to the Académie des Sciences in Paris about a talking dog. Leibniz describes the dog as a common middle-sized dog owned by a peasant. According to Leibniz, a young girl who heard the dog make noises resembling German words decided to teach it to speak. After much time and effort, it learned to pronounce approximately thirty words, including "thé", "caffé", "chocolat", and "assemblée"- French words which had passed into German unchanged. Leibniz also adds the crucial observation that the dog speaks only "as an echo", that is. after the master has pronounced the word; "it seems that the dog speaks only by force, though without ill-treatment". (The Cambridge Companion to Leibniz)

Figure 19.13: Leibniz' calculator.

20

The Eigenvalue Problem for an Elliptic Operator

> For since the fabric of the universe is most perfect and the work
> of a most wise Creator, nothing at all takes place in the universe
> in which some rule of maximum or minimum does not appear.
> (Euler)

In this chapter, we briefly consider the *eigenvalue problem* of finding
non-zero functions φ and real numbers $\lambda \in \mathbb{R}$ such that

$$\begin{cases} -\nabla \cdot (a\nabla\varphi) + c\varphi = \lambda\varphi & \text{in } \Omega, \\ \varphi = 0 & \text{on } \Gamma, \end{cases} \tag{20.1}$$

where $\Omega \subset \mathbb{R}^d$, Γ is the boundary of Ω, and $a = a(x) > 0$ and $c = c(x)$
are given coefficients. We refer to φ as an *eigenfunction* corresponding
to the *eigenvalue* λ. Recall that we discussed the eigenvalue problem in
reference to Fourier's method in Chapter 6. It turns out that the eigen-
values of (20.1) may be arranged as a sequence $\lambda_1 \leq \lambda_2 \leq \cdots \leq \lambda_n \to \infty$
with one eigenfunction φ_n corresponding to each eigenvalue λ_n. The
eigenfunctions corresponding to different eigenvalues are orthogonal with
respect to the $L_2(\Omega)$ scalar product and the eigenfunctions corresponding to the same eigenvalue form (together with the zero function) a finite dimensional vector space called the *eigenspace*. The eigenfunctions
$\{\varphi_n\}_{n=1}^{\infty}$ may be chosen as an orthonormal basis in $L_2(\Omega)$. In particular, any function $v \in L_2(\Omega)$ can be represented as a series $v = \sum_n v_n\varphi_n$,
where $v_n = \int_\Omega v\,\varphi_n\,dx$ are called the *generalized Fourier coefficients*. See
Strauss ([18]) for more information on these results.

With $a = 1$ and $c = 0$, we obtain the eigenvalue problem for the Laplace operator with Dirichlet boundary conditions

$$\begin{cases} -\Delta\varphi = \lambda\varphi & \text{in } \Omega, \\ \varphi = 0 & \text{on } \Gamma. \end{cases} \tag{20.2}$$

We recall that in the corresponding problem in one dimension with $\Omega = (0, \pi)$, the eigenfunctions are (modulo normalization) $\varphi_n(x) = \sin(nx)$ corresponding to eigenvalues $\lambda_n = n^2$, $n = 1, 2, ...$ In the case $d = 2$ and $\Omega = (0, \pi) \times (0, \pi)$, the eigenfunctions are $\varphi_{nm}(x_1, x_2) = \sin(nx_1)\sin(mx_2)$, $n, m = 1, 2, ...$, with eigenvalues $\lambda_{nm} = n^2 + m^2$. In the first case, all of the eigenspaces have dimension one, but in higher dimensions, all of the eigenspaces except for the eigenspace corresponding to the smallest eigenvalue have dimension larger than one.

Problem 20.1. Prove that eigenvalues of (20.2) are positive and that eigenfunctions corresponding to different eigenvalues are orthogonal in $L_2(\Omega)$.

The drum and the guitar

The motion of an elastic membrane supported at the edge along a curve Γ in the plane bounding the domain Ω, is described by the homogeneous wave equation

$$\begin{cases} \ddot{u} - \Delta u = 0 & \text{in } \Omega \times (0, T), \\ u = 0 & \text{on } \Gamma \times (0, T), \\ u(0) = u_0, \ \dot{u}(0) = \dot{u}_0 & \text{in } \Omega, \end{cases} \tag{20.3}$$

where $u(x, t)$ represents the transversal deflection of the membrane. If φ_n is an eigenfunction with corresponding eigenvalue λ_n of the eigenvalue problem (20.2), then the functions $\sin(\sqrt{\lambda_n}t)\varphi_n(x)$ and $\cos(\sqrt{\lambda_n}t)\varphi_n(x)$ satisfy the homogeneous wave equation (20.3) with specific initial data. These functions are called the *normal modes* of vibration of the membrane. A general solution u of the homogeneous wave equation (20.3) with initial values $u(0)$ and $\dot{u}(0)$ can be expressed as a linear combination of the normal modes $\sin(\sqrt{\lambda_n}t)\varphi_n(x)$ and $\cos(\sqrt{\lambda_n}t)\varphi_n(x)$. This is the Fourier's solution of the wave equation on $\Omega \times (0, T)$, which is analogous to the solution in one dimension given by (17.11).

If Ω is a circular disc, then (20.3) describes the vibrations of a drum head. The smallest eigenvalue corresponds to the basic tone of the drum.

This can be changed by changing the tension of the drum head, which corresponds to changing the coefficient a in the generalization (20.1).

In Fig. 20.1, we show contour plots for the first four eigenfunctions, corresponding to $\lambda_1 \approx 38.6$, $\lambda_2 \approx 83.2$, $\lambda_3 \approx 111.$, and $\lambda_4 \approx 122.$, computed using Femlab in a case where (20.2) describes the vibrations of the lid of a guitar with Dirichlet boundary conditions on the outer boundary, described as an ellipse, and Neumann boundary conditions at the hole in the lid, described as a circle.[1] The distribution of the eigenvalues determine the sound produced by the guitar lid and the computational results could be used to find good shapes of a guitar lid.

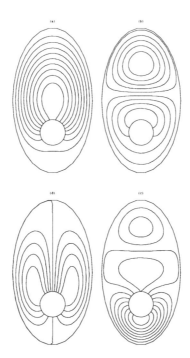

Figure 20.1: Contour plots of the first four eigenfunctions of the guitar lid corresponding to (a) $\lambda_1 \approx 38.6$, (b) $\lambda_2 \approx 83.2$, (c) $\lambda_3 \approx 111.$, and (d) $\lambda_4 \approx 122.$. These were computed with Femlab with a fixed mesh size of diameter .02.

---

[1]Computations provided courtesy of Marten Levenstam. The eigenvalues were computed in Femlab using a filtered k-step Arnoldi method as described in D.C. Sorensen, SIAM J. Matrix Anal. Appl. 13 (1992), pp. 357–385.

Often the smaller eigenvalues are the most important in consideration of design. This is the case for example in designing suspension bridges, which must be built so that the lower eigenvalues of vibrations in the bridge are not close to possible wind-induced frequencies. This was not well understood in the early days of suspension bridges which caused the famous collapse of the Tacoma bridge in 1940.

20.0.4. The Rayleigh quotient

The variational form of the eigenvalue problem (20.1) is to find $\lambda \in \mathbb{R}$ and a non-zero $\varphi \in V$ such that

$$(a\nabla\varphi, \nabla\psi) + (c\varphi, \psi) = \lambda(\varphi, \psi) \quad \text{for all } \psi \in V, \tag{20.4}$$

where

$$V = \left\{ v : \int_\Omega (a|\nabla v|^2 + v^2)\, dx < \infty,\ v = 0 \text{ on } \Gamma \right\},$$

and $(\cdot, \cdot)$ as usual denotes $L_2(\Omega)$ inner product. Setting $\psi = \varphi$ gives a formula for the eigenvalue corresponding to φ,

$$\lambda = \frac{(a\nabla\varphi, \nabla\varphi) + (c\varphi, \varphi)}{(\varphi, \varphi)}.$$

Introducing the *Rayleigh quotient*

$$RQ(\psi) = \frac{(a\nabla\psi, \nabla\psi) + (c\psi, \psi)}{(\psi, \psi)} \quad \text{for } \psi \in V,$$

the previous equality can be rewritten as $\lambda = RQ(\varphi)$, or in words: the Rayleigh quotient of an eigenfunction is equal to the corresponding eigenvalue.

We can turn this argument around and consider how $RQ(\psi)$ varies as ψ varies in V. In particular, there is a function $\varphi_1 \in V$ that minimizes the Rayleigh quotient over all functions in V and this function is the eigenfunction corresponding to the smallest eigenvalue λ_1:

$$\lambda_1 = \min_{\psi \in V} RQ(\psi) = RQ(\varphi_1). \tag{20.5}$$

More generally, the eigenfunction φ_j minimizes the Rayleigh quotient over all functions in V orthogonal to the eigenfunctions φ_i, $i = 1, 2, ...j - 1$, and $\lambda_j = RQ(\varphi_j)$.

Problem 20.2. State and prove the analog of the Rayleigh quotient minimum principle for a diagonal matrix.

Problem 20.3. Suppose $\varphi_1 \in V$ minimizes the Rayleigh quotient. Prove that φ_1 is the eigenfunction corresponding to a smallest eigenvalue λ_1 satisfying (20.4) with $\lambda = \lambda_1$. Hint: Define the function $f(\epsilon) = RQ(\varphi + \epsilon\psi)$, where $\psi \in V$ and $\epsilon \in \mathbb{R}$, and use that $f'(0) = 0$.

Problem 20.4. Consider the problem of finding the smallest interpolation constant C_i in an error estimate of the form $\|v - \pi v\|_{L_2(0,1)} \leq C_i \|v'\|_{L_2(0,1)}$, where $\pi v \in \mathcal{P}^1(0,1)$ interpolates $v(x)$ at $x = 0, 1$. Hint: show first that it suffices to consider the case $v(0) = v(1) = 0$ with $\pi v = 0$. Then rewrite this problem as a problem of determining the smallest eigenvalue and show that $C_i = 1/\pi$. Show similarly that the best constant C_i in the estimate $\|v - \pi v\|_{L_2(0,1)} \leq C_i \|v''\|_{L_2(0,1)}$ is equal to $1/\pi^2$.

20.1. Computation of the smallest eigenvalue

We consider the computation of the smallest eigenvalue in the eigenvalue problem (20.2) for the Laplacian by minimizing the Rayleigh quotient over the usual finite element subspace $V_h \subset V$ consisting of continuous piecewise linear functions vanishing on Γ,

$$\lambda_1^h = \min_{\psi \in V_h} RQ(\psi). \tag{20.6}$$

The difference between λ_1 given by (20.5) and λ_1^h given by (20.6) is that the minimization in (20.6) is over the finite dimensional vector space V_h instead of V. Since $V_h \subset V$, we must have $\lambda_1^h \geq \lambda_1$. The question is thus how much larger λ_1^h is than λ_1. To answer this question, we prove an a priori error estimate showing the error in the smallest eigenvalue is bounded by the square of the energy norm interpolation error of the eigenfunction φ_1. This result extends to approximation of larger eigenvalues λ_j with $j > 1$, but the proof is more subtle in this case. We comment on computation of larger eigenvalues in the next section and in the companion volume.

Theorem 20.1. *There is a constant C_i such that for h sufficiently small,*

$$0 \leq \lambda_1^h - \lambda_1 \leq C_i \|hD^2\varphi_1\|^2. \tag{20.7}$$

Proof. Assume φ satisfies (20.2) with $\|\varphi\| = 1$ with corresponding eigenvalue $\lambda = RQ(\varphi) = \|\nabla\varphi\|^2$. We shall use the following identity for all $v \in V$ with $\|v\| = 1$, which follows from the definition of φ

$$\|\nabla v\|^2 - \lambda = \|\nabla(\varphi - v)\|^2 - \lambda\|\varphi - v\|^2.$$

Problem 20.5. Prove this identity.

Using this identity with $v \in V_h$, $\lambda = \lambda_1$ and $\varphi = \varphi_1$, and recalling the characterization (20.6), we obtain

$$\lambda_1^h - \lambda_1 \le \|\nabla v\|^2 - \lambda_1 \le \|\nabla(\varphi_1 - v)\|^2. \tag{20.8}$$

We now take $v \in V_h$ to be a suitable approximation of φ_1 such that $\|\nabla(\varphi_1 - v)\| \le C\|hD^2\varphi_1\|$, which may put a condition on the size of h because of the restriction $\|v\| = 1$, and the desired result follows. ∎

Problem 20.6. Verify that it is possible to find the approximation v to φ_1 required in the proof of (20.7).

Problem 20.7. Derive an a posteriori error error estimate for $\lambda_1^h - \lambda_1$. Hint: multiply the equation $-\Delta\varphi_1 - \lambda_1\varphi_1 = 0$ satisfied by the continuous eigenfunction φ_1 corresponding to λ_1, by the discrete eigenfunction $\Phi_1 \in V_h$ satisfying $(\nabla\Phi_1, \nabla v) = \lambda_1^h(\Phi_1, v)$ for all $v \in V_h$, to get

$$(\lambda_1 - \lambda_1^h)(\varphi_1, \Phi_1) = (\nabla\Phi_1, \nabla(\varphi_1 - \pi_h\varphi_1)) - \lambda_1^h(\Phi_1, \varphi_1 - \pi_h\varphi_1),$$

where φ_1 and Φ_1 are normalized to have L_2 norm equal to one. Assuming that $(\varphi_1, \Phi_1) \ge c > 0$, where c is a positive constant, derive an a posteriori error estimate in the usual way. (see M. Larsson, A posteriori error estimates for eigenvalue problems, to appear).

20.2. On computing larger eigenvalues

We give an example illustrating the approximation of the larger eigenvalues. In principle, larger eigenvalues and their associated eigenfunctions could be computed in the same fashion as the first eigenpair by finding the stationary points of the Rayleigh quotient over the appropriate finite element space. However, since the eigenfunctions corresponding to larger eigenvalues generally oscillate at larger frequencies, we expect the accuracy of the approximations on a fixed mesh to deteriorate with increasing eigenvalues. In fact, the eigenvalues of the continuous problem tend to infinity, while those of the finite element approximation are finite, so some of the eigenvalues of the continuous problem cannot be captured in the approximation no matter how small we choose the mesh size.

As an example, we consider a finite element discretization of the weak form of the eigenvalue problem (20.2) with $\Omega = (0, \pi)$, which reads: compute $\Phi \in V_h$ and $\lambda_h \in \mathbb{R}$ such that

$$(\Phi', \psi') = \lambda_h (\Phi, \psi) \quad \text{for all } \psi \in V_h, \tag{20.9}$$

where V_h is the space of continuous piecewise linear functions, vanishing at $x = 0, \pi$, on a uniform discretization of $(0, \pi)$ into $M + 1$ elements with meshsize $h = \pi/(M + 1)$ and nodes $x_j = jh$. We also use lumped mass quadrature to evaluate the integral on the right-hand side of (20.9). This gives the matrix eigenvalue problem

$$A\xi = \lambda\xi, \tag{20.10}$$

where ξ denotes the vector of nodal values of Φ and the coefficient matrix A is the product of the inverse of the diagonal lumped mass matrix and the stiffness matrix; cf. Section 15.1.5. Let ξ_n, $n = 1, 2, ..., M$ be the eigenvectors of (20.10) and Φ_n the corresponding finite element approximations.

Problem 20.8. Compute the finite element approximation of (20.9) using lumped mass quadrature and derive (20.10).

We know that the eigenvalues of the continuous problem are n^2, $n = 1, 2, ...$, with corresponding eigenfunctions $\varphi_n(x) = \sin(nx)$. It turns out in this very special case that the nodal values of the discrete eigenfunctions Φ_n agree with the nodal values of the exact eigenfunctions $\sin(nx)$ for $n = 1, ..., M$, that is $\Phi_n(jh) = \sin(njh)$, $n, j = 1, 2, ..., M$.

Problem 20.9. Prove by substitution that Φ_n is an eigenvector satisfying (20.10) with eigenvalue $\lambda_n^h = 2(1 - \cos(nh))/h^2$ for $n = 1, 2, ..., N$.

When n is small the discrete eigenvalue λ_n^h is a good approximation of the continuous eigenvalue λ_n since by Taylor's theorem

$$\frac{2(1 - \cos(nh))}{h^2} \approx n^2 + O(n^4 h^2), \tag{20.11}$$

However, despite the interpolation property of the discrete eigenfunctions the L_2 norm of the error $\|\Phi_n - \varphi_n\|$, or even worse the energy norm of the error $\|\Phi'_n - \varphi'_n\|$, becomes large when n gets close to M, see Fig. 20.2. In this case,

$$\frac{2(1 - \cos(nh))}{h^2} \approx \frac{4}{h^2} - \frac{(M - n)^2}{2}, \tag{20.12}$$

which is not close to n^2. In Fig. 20.3 we show the first 100 continuous and discrete eigenvalues for $M = 100$. We conclude that eigenvalues corresponding to eigenfunctions that oscillate with a wavelength on the order of the meshsize and smaller are not well approximated.

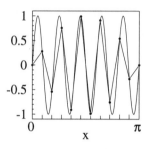

Figure 20.2: $\sin(10x)$ and $\Phi_{10}(x)$ for $M = 10$.

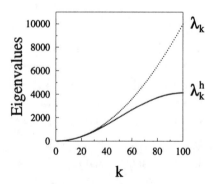

Figure 20.3: The continuous and discrete eigenvalues with $M = 100$.

Problem 20.10. Verify estimates (20.11) and (20.12).

Problem 20.11. Define $f(x) = 100^2 \sin(100x)$ (i.e. $f(x) = \lambda_{100} \varphi_{100}$). Use Femlab to solve $-u'' = f$ on $(0, \pi)$ together with $u(0) = u(\pi) = 1$. Plot the approximation together with the true solution. How many elements did Femlab use to get an accurate approximation? Explain the significance of this for the discussion above.

This can have a strong consequences for the time behavior of a discrete approximation to a time dependent problem such as the heat equation or the wave equation. The following problem is an interesting illustration.

Problem 20.12. Consider the initial-boundary value problem for the wave equation:

$$\begin{cases} \ddot{u} - u'' = 0, & x \in [0, \pi], t > 0, \\ u(0, t) = u(\pi, t) = 0, & t > 0, \\ u(x, 0) = u_0(x), \dot{u}(x, 0) = 0, & x \in [0, \pi]. \end{cases} \tag{20.13}$$

Let U denote the continuous piecewise linear semi-discrete finite element approximation computed on a uniform mesh on $[0, \pi]$. Compare the time behavior of U to that of u when the initial data u_0 is nonsmooth. Can you say something about the time behavior of a finite element approximation that is discrete in time and space? Hint: discretize (20.13) in space using the finite element method on a uniform mesh as indicated. Now use separation of variables to get a scalar ordinary differential equation in time and a matrix eigenvalue problem in space, then solve both problems. Nonsmooth functions are characterized by large Fourier coefficients in the higher modes, so choose the data to be the discrete eigenfunction Φ_M. Compare the solution of (20.13) to the solution of the system of ordinary differential equations as time passes. Plot the two solutions.

Problem 20.13. Consider the finite element approximation of (20.2) with $\Omega = (0, \pi) \times (0, \pi)$ computed using the standard triangulation and continuous piecewise linear functions. (a) Compute the discrete eigenvalues and eigenfunctions. Hint: use separation of variables and Problem 20.9. (b) Estimate the convergence rate of the Jacobi iterative method for solving $A\xi = b$.

20.3.   The Schrödinger equation for the hydrogen atom

It does not require much imagination to see an analogy between the mirroring activity of the Leibniz monad, which appears to our

confused vision like a casual activity, emanating from one monad
and impinging on the other, and the modern view in which the
chief activity of the electrons consists in radiating to one another.
(Wiener)

The quantum mechanical model of a hydrogen atom consisting of one
electron orbiting around one proton at the origin, takes the form of the
following eigenvalue problem in $\Omega = \mathbb{R}^3$:

$$\begin{cases} -\Delta\varphi - \frac{2}{r}\varphi = \lambda\varphi & \text{in } \Omega, \\ \int_\Omega \varphi^2 \, dx = 1. \end{cases} \qquad (20.14)$$

The eigenfunction φ is a *wave function* for the electron describing the
position of the electron in the sense that the integral $\int_\omega \varphi^2 dx$ represents
the probability that the electron is in the domain $\omega \subset \mathbb{R}^3$. In fact, (20.14)
is the eigenvalue problem associated with the *Schrödinger equation*

$$i\dot{\varphi} - \Delta\varphi - \frac{2}{r}\varphi = 0$$

describing the motion of the electron.

The Rayleigh quotient for the eigenvalue problem (20.14) is given by

$$RQ(\psi) = \frac{\int_\Omega |\nabla\psi|^2 \, dx - 2 \int_\Omega \psi^2/r \, dx}{\int_\Omega \psi^2 \, dx}, \qquad (20.15)$$

and is defined for $\psi \in V = \{\psi : \int_{\mathbb{R}^3}(|\nabla\psi|^2 + \psi^2/r)dx < \infty\}$. The
quantity $\int_\Omega |\nabla\psi|^2 dx$ represents the kinetic energy of an electron with
wave function ψ and $-2 \int_\Omega \psi^2/r \, dx$ represents the potential energy corresponding to the attractive Coulomb force between the proton and elec-
tron. The equation (20.14) is one of the few equations of quantum mechanics that can be solved analytically and this is due to the spherical
symmetry. The eigenvalues are $\lambda_n = -1/n^2$, for integers $n \geq 1$, called
the *principal quantum number* and represent energy levels. There are
n^2 eigenfunctions corresponding to each energy level λ_n, of which one
depends only on the radius, see Strauss ([18]). The eigenfunctions are
called the *bound states* and the unique eigenfunction corresponding to
the smallest eigenvalue is called the *ground state* since it is the bound
state "closest" to the proton with the smallest energy. As soon as more
than one electron or proton are involved, that is for all atoms except the

hydrogen atom, analytical solution of Schrödinger's equation is practically impossible and a variety of approximate solution methods have been developed.

Among other things, the model (20.14) predicts that the electron may jump from one state with eigenvalue λ_i to another with eigenvalue λ_j by emitting or absorbing a corresponding "quantum" of energy $\lambda_i - \lambda_j$ as was observed in the famous experiments of Bohr. Note that the fact that $\lambda_i \geq -1$ implies that the hydrogen atom is stable in the sense that the electron does not fall into the proton.

We note that the domain in (20.14) is the whole of $\mathbb{R}^3$. Looking for solutions in a space V of functions that are square integrable functions means that we exclude certain oscillating solutions of the Schrödinger equation corresponding to free states of the electron. This is related to the existence of solutions $u(x,t)$ of the problem $i\dot{u} - u'' = 0$ in $\mathbb{R} \times \mathbb{R}$ of the form $u(x,t) = \exp(i\lambda^2 t)\exp(i\lambda x)$ for any $\lambda \geq 0$. The value λ belongs to the "continuous spectrum" for which the corresponding "eigen-functions" are not square integrable. The eigenvalues with eigenfunctions in V belong to the "discrete spectrum".

To discretize the Schrödinger eigenvalue problem (20.14) in $\mathbb{R}^3$, we generally truncate the domain to be finite, say $\{x : |x| < R\}$ for some $R > 0$, and impose suitable boundary conditions, such as Dirichlet boundary conditions, on the boundary $\{x : |x| = R\}$. The relevant choice of R is related to the eigenvalue/eigenfunction being computed and the tolerance level.

Problem 20.14. (For amateur quantum physicists) Prove that the hydrogen atom is stable in the sense that the Rayleigh quotient (20.15) satisfies

$$\min_{\psi \in V} RQ(\psi) \geq -4,$$

showing that the electron does not fall into the proton. Hint: estimate $\int_\Omega \psi \frac{\psi}{r}$ using Cauchy's inequality and the following Poincaré inequality for functions $\psi \in V$:

$$\int_\Omega \frac{\psi^2}{r^2}\, dx \leq 4 \int_\Omega |\nabla \psi|^2\, dx. \tag{20.16}$$

This shows that the potential energy cannot outpower the kinetic energy in the Rayleigh quotient. To prove the last inequality, use the representation

$$\int_\Omega \frac{\psi^2}{r^2}\, dx = -\int_\Omega 2\psi\nabla\psi \cdot \nabla \ln(|x|)\, dx.$$

resulting from Green's formula, together with Cauchy's inequality.

Problem 20.15. (a) Show that the eigenvalue problem (20.14) for the hydrogen atom for eigenfunctions with radial dependence only, may be formulated as the following one-dimensional problem

$$- \varphi_{rr} - \frac{2}{r}\varphi_r - \frac{2}{r}\varphi = \lambda\varphi, \quad r > 0, \quad \varphi(0) \text{ finite}, \quad \int_{\mathbb{R}} \varphi^2 r^2 \, dr < \infty, \tag{20.17}$$

where $\varphi_r = \dfrac{d\varphi}{dr}$. (b) Show that $\psi(r) = \exp(-r)$ is an eigenfunction corresponding to the eigenvalue $\lambda = -1$. (b) Is this the smallest eigenvalue? (c) Determine λ_2 and the corresponding eigenfunction by using a change of variables of the form $\varphi(r) = v(r)\exp(-\frac{r}{2})$. (d) Solve (20.17) using Femlab.

Problem 20.16. Formulate a two-dimensional analog of (20.14) of physical significance and compute approximate solutions using Femlab.

20.4. The special functions of mathematical physics

The one-dimensional analog of (20.1) is called a *Sturm-Liouville problem*. Such eigenvalue problems occur for example when separation of variables is used in various coordinate systems, and the corresponding eigenfunctions are the classical *special functions* of mathematical physics. We list some of these functions below together with the corresponding Sturm-Liouville problem.

Bessel's equation

The eigenfunctions $u_n(x)$ and eigenvalues λ_n of Bessel's equation

$$\begin{cases} -(xu')' + x^{-1}m^2u = \lambda xu & \text{for } 0 < x < 1, \\ |u(0)| < \infty, \quad u(0) = 1, \end{cases} \tag{20.18}$$

are given by $u_n(x) = J_m(\lambda_n^{1/2}x)$ and $\lambda_n = \mu^2$, where μ is a zero of the Bessel function J_m satisfying (20.18) with $\lambda = 1$ for $x \in \mathbb{R}$ and $|u(0)| < \infty$.

Legendre's equation

The eigenfunctions $u_n(x)$ and eigenvalues λ_n of Legendre's equation

$$\begin{cases} -((1-x^2)u')' + (1-x^2)^{-1}m^2 u = \lambda u & \text{for } 0 < x < 1, \\ |u(-1)| < \infty, \quad |u(1)| < \infty, \end{cases} \tag{20.19}$$

are given by $\lambda_n = n(n+1)$ and

$$u_n(x) = \frac{1}{2^n n!}(1-x^2)^{m/2} \frac{d^{m+n}((x^2-1)^n)}{dx^{m+n}}.$$

Tchebycheff's equation

The eigenfunctions $u_n(x)$ and eigenvalues λ_n of Tchebycheff's equation

$$\begin{cases} -((1-x^2)^{1/2}u')' = (1-x^2)^{-1/2}\lambda u & \text{for } 0 < x < 1, \\ |u(-1)| < \infty, \quad |u(1)| < \infty, \end{cases} \tag{20.20}$$

are given by $\lambda_n = n^2$ and $u_n(x) = 2^{-(n-1)}\cos(n\cos^{-1}x)$.

Problem 20.17. Use the method of separation of variables to solve the Poisson equation on the disc $\{x \in \mathbb{R}^2 : |x| < 1\}$ with homogeneous Dirichlet boundary conditions. Hint: use polar coordinates and the eigenfunctions of Bessel's equation with $m = 0$.

> Plowhand has been my name
> seems like a thousand years or more
> I ain't gonna pick no more cotton,
> I declare I ain't gonna plant no more corn.
> If a mule wants to run away with the world
> oooh Lord, I'll let it go it on.
> I wouldn't tell a mule to get up,
> Naah, if he sit down in my lap.
> I'm through with plowin'
> cause it killed my old grandpap. (R. Howard)

21

The Power of Abstraction

Maybe in order to understand mankind, we have to look at the
word itself. *Mankind.* Basically, it's made up of two separate
words - "mank" and "ind". What do these word mean? It's a
mystery, and that's why, so is mankind. (Jack Handley)

The use of mathematical symbolism eliminates the waste of mental
energy on trivialities, and liberates this energy for deployment
where it is needed, to wit, on the chaotic frontiers of theory and
practice. It also facilitates reasoning where it is easy, and restrains
it where it is complicated. (Whitehead)

Up until now we have considered a set of specific examples spanning
the fundamental models in science. In this chapter, we consider an "abstract" linear elliptic problem, concentrating on the basic questions of
existence, uniqueness, and stability of solutions together with the basic approximation properties of the Galerkin method. After that, we
apply the abstract theory to specific problems including Poisson's equation with various boundary conditions, a model of linear elasticity, and
Stoke's equations for creeping fluid flow. The abstract framework we
describe is the result of a long development of variational methods initiated by Euler and Lagrange, continued by Dirichlet, Riemann, Hilbert,
Rayleigh, Ritz, Galerkin, and continuing at the present time partly because of the modern interest in the finite element method. The advantage of considering a problem in abstract form is that we can emphasize
the essential ingredients and moreover we can apply results for the abstract problem to specific applications as soon as the assumptions of the
abstract problem are satisfied without having go through the same type

of argument over and over again. This is the real "power" of abstraction. We focus on linear elliptic problems, since setting up an abstract framework is easist in this case. The framework may be extended naturally to a class of nonlinear elliptic problems related to convex minimization problem and to the related parabolic problems. An abstract framework for hyperbolic problems is less developed; see the advanced companion book for details. We keep the presentation in this chapter short, and give more details in the advanced companion volume. The idea is to indicate a framework, not to develop it in detail.

We recall from Chapters 8 and 15 that we started by rewriting a given boundary value problem in variational form. We then applied Galerkin's method to compute an approximate solution in a subspace of piecewise polynomials and we proved energy norm error estimates using the Galerkin orthogonality. The abstract elliptic problem we consider is formulated in variational terms and has stability and continuity properties directly related to the energy norm. The basic theorem on the the existence, uniqueness, and stability of the solution of the abstract elliptic problem is the *Lax-Milgram theorem*. We also give a related result stating that Galerkin's method is optimal in the energy norm. These results guarantee that some of the basic models of science including Poisson's equation and the equations for linear elasticity and Stokes flow have a satisfactory mathematical form and may be solved approximately using Galerkin's method. This is a cornerstone in science and engineering.

21.1. The abstract formulation

The ingredients of the abstract formulation are

(i) a Hilbert space V where we look for the solution, with norm $\|\cdot\|_V$ and scalar product $(\cdot,\cdot)_V$,

(ii) a bilinear form $a : V \times V \to \mathbb{R}$ that is determined by the underlying differential equation,

(iii) a linear form $L : V \to \mathbb{R}$ that is determined by the data.

A *Hilbert space* is a vector space with a scalar product that is *complete*, which means that any Cauchy sequence in the space converges to a limit in the space. Recall that we discussed the importance of using a space with this property in Chapter 3, where we used the completeness of the continuous functions on an interval to prove the Fundamental Theorem of Calculus. A *bilinear form* $a(\cdot, \cdot)$ is a function that takes $V \times V$ into the real numbers, i.e. $a(v, w) \in \mathbb{R}$ for all $v, w \in V$, such that $a(v, w)$ is linear in each argument v and w, that is $a(\alpha_1 v_1 + \alpha_2 v_2, w_1) = \alpha_1 a(v_1, w_1) + \alpha_2 a(v_2, w_1)$ and $a(v_1, \alpha_1 w_1 + \alpha_2 w_2) = \alpha_1 a(v_1, w_1) + \alpha_2 a(v_1, w_2)$ for $\alpha_i \in \mathbb{R}$, v_i, $w_i \in V$. Finally, a *linear form* $L(\cdot)$ is a function on V such that $L(v) \in \mathbb{R}$ for all $v \in V$ and $L(v)$ is linear in v.

The abstract problem reads: find $u \in V$ such that

$$a(u, v) = L(v) \quad \text{for all } v \in V. \tag{21.1}$$

We make some assumptions on $a(\cdot, \cdot)$ and $L(\cdot)$, which gives an abstract definition of a linear "elliptic" problem. We first assume that $a(\cdot, \cdot)$ is *V-elliptic* or *coercive*, which means that there is a positive constant κ_1 such that for all $v \in V$,

$$a(v, v) \geq \kappa_1 \|v\|_V^2. \tag{21.2}$$

We also require that $a(\cdot, \cdot)$ is *continuous* in the sense that there is a constant κ_2 such that for all $v, w \in V$

$$|a(v, w)| \leq \kappa_2 \|v\|_V \|w\|_V. \tag{21.3}$$

We finally require that the linear form $L(\cdot)$ is *continuous* in the sense that there is a constant κ_3 such that for all $v \in V$,

$$|L(v)| \leq \kappa_3 \|v\|_V. \tag{21.4}$$

The reason that we say that L is continuous if (21.4) holds is because by linearity $|L(v) - L(w)| \leq \kappa_3 \|v - w\|_V$, which shows that $L(v) \to L(w)$ if $\|v - w\|_V \to 0$, i.e., if $v \to w$ in V. Assumption (21.3) similarly implies that $a(\cdot, \cdot)$ is continuous in each variable. Further, we define the *energy norm* $\| \cdot \|_a$ by $\|v\|_a = \sqrt{a(v, v)}$, noting that (21.2) in particular guarantees that $a(v, v) \geq 0$. By (21.2) and (21.3), we have $\kappa_1 \|v\|_V^2 \leq \|v\|_a^2 \leq \kappa_2 \|v\|_V^2$. In other words, if a quantity is small in the energy norm $\| \cdot \|_a$ then it is small in the norm $\| \cdot \|_V$ and vica versa. We refer to this situation by saying that $\| \cdot \|_a$ and $\| \cdot \|_V$ are *equivalent norms*. Thus, without changing anything qualitatively, we could choose the norm in the Hibert space V to be the energy norm $\| \cdot \|_a$ related to the bilinear form a, in which case $\kappa_1 = \kappa_2 = 1$. In this sense, the energy norm is a natural choice to use to analyze the bilinear form a. In applications, the energy norm fits with the notion of energy in mechanics and physics.

Problem 21.1. Determine a and L for (8.2), (8.1), and (15.18).

21.2.   The Lax-Milgram theorem

We now state and prove the basic Lax-Milgram theorem.

Theorem 21.1. *Suppose $a(\cdot,\cdot)$ is a continuous, V-elliptic bilinear form on the Hilbert space V and L is a continuous linear functional on V. Then there is a unique element $u \in V$ satisfying (21.1). Moreover, the following stability estimate holds*

$$\|u\|_V \leq \frac{\kappa_3}{\kappa_1}. \tag{21.5}$$

Recall that the bilinear forms a associated to the two-point boundary value problem (8.2) and to Poisson's equation (15.18) are *symmetric*, i.e.

$$a(v, w) = a(w, v) \quad \text{for all } v, w \in V.$$

Symmetric problems have additional structure that make the proof of the Lax-Milgram theorem easier, and this is the case we consider now. We treat the non-symmetric case in the companion volume, see also Renardy and Rogers ([14]).

If a is symmetric, then the variational problem (21.1) is equivalent to the minimization problem: find $u \in V$ such that

$$F(u) \leq F(v) \quad \text{for all } v \in V, \tag{21.6}$$

where $F(v) = a(v, v)/2 - L(v)$. We state and prove this equivalence in the following theorem.

Theorem 21.2. *An element $u \in V$ satisfies (21.1) if and only if u satisfies (21.6).*

Proof. Assume first that $u \in V$ satisfies (21.6). Choose $v \in V$ and consider the function $g(\epsilon) = F(u + \epsilon v)$ for $\epsilon \in \mathbb{R}$. By (21.6) we know that $g(\epsilon) \geq g(0)$ for $\epsilon \in \mathbb{R}$, so that $g'(0) = 0$ if $g'(0)$ exists. But, differentiating the expression $g(\epsilon) = (a(u, u) + \epsilon(a(u, v) + a(v, u)) + \epsilon^2 a(v, v))/2 - L(u) - \epsilon L(v)$ with respect to ϵ and setting $\epsilon = 0$, gives $a(u, v) - L(v) = 0$, and (21.1) follows. Note that the symmetry of $a(\cdot,\cdot)$ is crucial to this argument.

Conversely, if (21.1) is satisfied, then for all $w \in V$,

$$F(u + w) = \frac{1}{2}a(u, u) + a(u, w) + \frac{1}{2}a(w, w) - L(u) - L(w)$$

$$= F(u) + \frac{1}{2}a(w, w) \geq F(u),$$

with equality only if $w = 0$, which proves (21.6).   ■

We now prove the Lax-Milgram theorem for symmetric $a(\cdot, \cdot)$ by using the equivalence of (21.1) and (21.6).

Proof. Since we have assumed that the energy norm and the norm of V are equivalent in (21.2) and (21.3), without loss of generality, we can take $(\cdot, \cdot)_V$ to be $a(\cdot, \cdot)$, so that $a(v, v) = \|v\|_V^2$, and $\kappa_1 = \kappa_2 = 1$.

We consider the set of real numbers that can be obtained as the limit of sequences $\{F(u_j)\}$ with $u_j \in V$. We observe that this set is bounded below by $-1/2$ since $F(v) \geq \|v\|_V^2/2 - \|v\|_V \geq -1/2$ for all $v \in V$. We claim that the set of limits of $\{F(u_j)\}$ contains a smallest real number and we denote this number by β. Clearly, $\beta \geq -1/2$ if β exists. Now, the existence of β follows from the basic property of the real numbers that a set of real numbers that is bounded below has a greatest lower bound. In other words, there is a largest real number that is smaller or equal to all numbers in the set, which in our case is the number β. This property is equivalent to the property of convergence of a Cauchy sequence of real numbers. As another example, the set of positive real numbers ξ such that $\xi^2 > 2$ is clearly bounded below and its largest lower bound is nothing but $\sqrt{2}$. See Rudin ([15]) for more details.

Accepting the existence of β, we also know that $\beta \leq F(0) = 0$, and thus $-1/2 \leq \beta \leq 0$. Now let $\{u_j\}$ be a *minimizing sequence* for the minimization problem (21.6), i.e. a sequence such that

$$F(u_j) \to \beta \quad \text{as } j \to \infty. \tag{21.7}$$

We prove that $\{u_j\}$ is a *Cauchy sequence* in the sense that for any given $\epsilon > 0$ there is a natural number N_ϵ such that

$$\|u_i - u_j\|_V < \epsilon \quad \text{if } i, j \geq N_\epsilon. \tag{21.8}$$

Since V is complete, there is a $u \in V$ such that $\|u - u_j\| \to 0$ as $j \to \infty$. By the continuity properties of F, it follows that $F(u) = \beta$. and thus

$u \in V$ is a solution of (21.6) and therefore (21.1). The uniqueness follows from the last inequality of the proof of Theorem 21.2 above.

To prove that the minimizing sequence is a Cauchy sequence, we note that (21.7) implies that for any $\epsilon > 0$ there is a N_ϵ such that

$$F(u_j) \le \beta + \frac{\epsilon^2}{8} \quad \text{if } j \ge N_\epsilon. \tag{21.9}$$

We use the parallelogram law

$$\|u_i - u_j\|_V^2 = 2\|u_i\|_V^2 + 2\|u_j\|_V^2 - \|u_i + u_j\|_V^2,$$

together with the definition of $F(v)$, the definition of β, and (21.9), to argue

$$\frac{1}{4}\|u_i - u_j\|_V^2 = F(u_i) + F(u_j) - 2F\left(\frac{1}{2}(u_i + u_j)\right)$$

$$\le F(u_i) + F(u_j) - 2\beta \le \frac{\epsilon^2}{4},$$

proving (21.8).

Finally, the stability estimate follows immediately after taking $v = u$ in (21.1) and using the V-ellipticity and the continuity of L. ∎

21.3. The abstract Galerkin method

We consider Galerkin's method in abstract form applied to the problem (21.1): given a finite dimensional space $V_h \subset V$, find $U \in V_h$ such that

$$a(U, v) = L(v) \quad \text{for all } v \in V_h. \tag{21.10}$$

This leads to a linear system of equations whose size is determined by the dimension of V_h. We could for example choose V_h to be the space of polynomials of a fixed degree or less, the space of trigonometric functions with integer frequencies up to a fixed maximum, or in the case of the finite element method, the space of piecewise polynomial functions. We note that since $V_h \subset V$, we have the familiar Galerkin orthogonality:

$$a(u - U, v) = 0 \quad \text{for all } v \in V_h. \tag{21.11}$$

The basic a priori error estimate reads:

Theorem 21.3. *If u and U satisfy (21.1) and (21.10) then for all $v \in V_h$,*

$$\|u - U\|_V \le \frac{\kappa_2}{\kappa_1}\|u - v\|_V.$$

If the norm $\|\cdot\|_V$ is equal to the energy norm $\|\cdot\|_a$, then

$$\|u - U\|_a \le \|u - v\|_a, \tag{21.12}$$

which expresses the optimality of Galerkin's method in the energy norm.

Proof. The V-ellipticity and continuity of a together with Galerkin orthogonality implies that for all $v \in V_h$,

$$\kappa_1\|u - U\|_V^2 \le a(u - U, u - U) = a(u - U, u - U) + a(u - U, U - v)$$
$$= a(u - U, u - v) \le \kappa_2\|u - U\|_V\|u - v\|_V,$$

which proves the desired result.   ∎

> **Problem 21.2.**   Prove (21.12). Prove that the solution U of (21.10) satisfies the following analog of (21.5): $\|U\|_V \le \kappa_3/\kappa_1$.

21.4.   Applications

We now present some basic applications of the Lax-Milgram theorem. In each case, we need to specify a, L and V and show that the assumptions of the Lax-Milgram theorem are satsified. Usually, the main issue is to verify the V-ellipticity of the bilinear form a. We illustrate some tools for this purpose in a series of examples.

21.4.1.   A problem with Neumann boundary conditions

As a first example, we consider Poisson's equation with an absorption term together with Neumann boundary conditions as given in Problem 21.45,

$$\begin{cases} -\Delta u + u = f & \text{in } \Omega, \\ \partial_n u = 0 & \text{on } \Gamma, \end{cases} \tag{21.13}$$

where Ω is a bounded domain in $\mathbb{R}^d$ with boundary Γ. This problem takes the variational form (21.1) with

$$a(v, w) = \int_\Omega (\nabla v \cdot \nabla w + vw)\, dx, \quad L(v) = \int_\Omega fv\, dx, \tag{21.14}$$

and

$$V = \left\{ v : \int_\Omega \left(|\nabla v|^2 + v^2 \right) dx < \infty \right\}. \tag{21.15}$$

The issue is to verify that the assumptions of the Lax-Milgram theorem are satisfied with these choices.

Clearly, V has natural scalar product and norm

$$(v, w)_V = \int_\Omega \left(\nabla v \cdot \nabla w + vw \right) dx, \quad \|v\|_V = \left(\int_\Omega \left(|\nabla v|^2 + v^2 \right) dx \right)^{1/2}.$$

It turns out that V is complete, a fact ultimately based the completeness of $\mathbb{R}$, and therefore V is a Hilbert space. Further, we note that (21.2) and (21.3) trivially hold with $\kappa_1 = \kappa_2 = 1$. Finally, to show (21.4), we note that

$$|L(v)| \le \|f\|_{L_2(\Omega)} \|u\|_{L_2(\Omega)} \le \|f\|_{L_2(\Omega)} \|u\|_V,$$

which means that we may take $\kappa_3 = \|f\|_{L_2(\Omega)}$ provided we assume that $f \in L_2(\Omega)$. We conclude that the Lax-Milgram theorem applies to (21.13).

21.4.2.　The spaces $H^1(\Omega)$ and $H_0^1(\Omega)$

The space V defined in (21.15) naturally occurs in variational formulations of second order elliptic differential equations and it has a special notation:

$$H^1(\Omega) = \left\{ v : \int_\Omega \left(|\nabla v|^2 + v^2 \right) dx < \infty \right\}, \tag{21.16}$$

while the scalar product and norm are denoted by

$$(v, w)_{H^1(\Omega)} = \int_\Omega \left(\nabla v \cdot \nabla w + vw \right) dx,$$

and the associated norm

$$\|v\|_{H^1(\Omega)} = \left(\int_\Omega \left(|\nabla v|^2 + v^2 \right) dx \right)^{1/2}.$$

The space $H^1(\Omega)$ is the *Sobolev space* of functions on Ω that are square integrable together with their gradients, named after the Russian mathematician Sobolev (1908-1994). The index one refers to the fact that we require first derivatives to be square integrable.

We also use the subspace $H_0^1(\Omega)$ of $H^1(\Omega)$ consisting of the functions in $H^1(\Omega)$ that vanish on the boundary Γ of Ω:

$$H_0^1(\Omega) = \{v \in H^1(\Omega) : v = 0 \text{ on } \Gamma\}.$$

We motivate below why this is a Hilbert space with the same norm and scalar product as $H^1(\Omega)$.

Problem 21.3. (a) Find r such that $x^r \in H^1(0,1)$ but $x^s \notin H^1(0,1)$ for any $s < r$. (b) With $\Omega = \{x : |x| \leq 1\}$ denoting the unit disk, find conditions on r such that $|x|^r \in H^1(\Omega)$ but $|x|^s \notin H^1(\Omega)$ for any $s < r$.

Problem 21.4. Define $H^2(\Omega)$ and find a function that is in $H^1(\Omega)$ but not in $H^2(\Omega)$ where Ω is the unit disk.

21.4.3.   Poisson's equation with Dirichlet boundary conditions

The first elliptic problem in several dimensions we studied was Poisson's equation with homogeneous Dirichlet boundary conditions posed on a bounded domain $\Omega \subset \mathbb{R}^2$ with boundary Γ:

$$\begin{cases} -\Delta u = f & \text{in } \Omega, \\ u = 0 & \text{on } \Gamma. \end{cases}$$

This problem has the variational formulation (21.1) with $V = H_0^1(\Omega)$ and

$$a(v,w) = \int_\Omega \nabla v \cdot \nabla w \, dx, \quad L(v) = \int_\Omega fv \, dx.$$

In this case the V-ellipticity of a does not follow automatically from the definition of the norm in $V = H_0^1(\Omega)$ as above, because the bilinear form $a(v,v)$ in this case does not contain the term $\int_\Omega v^2 \, dx$ contained in the squared V norm. Further, we need to show that it makes sense to impose the boundary condition $v = 0$ on Γ for functions v in $V = H^1(\Omega)$, which is the essential issue in proving that $H_0^1(\Omega)$ is a Hilbert space.

To verify the V-ellipticity we use the *Poincaré-Friedrichs inequality*, which states that the $L_2(\Omega)$-norm of a function $v \in H^1(\Omega)$ can be estimated in terms of the $L_2(\Omega)$-norm of the gradient ∇v plus the $L_2(\Gamma)$-norm of the restriction of v to the boundary Γ. The corresponding theorem in one dimension for an interval $(0,1)$ states that

$$\|v\|_{L_2(0,1)}^2 \leq 2\big(v(0)^2 + \|v'\|_{L_2(0,1)}^2\big). \tag{21.17}$$

This inequality is proved easily by integrating the inequality

$$v^2(x) = \left(v(0) + \int_0^x v'(y)\, dy\right)^2 \leq 2\left(v^2(0) + \int_0^1 (v'(y))^2\, dy\right)$$

for $0 \leq x \leq 1$, which is obtained by using Cauchy's inequality and the fact that $(a+b)^2 \leq 2(a^2 + b^2)$. The result for higher dimensions is

Theorem 21.4. *There is a constant C depending on Ω such that for all $v \in H^1(\Omega)$,*

$$\|v\|^2_{L_2(\Omega)} \leq C\left(\|v\|^2_{L_2(\Gamma)} + \|\nabla v\|^2_{L_2(\Omega)}\right). \qquad (21.18)$$

Problem 21.5. (a) Prove (21.18). Hint: Take $\varphi = |x|^2/(2d)$ where $\Omega \subset \mathbb{R}^d$, so $\Delta\varphi = 1$ and use the fact that

$$\int_\Omega v^2 \Delta\varphi\, dx = \int_\Gamma v^2 \partial_n\varphi\, ds - \int_\Omega 2v\nabla v \cdot \nabla\varphi\, dx. \qquad (21.19)$$

(b) Give a different proof for square domains of the form $\{x \in \mathbb{R}^2 : |x_i| \leq 1\}$ analogous to the proof in one dimension by directly representing u in Ω through line integrals starting at Γ.

For functions $v \in H^1(\Omega)$ with $v = 0$ on Γ, i.e., $\|v\|_{L_2(\Gamma)} = 0$, Poincaré-Friedrichs' inequality implies

$$\|v\|^2_{H^1(\Omega)} = \|\nabla v\|^2_{L_2(\Omega)} + \|v\|^2_{L_2(\Omega)} \leq (1+C)\|\nabla v\|^2_{L_2(\Omega)} = (1+C)a(v,v),$$

which proves the V-ellipticity (21.2) with $\kappa_1 = (1+C)^{-1} > 0$.

Since (21.3) and (21.4) follow exactly as in the case of Neumann boundary conditions considered above, it now remains to show that the space $H_0^1(\Omega)$ is a well defined Hilbert space, that is, we need to show that a function in $H_0^1(\Omega)$ has well defined values on the boundary Γ. We start noting that it is in general impossible to uniquely define the boundary values of a function v in $L_2(\Omega)$. This is because by changing a function $v \in L^2(\Omega)$ only very close to the boundary, we can significantly change the boundary values of v without much changing the $L_2(\Omega)$ norm. This is reflected by the fact that there is no constant C such that $\|v\|_{L_2(\Gamma)} \leq C\|v\|_{L_2(\Omega)}$ for all functions $v \in L_2(\Omega)$. However, if we change $L_2(\Omega)$ to $H^1(\Omega)$, such an equality holds, and therefore a function v in $H^1(\Omega)$ has well defined boundary values, i.e., the *trace* of $v \in H^1(\Omega)$ on the boundary Γ is well defined. This is expressed in the following *trace inequality:*

Theorem 21.5. *If Ω is a bounded domain with boundary Γ, then there is a constant C such that for all $v \in H^1(\Omega)$,*

$$\|v\|_{L_2(\Gamma)} \leq C\|v\|_{H^1(\Omega)}. \tag{21.20}$$

Problem 21.6. Prove this. Hint: choose φ such that $\partial\varphi = 1$ on Γ and use (21.19).

Problem 21.7. Prove that there is no constant C such that $\|v\|_{L_2(\Gamma)} \leq C\|v\|_{L_2(\Omega)}$ for all $v \in L_2(\Omega)$.

The trace inequality shows that a function v in $H^1(\Omega)$ has well defined boundary values and in particular the boundary condition $v = 0$ on Γ makes sense, and it follows that $H_0^1(\Omega)$ is a Hilbert space.

Note that (21.18) implies that we may use the energy norm $\|\nabla v\|_{L_2(\Omega)} = \sqrt{a(v, v)}$ as an equivalent norm on $V = H_0^1(\Omega)$. As we said, choosing this norm, (21.2) and (21.3) hold with $\kappa_1 = \kappa_2 = 1$.

Problem 21.8. Verify that the assumptions of the Lax-Milgram theorem are satisfied for the following problems with appropriate assumptions on α and f:

(a) $\begin{cases} -u'' + \alpha u = f & \text{in } (0,1), \\ u(0) = u'(1) = 0, & \alpha = 0 \text{ and } 1. \end{cases}$

(b) $\begin{cases} -u'' = f & \text{in } (0,1), \\ u(0) - u'(0) = u(1) + u'(1) = 0. \end{cases}$

Problem 21.9. Verify that the assumptions of the Lax-Milgram theorem are satisfied for the beam problem:

$$\frac{d^4 u}{dx^4} = f \quad \text{in } (0,1),$$

with the boundary conditions; (a) $u(0) = u'(0) = u(1) = u'(1) = 0$, (b) $u(0) = u''(0) = u'(1) = u'''(1) = 0$, (c) $u(0) = -u''(0) + u'(0) = 0$, $u(1) = u''(1) + u'(1) = 0$; under appropriate assumptions on f. Give mechanical interpretations of the boundary conditions.

Remark 21.4.1. We saw earlier that if $f \in L_2(\Omega)$ then (21.4) holds with $V = H^1(\Omega)$ and $\kappa_3 = \|f\|_{L_2(\Omega)}$. We may ask what is the weakest assumption on the right-hand side f that allows (21.4) to hold with

$\kappa_3 < \infty$. In true mathematical style, we answer this by defining a weak $H^{-1}(\Omega)$ norm of f,

$$\|f\|_{H^{-1}(\Omega)} = \sup_{v \in H_0^1(\Omega)} \frac{(f, v)}{\|v\|_V},$$

where $V = H_0^1(\Omega)$ using the equivalent norm $\|v\|_{H_0^1(\Omega)} = \|\nabla v\|_{L_2(\Omega)}$. By definition, (21.4) holds with $\kappa_3 = \|f\|_{H^{-1}(\Omega)}$. By (21.18), the norm $\|\cdot\|_{H^{-1}(\Omega)}$ may be dominated by the $L_2(\Omega)$ norm:

$$\|f\|_{H^{-1}(\Omega)} \leq \frac{\|f\|_{L_2(\Omega)} \|v\|_{L_2(\Omega)}}{\|\nabla v\|_{L_2(\Omega)}} \leq \sqrt{C} \|f\|_{L_2(\Omega)};$$

In fact, the $H^{-1}(\Omega)$ norm is weaker than the $L_2(\Omega)$ norm, which allows us to use right-hand sides $f(x)$ in Poisson's equation that do not belong to $L_2(\Omega)$, such as the "near point load" used in the tent problem considered in Chapter 15.

Problem 21.10. Show that Lax-Milgram applies to problem (21.13) with $\Omega = \{x \in \mathbb{R}^2 : |x| < 1\}$ and $f(x) = |x|^{-1}$, although in this case $f \notin L_2(\Omega)$.

21.4.4. Non-homogeneous boundary data

Generally, nonhomogeneous boundary data is incorporated into the linear form L along with the right-hand side f. For example, recalling the discussion on Neumann/Robin boundary conditions in Chapter 15, we see that the problem $-\Delta u + u = f$ in Ω posed with nonhomogeneous Neumann conditions $\partial_n u = g$ on Γ takes the variational form (21.1) with $V = H^1(\Omega)$, $a(u, v)$ defined as in (21.14) and

$$L(v) = \int_\Omega fv \, dx + \int_\Gamma gv \, ds.$$

The continuity of $L(\cdot)$ follows assuming $f \in H^{-1}(\Omega)$ and $g \in L_2(\Gamma)$.

Problem 21.11. Prove the last claim.

Problem 21.12. Formulate the variational problem associated to Poisson's equation with non-homogeneous Dirichlet boundary conditions given by g on Γ.

Problem 21.13. Show that the Lax-Milgram theorem applies to the problem $-\Delta u + \alpha u = f$ in Ω, $\partial_n u + \sigma u = g$ on Γ, for (a) $\alpha = 1$ and $\sigma = 0$, (b) $\alpha = 0$ and $\sigma = 1$. What can be said in the case $\alpha = \sigma = 0$.

21.4.5. A diffusion dominated convection-diffusion problem

The convection-diffusion problem

$$\begin{cases} -\epsilon\Delta u + \beta \cdot \nabla u + \alpha u = f & \text{in } \Omega, \\ u = 0 & \text{on } \Gamma, \end{cases} \tag{21.21}$$

where Ω is domain in $\mathbb{R}^d$ with boundary Γ, $\epsilon > 0$ is constant, and $\beta(x)$ and $\alpha(x)$ are given coefficients, takes the variational form (21.1) with $V = H_0^1(\Omega)$ and

$$a(u, v) = \int_\Omega \left(\epsilon\nabla u \cdot \nabla v + \beta \cdot \nabla u \, v + \alpha u \, v \right) dx, \quad L(v) = \int_\Omega fv \, dx.$$

In this case, $a(\cdot, \cdot)$ is not symmetric because of the convection term. To guarantee ellipticity we assume recalling (18.14) that $-\frac{1}{2}\nabla \cdot \beta + \alpha \geq 0$ in Ω, which by (18.18) guarantees that for all $v \in H_0^1(\Omega)$,

$$\int_\Omega \left(\beta \cdot \nabla v \, v + \alpha v^2 \right) dx \geq 0.$$

It follows that $a(v, v) \geq 0$ and the assumptions of the Lax-Milgram theorem hold, but the stability estimate degrades with decreasing ϵ so that the theorem is mostly relevant for diffusion-dominated problems.

Problem 21.14. Prove the preceding statement with specific focus on the dependence of the constants on ϵ.

21.4.6. Linear elasticity in $\mathbb{R}^3$

> No body is so small that it is without elasticity. (Leibniz)

As an example of a problem in $\mathbb{R}^3$, we let Ω be a bounded domain in $\mathbb{R}^3$ with boundary Γ split into two parts Γ_1 and Γ_2 and consider the basic problem of linear elasticity modeled by *Cauchy-Navier's elasticity equations*: find the *displacement* $u = (u_i)_{i=1}^3$ and the *stress tensor* $\sigma = (\sigma_{ij})_{i,j=1}^3$ satisfying

$$\begin{cases} \sigma = \lambda \operatorname{div} u \, I + 2\mu\epsilon(u) & \text{in } \Omega, \\ -\operatorname{div} \sigma = f & \text{in } \Omega \\ u = 0 & \text{on } \Gamma_1, \\ \sigma \cdot n = g & \text{on } \Gamma_2, \end{cases} \tag{21.22}$$

where λ and μ are positive constants called the *Lamé coefficients*, $\epsilon(u) = (\epsilon_{ij}(u))_{i,j=1}^3$ is the *strain tensor* with components

$$\epsilon_{ij}(u) = \frac{1}{2}\left(\frac{\partial u_i}{\partial x_j} + \frac{\partial u_j}{\partial x_i}\right),$$

$$\mathrm{div}\ \sigma = \left(\sum_{j=1}^3 \frac{\partial \sigma_{ij}}{\partial x_j}\right)_{i=1}^3 \quad \text{and} \quad \mathrm{div}\ u = \sum_{i=1}^3 \frac{\partial u_i}{\partial x_i},$$

$I = (\delta_{ij})_{i,j=1}^3$ with $\delta_{ij} = 1$ if $i = j$ and $\delta_{ij} = 0$ if $i \neq j$, $f \in [L_2(\Omega)]^3$ and $g \in [L_2(\Gamma_1)]^3$ are given loads, $n = (n_j)$ is the outward unit normal to Γ_1, and $(\sigma \cdot n)_i = \sum_{j=1}^3 \sigma_{ij} n_j$. For simplicity, we assume that λ and μ are constant. The equations (21.22) express *Hooke's law* connecting stresses and strains and the *equilibrium equation* stating equilibrium of external and internal forces.

The problem has the variational form (21.1) with the choices:

$$V = \left\{v \in [H^1(\Omega)]^3 : v = 0 \text{ on } \Gamma_1\right\},$$

$$a(u,v) = \int_\Omega (\lambda\, \mathrm{div}\ u\, \mathrm{div}\ v + 2\mu\epsilon(u) : \epsilon(v))\ dx,$$

$$L(v) = \int_\Omega f \cdot v\, dx + \int_{\Gamma_1} g \cdot v\, ds,$$

where $\epsilon(u) : \epsilon(v) = \sum_{i,j=1}^3 \epsilon_{ij}(u)\epsilon_{ij}(v)$. We note that the bilinear form a has the form of "virtual work",

$$a(u,v) = \int_\Omega \sigma(u) : \epsilon(v)\, dx,$$

where $\sigma(u) = \lambda \mathrm{div}\ u\, I + 2\mu\epsilon(u)$. To prove V-ellipticity, we use *Korn's inequality*. For simplicity, we assume that $\Gamma_1 = \Gamma$.

Theorem 21.6. *There is a constant c such that for all $v \in [H_0^1(\Omega)]^3$,*

$$\sum_{i,j=1}^3 \int_\Omega \epsilon_{ij}(v)\epsilon_{ij}(v)\, dx \geq c \sum_{i=1}^3 \|v_i\|_{H^1(\Omega)}^2.$$

Proof. Using the notation $v_{i,j} = \partial v_i / \partial x_j$, $v_{i,jl} = \partial^2 v_i / \partial x_j \partial x_l$, etc.,

$$\sum_{i,j=1}^{3} \epsilon_{ij}(v)\epsilon_{ij}(v) = \sum_{i,j=1}^{3} \frac{1}{2} v_{i,j} v_{i,j} + \sum_{i,j=1}^{3} \frac{1}{2} v_{i,j} v_{j,i}.$$

Integrating the second term on the right and then using integration by parts, we get

$$\sum_{i,j=1}^{3} \int_{\Omega} v_{i,j} v_{j,i} \, dx = \int_{\Gamma} v_{i,j} v_j n_i \, ds - \int_{\Omega} v_{i,ji} v_j \, dx$$

$$= \sum_{i,j=1}^{3} \int_{\Gamma} v_{i,j} v_j n_i \, ds - \int_{\Gamma} v_{i,i} v_j n_j \, ds + \int_{\Omega} v_{i,i} v_{j,j} \, dx$$

$$= \sum_{i,j=1}^{3} \int_{\Omega} v_{i,i} v_{j,j} \, dx,$$

since $v = 0$ on Γ. We conclude that

$$\sum_{i,j=1}^{3} \int_{\Omega} \epsilon_{ij}(v)\epsilon_{ij}(v) \, dx = \frac{1}{2} \sum_{i,j=1}^{3} \int_{\Omega} (v_{i,j})^2 \, dx + \frac{1}{2} \int_{\Omega} \left(\sum_{i=1}^{} v_{i,i}\right)^2 dx.$$

The desired inequality follows using Poincaré's inequality to bound the L_2 norm of v_i in terms of the L_2 norm of ∇v_i. ∎

Problem 21.15. Provide the last details.

Problem 21.16. Solve the Cauchy-Navier elasticity equations for the cantilever beam in two dimensions using Femlab. Compare with analytic solutions of the beam equation.

21.4.7. The Stokes equations

The Stokes equations for stationary incompressible creeping fluid flow with zero velocity boundary conditions read: find the *velocity* $u = (u_i)_{i=1}^{3}$, total *stress* $\sigma = (\sigma_{ij})_{i,j=1}^{3}$, and the *pressure* p such that

$$\begin{cases} \sigma = -pI + 2\mu\epsilon(u) & \text{in } \Omega, \\ -\text{div } \sigma = f & \text{in } \Omega, \\ \text{div } u = 0 & \text{in } \Omega, \\ u = 0 & \text{on } \Gamma, \end{cases}$$

Eliminating the stress σ gives

$$\begin{cases} -\mu\Delta u + \nabla p = f & \text{in } \Omega, \\ \text{div } u = 0 & \text{in } \Omega, \\ u = 0 & \text{on } \Gamma. \end{cases} \qquad (21.23)$$

This can be formulated in variational form (21.1) with

$$V = \{v \in [H^1(\Omega)]^3 : \text{div}\, u = 0 \text{ in } \Omega\},$$

$$a(u, v) = \int_\Omega \sum_{i=1}^3 \nabla u_i \cdot \nabla v_i \, dx, \text{ and } L(v) = \int_\Omega f \cdot v \, dx.$$

The picture on the cover of the book shows streamlines of Stokes flow around a sphere.

Problem 21.17. Prove that the assumptions of the Lax-Milgram theorem hold in this case. this.

Problem 21.18. Extend the mechanical models of Section 13.7 to several dimensions.

Note that the stationary Navier-Stokes equations are obtained by adding the term $(\nabla \cdot u)u$ to the first equation in (21.23).

21.5. A strong stability estimate for Poisson's equation

We conclude this chapter by proving the strong stability estimate (15.6) for solutions to Poisson's equation that we used in the proofs of the L_2 error estimates for elliptic and parabolic problems. The estimate shows that the $L_2(\Omega)$ norm of all second derivatives of a function v vanishing on the boundary of a convex domain are bounded by the $L_2(\Omega)$ norm of the particular combination of second derivatives given by the Laplacian. For simplicity, we consider the case of a convex domain in the plane with smooth boundary.

Theorem 21.7. *If Ω is a bounded domain in $\mathbb{R}^2$ with smooth boundary Γ then for all smooth functions v with $v = 0$ on Γ,*

$$\sum_{i,j=1}^3 \int_\Omega (D^2 v)^2 \, dx + \int_\Gamma \frac{1}{R}\left(\frac{\partial v}{\partial n}\right)^2 ds = \int_\Omega (\Delta v)^2 \, dx,$$

where $R(x)$ is the radius of curvature of Γ at $x \in \Gamma$ with $R(x) \geq 0$ if Ω is convex, see Fig. 21.1.

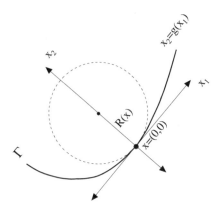

Figure 21.1: The radius of curvature and the local coordinate system near a point x on Γ.

Proof. We use the notation $v_{(i)} = \partial v / \partial x_i$, $v_{(ij)} = \partial^2 v / \partial x_i \partial x_j$, etc.. Assuming that v is smooth with $v = 0$ on Γ, integration by parts gives

$$\int_\Omega \Delta v \Delta v \, dx = \sum_{i,j=1}^{3} \int_\Omega v_{(ii)} v_{(jj)} \, dx$$

$$= \sum_{i,j=1}^{3} \int_\Gamma v_{(i)} v_{(jj)} n_i \, ds - \sum_{i,j=1}^{3} \int_\Omega v_{(i)} v_{(ijj)} \, dx$$

$$= \sum_{i,j=1}^{3} \int_\Gamma \left(v_{(i)} v_{(jj)} n_i - v_{(i)} v_{(ij)} n_j \right) ds + \sum_{i,j=1}^{3} \int_\Omega v_{(ij)} v_{(ij)} \, dx.$$

Recalling the definition of $D^2 v$ from Chapter 14

$$\int_\Omega \left((\Delta v)^2 - (D^2 v)^2 \right) dx = \sum_{i,j=1}^{3} \int_\Gamma \left(v_{(i)} v_{(jj)} n_i - v_{(i)} v_{(ij)} n_j \right) ds.$$

To evaluate the integrand on the right at a point $x \in \Gamma$, we use the fact that the integrand is invariant under orthogonal coordinate transformations. We may assume that $x = (0, 0)$ and that in a neighborhood of x, the graph of Γ is described by the equation $x_2 = g(x_1)$ in a local coordinate system, see Fig. 21.1. Since $v = 0$ on Γ, we have $v(x_1, g(x_1)) = 0$ for x_1 in some neighborhood of 0 and thus by differentiation with respect to x_1, we find that

$$v_{(1)} + v_{(2)}g'(x_1) = 0,$$
$$v_{(11)} + 2v_{(12)}g'(x_1) + v_{(22)}(g'(x_1))^2 + v_{(2)}g''(x_1) = 0.$$

Since $g'(0) = 0$ and, by the definition of the radius of curvature, $g''(0) = 1/R(0)$, we conclude that

$$v_{(1)}(0,0) = 0$$
$$v_{(11)}(0,0) = -v_{(2)}(0,0)/R(0).$$

At $x = (0,0)$, since $n = (0,-1)^\top$ at that point,

$$\sum_{i,j=1}^{3} \left(v_{(i)}v_{(jj)}n_i - v_{(i)}v_{(ij)}n_j \right) = -v_{(2)}(v_{(1)1} + v_{(22)}) + v_{(2)}v_{(22)}$$
$$= -v_{(2)}v_{(11)} = \left(v_{(2)}\right)^2/R = \left(\partial v/\partial n\right)^2/R.$$

and the statement of the theorem follows. ∎

Problem 21.19. (A maximum principle). Prove that if u is continuous in $\Omega \cup \Gamma$, where Ω is a domain with boundary Γ, and $\Delta u(x) \geq 0$ for $x \in \Omega$, then u attains its maximum on the boundary Γ. Hint: consider first the case that $\Delta u(x) > 0$ for $x \in \Omega$ and arrive at a contradiction by assuming a maximum is attained in Ω that is not on Γ by using the fact that at such a point, the second derivatives with respect to x_i cannot be positive. Extend this to the case $\Delta u(x) \geq 0$ by considering the function $u_\epsilon(x) = u(x) + \epsilon|x - \bar{x}|^2$, which for $\epsilon > 0$ sufficiently small also has an interior maximum.

Problem 21.20. Consider the problem

$$\begin{cases} -(u_{(11)} - u_{(12)} + 2u_{(22)}) + u_{(1)} + u = f & \text{in } \Omega, \\ u = 0 & \text{on } \Gamma_1, \\ u_{(1)}n_1 - \tfrac{1}{2}u_{(1)}n_2 - \tfrac{1}{2}u_{(2)}n_1 + u_{(2)}n_2 + u = g & \text{on } \Gamma_2. \end{cases}$$

Give a variational formulation of this problem and show that the conditions in the Lax-Milgram lemma (except symmetry) are satisfied.

Figure 21.2: Queen Sophie Charlotte von Brandenburg, gifted student
of Leibniz's philosophy.

Ein jeder Geist steht vor den ganzen Bau der Dinge,
Als ob die Fernung sich in einen Spiegel bringe,
Nach jeden Augenpunct, verdunckelt oder klar,
Er ist ein Bild, wie er ein Zweck der Schöpfung war
(Leibniz, at the funeral of Queen Sophie Charlotte, 1705)

Bibliography

[1] V. I. ARNOLD, *Huygens and Barrow, Newton and Hooke*, Birkhäuser, New York, 1990.

[2] K. ATKINSON, *An Introduction to Numerical Analysis*, John Wiley & Sons, Inc., New York, 1989.

[3] L. BERS, F. JOHN, AND M. SCHECTER, *Partial Differential Equations*, John Wiley & Sons, Inc., New York, 1964.

[4] S. BRENNER AND L. R. SCOTT, *The Mathematical Theory of Finite Element Methods*, Springer-Verlag, New York, 1994.

[5] P. CIARLET, *The Finite Element Method for Elliptic Problems*, North–Holland, New York, 1978.

[6] K. ERIKSSON, D. ESTEP, P. HANSBO, AND C. JOHNSON, *Introduction to adaptive methods for differential equations*, Acta Numerica, (1995), pp. 105–158.

[7] G. GOLUB AND C. V. LOAN, *Matrix Computations*, Johns Hopkins University Press, Maryland, 1983.

[8] P. HENRICI, *Discrete Variable Methods in Ordinary Differential Equations*, John Wiley & Sons, Inc., New York, 1962.

[9] E. ISAACSON AND H. KELLER, *Analysis of Numerical Methods*, John Wiley & Sons, Inc., New York, 1966.

[10] C. JOHNSON, *Numerical solution of partial differential equations by the finite element method*, Studentlitteratur and Cambridge University Press, Lund, Sweden, and Cambridge, England, 1987.

[11] N. JOLLEY, *The Cambridge Companion to Leibniz*, Cambridge University Press, Cambridge, England, 1995.

[12] M. KLINE, *Mathematical Thought from Ancient to Modern Times*, Oxford University Press, Oxford, England, 1972.

[13] G. W. LEIBNIZ, *Nova methodus pro maximis et minimis, itemque tangentibus, que nec fractas, nec irrationales quantitates moratur, et singulare pro illis calculi genus*, Acta Eruditorum, (1684), pp. 466–473.

[14] M. RENARDY AND R. ROGERS, *An Introduction to Partial Differential Equations*, Springer-Verlag, New York, 1993.

[15] W. RUDIN, *Principles of Mathematical Analysis, 3rd ed.*, McGraw-Hill, New York, 1976.

[16] E. STEIN AND A. HEINEKAMP, *Leibniz, Mathematiker, Physiker, Techniker*, Leibniz-Gesellschaft, Hannover, Germany, 1990.

[17] G. STRANG, *Introduction to Applied Mathematics*, Wellesley-Cambridge Press, Cambridge, Mass., 1986.

[18] W. A. STRAUSS, *Partial Differential Equations: An Introduction*, John Wiley & Sons, Inc., New York, 1992.

Index